T0211010

Communications
in Computer and Information Science   1263

Commenced Publication in 2007
Founding and Former Series Editors:
Simone Diniz Junqueira Barbosa, Phoebe Chen, Alfredo Cuzzocrea,
Xiaoyong Du, Orhun Kara, Ting Liu, Krishna M. Sivalingam,
Dominik Ślęzak, Takashi Washio, Xiaokang Yang, and Junsong Yuan

Editorial Board Members

More information about this series at http://www.springer.com/series/7899

Leonid Sokolinsky · Mikhail Zymbler (Eds.)

Parallel Computational Technologies

14th International Conference, PCT 2020
Perm, Russia, May 27–29, 2020
Revised Selected Papers

 Springer

Editors
Leonid Sokolinsky ⓘ
South Ural State University
Chelyabinsk, Russia

Mikhail Zymbler ⓘ
South Ural State University
Chelyabinsk, Russia

ISSN 1865-0929   ISSN 1865-0937 (electronic)
Communications in Computer and Information Science
ISBN 978-3-030-55325-8   ISBN 978-3-030-55326-5 (eBook)
https://doi.org/10.1007/978-3-030-55326-5

This Springer imprint is published by the registered company Springer Nature Switzerland AG
The registered company address is: Gewerbestrasse 11, 6330 Cham, Switzerland

Preface

This volume contains a selection of the papers presented at the 14th International Scientific Conference on Parallel Computational Technologies (PCT 2020). The PCT 2020 conference was supposed to be held in Perm, Russia, during March 31–April 2, 2020. However, considering the safety and well-being of all conference participants during the COVID-19 pandemic as a priority, the Organizing Committee took the hard decision to hold PCT 2020 online during May 27–29, 2020.

The PCT series of conferences aims at providing an opportunity to discuss the future of parallel computing, as well as to report the results achieved by leading research groups in solving both scientific and practical issues using supercomputer technologies. The scope of the PCT series of conferences includes all aspects of high-performance computing in science and technology such as applications, hardware and software, specialized languages, and packages.

The PCT series is organized by the Supercomputing Consortium of Russian Universities and the Federal Agency for Scientific Organizations. Originating in 2007 at the South Ural State University (Chelyabinsk, Russia), the PCT series of conferences has now become one of the most prestigious Russian scientific meetings on parallel programming and high-performance computing. PCT 2020 in Perm continued the series after Chelyabinsk (2007), St. Petersburg (2008), Nizhny Novgorod (2009), Ufa (2010), Moscow (2011), Novosibirsk (2012), Chelyabinsk (2013), Rostov-on-Don (2014), Ekaterinburg (2015), Arkhangelsk (2016), Kazan (2017), Rostov-on-Don (2018), and Kaliningrad (2019).

Each paper submitted to the conference was scrupulously evaluated by three reviewers on the relevance to the conference topics, scientific and practical contribution, experimental evaluation of the results, and presentation quality. The Program Committee of PCT selected the 24 best papers to be included in this CCIS proceedings volume.

We would like to thank the Russian Foundation for Basic Research for their continued financial support of the PCT series of conferences, as well as respected PCT 2020 sponsors, namely platinum sponsors, IBM and Intel, gold sponsors, Hewlett Packard Enterprise, RSC Group, and AMD, and silver sponsor, NVIDIA.

We would like to express our gratitude to every individual who contributed to the success of PCT 2020 amid the COVID-19 outbreak. Special thanks to the Program Committee members and the external reviewers for evaluating papers submitted to the conference. Thanks also to the Organizing Committee members and all the colleagues involved in the conference organization from Perm National Research Polytechnic University, the South Ural State University (national research university), and Moscow State University. We thank the participants of PCT 2020 for sharing their research and presenting their achievements online as well.

Finally, we thank Springer for publishing the proceedings of PCT 2020 in the *Communications in Computer and Information Science* series.

June 2020
Leonid Sokolinsky
Mikhail Zymbler

Organization

The 14th International Scientific Conference on Parallel Computational Technologies (PCT 2020) was organized by the Supercomputing Consortium of Russian Universities and the Ministry of Science and Higher Education of the Russian Federation.

Steering Committee

Berdyshev, V. I.	Krasovskii Institute of Mathematics and Mechanics, Russia
Ershov, Yu. L.	United Scientific Council on Mathematics and Informatics, Russia
Minkin, V. I.	South Federal University, Russia
Moiseev, E. I.	Moscow State University, Russia
Savin, G. I.	Joint Supercomputer Center, Russian Academy of Sciences, Russia
Sadovnichiy, V. A.	Moscow State University, Russia
Chetverushkin, B. N.	Keldysh Institute of Applied Mathematics, Russian Academy of Sciences, Russia
Shokin, Yu. I.	Institute of Computational Technologies, Russian Academy of Sciences, Russia

Program Committee

Sadovnichiy, V. A. (Chair)	Moscow State University, Russia
Dongarra, J. (Co-chair)	University of Tennessee, USA
Sokolinsky, L. B. (Co-chair)	South Ural State University, Russia
Voevodin, Vl. V. (Co-chair)	Moscow State University, Russia
Zymbler, M. L. (Academic Secretary)	South Ural State University, Russia
Ablameyko, S. V.	Belarusian State University, Belarus
Afanasiev, A. P.	Institute for Systems Analysis RAS, Russia
Akimova, E. N.	Krasovskii Institute of Mathematics and Mechanics UB RAS, Russia
Andrzejak, A.	Heidelberg University, Germany
Balaji, P.	Argonne National Laboratory, USA
Boldyrev, Y. Ya.	Saint Petersburg Polytechnic University, Russia
Carretero, J.	Carlos III University of Madrid, Spain
Gazizov, R. K.	Ufa State Aviation Technical University, Russia
Gergel, V. P.	Lobachevsky State University of Nizhny Novgorod, Russia

Glinsky, B. M.	Institute of Computational Mathematics and Mathematical Geophysics SB RAS, Russia
Goryachev, V. D.	Tver State Technical University, Russia
Il'in, V. P.	Institute of Computational Mathematics and Mathematical Geophysics SB RAS, Russia
Kobayashi, H.	Tohoku University, Japan
Kunkel, J.	University of Hamburg, Germany
Labarta, J.	Barcelona Supercomputing Center, Spain
Lastovetsky, A.	University College Dublin, Ireland
Ludwig, T.	German Climate Computing Center, Germany
Lykosov, V. N.	Institute of Numerical Mathematics RAS, Russia
Mallmann, D.	Jülich Supercomputing Centre, Germany
Malyshkin, V. E.	Institute of Computational Mathematics and Mathematical Geophysics SB RAS, Russia
Michalewicz, M.	A*STAR Computational Resource Centre, Singapore
Modorsky, V. Ya.	Perm Polytechnic University, Russia
Shamakina, A. V.	High Performance Computing Center in Stuttgart, Germany
Shumyatsky, P.	University of Brasília, Brazil
Sithole, H.	Centre for High Performance Computing, South Africa
Starchenko, A. V.	Tomsk State University, Russia
Sterling, T.	Indiana University, USA
Sukhinov, A. I.	Don State Technical University, Russia
Taufer, M.	University of Delaware, USA
Turlapov, V. E.	Lobachevsky State University of Nizhny Novgorod, Russia
Wyrzykowski, R.	Czestochowa University of Technology, Poland
Yakobovskiy, M. V.	Keldysh Institute of Applied Mathematics RAS, Russia
Yamazaki, Y.	Federal University of Pelotas, Brazil

Organizing Committee

Tashkinov, A. S. (Chair)	Perm Polytechnic University, Russia
Shevelev, N. A. (Co-chair)	Perm Polytechnic University, Russia
Korotaev, V. N. (Co-chair)	Perm Polytechnic University, Russia
Trushnikov, D. N. (Deputy Chair)	Perm Polytechnic University, Russia
Modorsky, V. Ya. (Deputy Chair)	Perm Polytechnic University, Russia
Seregina, M. A. (Secretary)	Perm Polytechnic University, Russia
Antonov, A. S.	Moscow State University, Russia
Antonova, A. P.	Moscow State University, Russia
Cherepanov, I. E.	Perm Polytechnic University, Russia
Faizrahmanov, R. A.	Perm Polytechnic University, Russia
Grents, A. V.	South Ural State University, Russia
Hudyakova, D. N.	Perm Polytechnic University, Russia

Ilyushin, P. Yu.	Perm Polytechnic University, Russia
Kalyulin, S. L.	Perm Polytechnic University, Russia
Kashevarova, G. G.	Perm Polytechnic University, Russia
Kashnikov, Yu. A.	Perm Polytechnic University, Russia
Kraeva, Ya. A.	South Ural State University, Russia
Maksimov, D. S.	Perm Polytechnic University, Russia
Masich, G. F.	Perm Polytechnic University, Russia
Mikrukov, A. O.	Perm Polytechnic University, Russia
Nikitenko, D. A.	Moscow State University, Russia
Perkova, A. I.	South Ural State University, Russia
Pervadchuk, V. P.	Perm Polytechnic University, Russia
Pisarev, P. V.	Perm Polytechnic University, Russia
Sakovskaya, V. I.	South Ural State University, Russia
Sidorov, I. Yu.	Moscow State University, Russia
Sobolev, S. I.	Moscow State University, Russia
Trufanov, A. N.	Perm Polytechnic University, Russia
Voevodin, Vad. V.	Moscow State University, Russia
Zymbler, M. L.	South Ural State University, Russia

Contents

Supercomputer Simulation

High Performance Architectures, Tools and Technologies

XP-COM Hybrid Software Package for Parallelization by Data

Anton Baranov[1]([✉])[ID], Andrey Fedotov[2], Oleg Aladyshev[1][ID], Evgeniy Kiselev[1], and Alexey Ovsyannikov[1][ID]

[1] Joint Supercomputer Center of the RAS, Moscow, Russia
antbar@mail.ru
[2] The Moscow Institute of Physics and Technology, Moscow, Russia
andrey_university@mail.ru

Abstract. The paper discusses the XP-COM hybrid software package. XP-COM is a combination of X-COM and Pyramid, two well-known packages for parallelization by data. The parallelization overhead introduced by X-COM and Pyramid has been presented in the previous studies. The X-COM software showed the best results in iterating over string values read from a file. The Pyramid software showed the least overhead when iterating through value combinations of several parameters. The authors used X-COM data transfer infrastructure as a basis for the XP-COM package development. The parameter value combination enumeration mechanism of the Pyramid software was used as the computational work separation procedure. The resulting XP-COM hybrid software was discovered to show better performance than both X-COM and Pyramid.

Keywords: HPC · Supercomputer · Parallelization by data · X-COM software · Pyramid software · Parallelization overhead · XP-COM

1 Software for Parallelization by Data

Parallelization by data is parallel computing technique often used to solve a number of important applied computing problems such as virtual screening, search for simple numbers, password strength checking, etc. In these problems, to all the input data set elements the same computing sequence (applied algorithm) is applied. The input data set is distributed between processes or threads of a parallel program. Each process (thread) performs an independent portion of the input data set. There is no information interconnect between the processes (threads) of the parallel program. Typically, parallelization by data is used in distributed environments of volunteer computing [1]. However, such problems are often solved using separate supercomputers.

Assume our algorithm can be implemented as a single sequential program (SSP), for which the input data portion is determined by one or more parameter values. The input data pool is formed as a set of all possible SSP param

© Springer Nature Switzerland AG 2020
L. Sokolinsky and M. Zymbler (Eds.): PCT 2020, CCIS 1263, pp. 3–15, 2020.
https://doi.org/10.1007/978-3-030-55326-5_1

eters values in all their combinations. The well-known HashCat password cracking utility [2] is a typical SSP example. During parallelizing by data on the supercomputer, one or several SSP instances with the different input parameters values are executed at each supercomputer node. The parallel computing in this case consists of launching the SSP instances on the whole available set of nodes. Such parallelization model is referred to as a SSP model.

In the previous studies [3] experimental comparison of Pyramid [4], BOINC [5], and X-COM [6,7] software packages was made. The experiments were carried out at the separate supercomputer, not in a distributed environment. The overhead costs introduced by each studied software were determined for a set of typical tests. X-COM and Pyramid showed the best results, significantly ahead of BOINC.

The Pyramid software package [4] is designed to operate on a computing facility with a hierarchical structure. The computing facility consists of a central server, cluster management servers, and computing nodes in clusters. The main goal of creating the Pyramid software was to relieve the user of organizing parallel computing with parallelization by data. The user implements the applied algorithm as a SSP, and Pyramid launches many SSP instances on computing nodes and distribute the computational work between the running instances.

Pyramid is reliable and fault-tolerant. The failure of one or more computing nodes or one or more clusters does not stop the computations but slows them down. When a previously disabled node or cluster is fixed and enabled again, Pyramid automatically begins to distribute the computing operations with the repaired node. At the same time, Pyramid saves checkpoints at a specified frequency to restore computation if the entire computer facility fails.

Designed by the specialists of Research Computing Center of Moscow State University, X-COM [6] is written in Perl programming language, which makes it one of the most lightweight means of parallelization by data. X-COM is based on client-server architecture. X-COM server is in charge of dividing the original task into blocks (jobs), distributing the jobs to the clients, coordinating all the clients, checking the integrity and accumulating the results. Any computational unit (workstation computer, cluster node, virtual machine) able to perform an instance of an application program can act as a client. The clients are computing the blocks of the application task (jobs received from the server), requesting jobs from the server, and transferring results to the server.

X-COM operates in a distributed environment, but can be run on a separate supercomputer. Pyramid operates on a separate supercomputer only. X-COM requires the user to develop server and client management programs. Pyramid requires the job passport only. With X-COM, checkpoint saving is the care up to the management program developer. Pyramid saves the checkpoints automatically.

The X-COM and Pyramid comparison [3] carried out at Joint Supercomputer Center of the Russian Academy of Sciences (JSCC RAS) provided the following results. X-COM shows many-fold better results with processing the strings read from a text file. Highly likely, X-COM has a more advanced data transfer subsystem. Pyramid has showed many-fold better results with processing several SSP parameters. This is natural, since Pyramid implements a comprehensive search of all possible combinations for several SSP parameters.

To combine the Pyramid and X-COM advantages, it was decided to develop a hybrid software package called XP-COM.

2   XP-COM Hybrid Software Package Architecture and Algorithms

To run multiple SSP instances, X-COM requires a developed server and client control programs. The server part provides the logic for dividing the computational work into portions, and the client part runs SPP with the parameters of the next portion. X-COM transfers the work portion from a server to a client and gets the results back. X-COM itself does not contain mechanisms for dividing the work and starting the SPP, the user has to develop the corresponding control programs as a programmer. The ability to divide computational work and start multiple SPP instances, to save the checkpoints for the SSP computing model as a part of the X-COM server and client control programs is the main idea of XP-COM hybrid software.

The following engines were adopted from Pyramid to implement XP-COM hybrid software:

– Computational work distribution (division) mechanism
– Automatic checkpoints saving
– Launching SSP with the specified parameters

The source codes of these engines were adopted from Pyramid and implemented as separate program modules:

1. The computational work distribution module called Divisio. Divisio operates at the X-COM server side. It assigns data chunks to computing clients at X-COM server request and saves checkpoints at the regular intervals.
2. The SSP launching module called Divisio_client. Divisio_client operates at the X-COM client side. It converts the data chunks received from X-COM server into the SSP actual parameters, and runs SSP.

These modules were embedded into X-COM through the server and client management programs developed by the authors in Perl language, as shown in Fig. 1. The management programs were developed in accordance with the rules described in the X-COM user manual [7].

The interaction of the XP-COM client and server parts is carried out according to the following algorithm.

1. Starting the server side and the Divisio module
2. Starting X-COM clients
3. Requesting data portion by the client from the server
4. Chipping data portion with Divisio, if not possible, the work finish

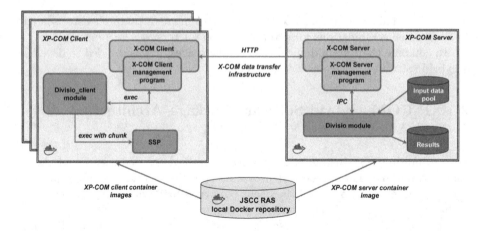

Fig. 1. The XP-COM software package architecture

5. Sending chipped data portion to the client from Divisio through the server
6. Processing received data portion by the client (Divisio_client launching)
7. Sending the processing results to the server
8. Recording the results received from the client
9. Go to step 3

Note the important detail of pairing the X-COM and Pyramid softwares. In Pyramid, the sharing and accounting of the computing work is carried out by the client threads' numbers. X-COM does not have client numbering. To enable the pairing, Divisio uses the dynamically allocated client numbers. The client is automatically assigned the lowest available number when requested the next data portion. The assigned number is recorded in the portion metadata and is considered occupied from this moment. Having completed the data portion processing, the client returns the result to the server. The result metadata includes the client number received earlier with the data portion. When the server receives the result, the number is freed.

XP-COM was implemented as the Docker containers to run in the supercomputer job management system. Two container images were formed – for the server and the client parts of XP-COM. The container images were placed in the JSCC RAS Docker local repository as considered in [8]. The Russian parallel jobs management system called SUPPZ [9] was used. The SUPPZ is the main job management system used on the JSCC RAS supercomputers. When the job starts, SUPPZ extracts XP-COM server and client images from the Docker local repository, copies them to the supercomputer nodes allocated for the job, launches container instances on the allocated nodes, connecting them to a separate virtual network. After that, XP-COM processes begin to run in the containers.

3   Methods the Performance Evaluating of Software Systems for Parallelization by Data

Various methods can be used to evaluate the performance and effectiveness of parallelization by data software systems. The paper [10] considers several methods to evaluate the X-COM effectiveness, such as evaluating the effectiveness by the number of redundant losses that are inevitable while ensuring reliability by the computing duplication method. The performance of parallelization by data can be assessed as a difference between the reference and the real resource allocation schedule as proposed in [11]. A benchmark suite which consists of a synthetic micro-benchmarks and real-world Hadoop applications was proposed in [12] to evaluate the performance using MapReduce. Both the job running time and the throughput (the number of tasks completed per minute) are evaluated. The MapReduce program completion time is estimated in [13] depending on the varying input dataset sizes and given cluster resources. The word count test is used in [14–16] to assess the software systems performance. The test that counts all the occurring RGB colors in a given bitmap image and the test that computes a line that best fits a file containing are used in [15]. The test that counts user visits to the Internet pages is used for performance evaluation [16].

The performance estimate approach proposed in [10] as the ratio of the time spent on organizing and supporting computations to the computing time itself seems to be the most suitable for us. In fact, it is proposed to evaluate the computing environment overhead costs. In relation to the SSP model, overhead costs can be determined as follows [3].

Let the SSP process the whole input data on one CPU core for a certain time T. If parallelization by data is applied at p CPU cores, the time T of processing the same size of input data will ideally be reduced by factor p. This will not happen in reality because of the overhead costs to the parallelization: time costs for data transfer among the computing nodes, for requesting the services (DBMS, web-server, scheduler), delays between receiving the data and starting its processing, between the end of processing and the start of the results transfer.

An elementary computational job was defined in [3] as processing of an indivisible (atomic, elementary) input data chunk, such as a string generated from input actual arguments or read from a text file, or a comprehensive search of values at a certain minimal range of the input data. Elementary job cannot be divided into smaller parts and, therefore, cannot be parallelized.

Let an elementary job be done for the time τ, and the entire size of the input data make N elementary chunks. Consequently, at one CPU core the entire size of the input data will be processed for the time $T_1 = N \cdot \tau$. In case of using p CPU cores the ideal time of the processing T_p will be

$$T_p = \frac{T_1}{p} = \frac{N \cdot \tau}{p}$$

Let the researched system process the input data at p CPU cores for the time $T_{exp}(p)$. Then the share of the overhead costs μ introduced by the researched system will make

$$\mu = 1 - \frac{T_p}{T_{exp}(p)} = 1 - \frac{N \cdot \tau}{p \cdot T_{exp}(p)}$$

As we can see, the overhead share depends on the parameters N, τ and p. The values of these parameters were varied in [3] to assess the dynamic pattern of the overhead costs.

Let us emphasize one feature of the experiments carried out in [3]. As already mentioned, X-COM is a framework for parallezation by data. It is up to the user to implement an SSP model in the X-COM framework as server and client control programs. Unlike the X-COM user, the Pyramid user does not have to be a programmer. In [3] the control programs implementing the SSP model for X-COM were developed from the position of an "amateur programmer", without any optimization, to put X-COM and Pyramid in the same comparative conditions. The time spent by the user on preparation for the calculation of the SSP model was approximately the same to compare the software systems. Thus, the overhead estimation for X-COM is relatively qualitative in its nature. In a way, the "amateur" control program for X-COM and the "professional" work dividing procedure in Pyramid were compared in [3].

For the experiments carried out in [3] three typical SSP test cases were used. The Opp_one test simulated the SSP operation with one input parameter, the actual value of which determined the input numerical data range. The Opp_three test simulated the SSP operation with three input parameters and enumerated over all combinations of their actual values. The Opp_file test simulated the processing of strings read from a text file. Each read string was the actual value of an input parameter. The SSP computations were modeled in [3] by the sleep() call, i.e. the test procedure "fell asleep" for a given time, simulating the computations. The md5 hash computed a certain number of times was used in this work instead the sleep() function call.

4   XP-COM Experimental Performance Assessment

The experiments were carried out at the JSCC RAS on the MVS-10P OP supercomputer (Broadwell subsystem). The tests were performed in the share mode under the SUPPZ control. The Broadwell subsystem of MVS-10P OP includes 136 nodes based on Intel Xeon E5-2697Av4 (Broadwell) processor. Each node has 2 processors of 16 cores (32 cores per node) and 128 GB RAM. Summary subsystem peak performance is 182 TFlops. Nodes are united by the Intel Omni-Path interconnect.

The test cases were formed as SUPPZ jobs submitted to the SUPPZ queue. Each time a job was run, the XP-COM server and client container images were downloaded from the local Docker repository, and the XP-COM container instances were launched at the nodes allocated by the SUPPZ. The allocated nodes were cleaned after the job was completed. One XP-COM client was launched per one CPU core.

Pyramid and X-COM test launched in the common job queue mode on the Broadwell subsystem showed that the overhead costs were no less than the results [3] obtained in exclusive access to the cluster. In this regard, the best results from [3] were used to compare the software systems.

Figure 2 shows the XP-COM overhead costs dependence for the test cases Opp_one, Opp_three and Opp_file with a variable cores number, with $N = 100,000$ elementary chunks, with the number of md5 hash calculations per chunk $1,300,000$ times (~ 1 s) and one-time chunk size per client in 98 elementary chunks.

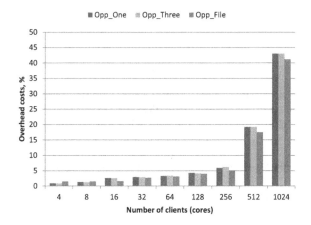

Fig. 2. The XP-COM overhead costs dependence for the test cases

The graph (Fig. 2) demonstrates that increasing the clients' number to 256 leads to a gradual increase in the overhead costs. The overhead does not exceed 6% with the clients number under 256. With a further increase in the clients' number, there is a sharp increase in the overhead costs.

Figure 3 shows a comparative diagram for XP-COM, X-COM and Pyramid overhead costs for the Opp_one test with a variable cores number, with a 100,000 elementary chunks, with the number of md5 hash calculations per chunk 1,300,000 times (~ 1 s). Overhead costs data for X-COM and Pyramid are taken from [3]. As you can see, the hybrid XP-COM software showed better performance.

Figure 4 shows the XP-COM overhead costs dependency graphs for the Opp_one test with a variable processing time of an elementary data chunk (the number of md5 hash calculations), with a data volume of 100,000 elementary chunks and the one-time chunk size per client in 98.

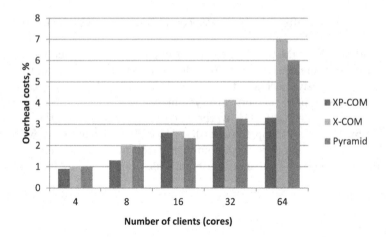

Fig. 3. The XP-COM, X-COM and Pyramid overhead costs for the Opp_one test

The graphs (Fig. 4) show that the overhead costs decrease as the elementary chunk processing time increases, regardless of the computing nodes number. This is primarily due to the fact that the XP-COM server accessing frequency is reduced, which in turn reduces the client wait time after the data portion request.

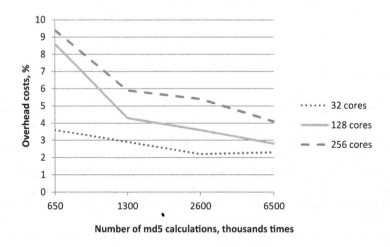

Fig. 4. The XP-COM overhead costs dependency graphs for the Opp_one test

Figure 5 shows the XP-COM, X-COM and Pyramid overhead costs diagram for the Opp_file test with a variable cores number, data volume of 100,000 elementary chunks, and the md5 hash calculations number per chunk 1,300,000 times (~1 s). The X-COM and Pyramid overhead costs data are taken from [3]. It is seen that the XP-COM hybrid software shows the best performance.

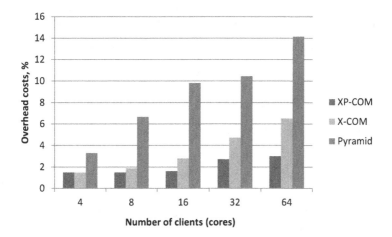

Fig. 5. The XP-COM, X-COM and Pyramid overhead costs for the Opp_file test

Figure 6 shows the XP-COM overhead costs dependency graphs for the Opp_file test with a variable chunk processing time (the md5 hash calculations number), with the data volume of 100,000 elementary chunks and the one-time chunk size per client in 98 elementary chunks.

From the graphs (Fig. 6) we can conclude that the overhead costs decrease with the increase in the elementary chunk processing time regardless of the computing nodes number.

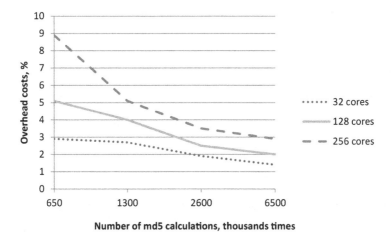

Fig. 6. The XP-COM overhead costs dependency graphs for the Opp_file test

Figure 7 shows the XP-COM, X-COM and Pyramid overhead costs diagram for the Opp_three test with a variable cores number, a data volume of 100,000 elementary chunks, and the md5 hash calculations number per chunk 1,300,000 times (∼1 s). The X-COM and Pyramid overhead costs data are taken from [3]. We can see that the XP-COM hybrid software shows the best performance.

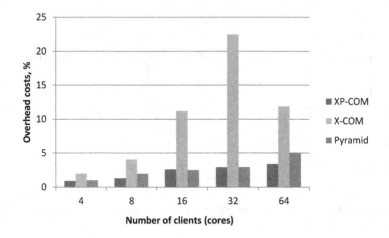

Fig. 7. The XP-COM, X-COM and Pyramid overhead costs diagram for the Opp_three test

Figure 8 shows the XP-COM overhead costs dependency graphs for the Opp_three test with a variable chunk processing time (the md5 hash calculations number), with the data volume of 100,000 elementary chunks and the one-time chunk size per client in 98 elementary chunks.

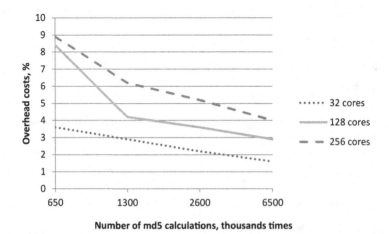

Fig. 8. The XP-COM overhead costs dependency graphs for the Opp_three test

From the graphs (Fig. 8) it can be concluded that the overhead costs decrease with the increase in the elementary chunk processing time regardless of the computing nodes number.

5   Conclusion

To combine the advantages of X-COM and Pyramid software systems for parallelization by data, the XP-COM hybrid system is proposed. The basis of the hybrid software was X-COM data transfer infrastructure. Effective engines for the computational work sharing and the checkpoints saving were borrowed from the Pyramid software. These engines were implemented as separate program modules called from the X-COM management programs.

The overall performance of the developed hybrid software was ahead of both X-COM and Pyramid. However, there was a sharp increase in the overhead costs with more than 256 clients in all tests. The problem may be due to the fact that XP-COM uses a two-rank structure with no intermediate servers, and the X-COM server fails to process all client requests. Clients that do not receive the next data portion in response to a request, send the request again after a timeout. During waiting, the client is idle and is unable to do computations.

In the future, the problem can be solved in one of the following ways:

1. Reducing the number of computing clients. If one XP-COM client on one node is run, the XP-COM server load will significantly decrease. A working client will have to launch several SSPs (in accordance with the job passport). The procedure for starting and monitoring the SSP instances can be implemented with the Divisio_client module.
2. Adding intermediate servers. The most powerful nodes are capable of hosting intermediate servers to accumulate a part of the total work and serve the downstream clients. This approach is directly recommended by the X-COM developers [17]. A tree structure will reduce the main server workload, since client requests will be evenly distributed between several servers. This approach involves making changes to the Divisio module, since the Divisio analogue should operate on each of the intermediate servers. The main server should arrange the computational work issuance to intermediate servers and the checkpoints saving.

Notes and Comments. The work was carried out within the RFBR grant (project No. 18-07-01325 A), the grant agreement with the Ministry of Science and Higher Education of the Russian Federation (No. 075-15-2019-1636), and the framework the JSCC RAS state assignment.

References

1. Tessema, M., Dunren, C.: Survey and taxonomy of volunteer computing. ACM Comput. Surv. **52**(3), 1–35. ACM, New York (2019). https://doi.org/10.1145/3320073

2. Binnie, C.: Password cracking with hashcat. In: Linux Server Security (2016). https://doi.org/10.1002/9781119283096.ch9
3. Baranov, A., Kiselev, E., Chernyaev, D.: Experimental comparison of performance and fault tolerance of software packages Pyramid, X-COM, and BOINC. Commun. Comput. Inf. Sci. **687**, 279–290 (2016). https://doi.org/10.1007/978-3-319-55669-7_22
4. Baranov, A., Kiselev, A., Kiselev, E., Korneev, V., Semenov, D.: The software package Pyramid for parallel computations arrangement with parallelization by data. In: Scientific Services & Internet: Supercomputing Centers and Applications: Proceedings of the International Supercomputing Conference, Novorossiysk, Russia, 20–25 September 2010, pp. 299–302. Publishing of Lomonosov Moscow State University, Moscow (2010). (in Russian)
5. Anderson, D.P.: BOINC: a system for public-resource computing and storage. In: 5th IEEE/ACM International Workshop on Grid Computing, pp. 4–10. IEEE/ACM, Pittsburgh (2004). https://doi.org/10.1109/GRID.2004.14
6. Filamofitsky, M.: The system X-Com for metacomputing support: architecture and tech-nology (in Russian). Numer. Methods Program. **5**, 123–137. Publishing of the Research Computing Center of Lomonosov Moscow State University, Moscow (2004)
7. X-Com: Distributed computing software. http://x-com.parallel.ru/sites/x-com.parallel.ru/files/download/X-Com_tutorial.pdf. Accessed 11 Nov 2019
8. Baranov, A.V., Savin, G.I., Shabanov, B.M., Shitik, A.S., Svadkovskiy, I.A., Telegin, P.N.: Methods of jobs containerization for supercomputer workload managers. Lobachevskii J. Math. **40**(5), 525–534 (2019). https://doi.org/10.1134/S1995080219050020
9. Savin, G.I., Shabanov, B.M., Telegin, P.N., Baranov, A.V.: Joint supercomputer center of the Russian Academy of Sciences: present and future. Lobachevskii J. Math. **40**(11), 1853–1862 (2019). https://doi.org/10.1134/S1995080219110271
10. Voevodin, V., Sobolev, S.: X-COM technology for distributed computing: abilities, problems, perspectives. Mech. Control Inf. IKI Semin. **5**, 183–191 (2011). (in Russian)
11. Khritankov, A.: Performance analysis of distributed computing systems using the X-Com system as an example. In: Scientific Services & Internet: Scalability. Parallelism, Efficiency: Proceedings of the International Supercomputing Conference, Novorossiysk, Russia, 21–26 September 2009, pp. 46–52. Publishing of Lomonosov Moscow State University, Moscow (2009). (in Russian)
12. Huang, S., Huang, J., Dai, J., Xie, T., Huang, B.: The HiBench benchmark suite: characterization of the MapReduce-based data analysis. In: Agrawal, D., Candan, K.S., Li, W.-S. (eds.) New Frontiers in Information and Software as Services. LNBIP, vol. 74, pp. 209–228. Springer, Heidelberg (2011). https://doi.org/10.1007/978-3-642-19294-4_9
13. Zhang, Z., Cherkasova, L., Loo, B.T.: Parameterizable benchmarking framework for designing a MapReduce performance model. Concurrency Comput. Pract. Experience **26**, 2005–2026 (2014). https://doi.org/10.1002/cpe.3229
14. Costa, F., Silva, L., Dahlin, M.: Volunteer cloud computing: MapReduce over the internet. In: 2011 IEEE International Parallel & Distributed Processing Symposium, pp. 1850–1857 (2011). http://hpcolsi.dei.uc.pt/hpcolsi/publications/mapreduce-pcgrid2011.pdf. Accessed 21 Nov 2019
15. Naegele, T.: Mapreduce framework performance comparison (2013). http://www.cs.ru.nl/bachelorscripties/2013/Thomas_Naegele___4031253___MapReduce_Framework_Performance_Comparison.pdf. Accessed 21 Nov 2019

16. Pavlo, A., Paulson, E., Rasin, A., et al.: A comparison of approaches to large-scale data analysis. https://doi.org/10.1145/1559845.1559865. Accessed 21 Nov 2019
17. Sobolev, S.: Hierarchical methods for improving the scalability and efficiency of distributed computing in the X-Com metacomputing system. In: Parallel Computational Technologies (PCT 2010), pp. 346–352 (2010). (in Russian)

Approach to Workload Analysis
of Large HPC Centers

Pavel Shvets⬤, Vadim Voevodin$^{(\boxtimes)}$⬤, and Dmitry Nikitenko⬤

Lomonosov Moscow State University, Leninskie Gory, 1, bld. 4, Moscow, Russia
pavel.shvets.srcc@gmail.com, vadim@parallel.ru, dan@parallel.ru

Abstract. In order to ensure high performance of a large supercomputer facility, its management must collect and analyze information on many different aspects of this complex operation in a timely manner. One of the most important directions in this area is to analyze the supercomputer workload. In particular, it is necessary to detect the subject areas being the most active in terms of consumed node-hours, the application packages using the GPU with the least efficiency and the reasons for that, the frequency of launching multi-process jobs etc. To do this, it is necessary to collect different types of data from various sources, integrate them within a single solution and develop methods for their analysis. In this article, we describe our approach to solving this problem and show real-life examples of results obtained on the Lomonosov-2 supercomputer.

Keywords: High performance computing · Supercomputers · Workload analysis · Job flow · Efficiency analysis · Usage statistics · System software

1 Introduction

Any modern large-scale supercomputer, e.g. from the Top500 list [10] of the most powerful systems in the world, is a highly sophisticated complex containing millions of different software and hardware components. The development of such system's architecture and its installation are very difficult and time-consuming tasks, requiring the participation of many highly qualified specialists from different fields, not to mention the huge financial costs. All these efforts are needed to create a fully functional high performance solution that will help scientists or industry workers to perform computations as quickly as possible, which, in turn, is needed to speed up their research.

Furthermore, it takes a lot of effort to maintain such system, so the developed supercomputer stays fully functional without any performance degradation. Any large-scale supercomputer has a dedicated support team that constantly monitors and controls the state and proper workload of the HPC system. One of the main tasks for this support team is to minimize the amount of idle or unavailable computational resources. Based on this, it is not surprising that there are a lot

© Springer Nature Switzerland AG 2020
L. Sokolinsky and M. Zymbler (Eds.): PCT 2020, CCIS 1263, pp. 16–30, 2020.
https://doi.org/10.1007/978-3-030-55326-5_2

of research aimed at studying the nature of failures that occur in supercomputer components to help system administrators to identify their root causes (e.g. [9]), minimize failure influence [14] and even predict when such failures could occur in the future [5]. To analyze and optimize the efficiency of a particular application, even more research is conducted: a wide spectrum of various performance analysis tools such as profilers, trace analyzers, debuggers etc. have been developed and are still being evolved.

Thus, obviously, significant efforts are made to reach and maintain the sustained supercomputer performance at the possible maximum. In light of this, it is somewhat surprising that another significant aspect of this problem is underconsidered. We are talking about the analysis of the HPC center workload. This aspect is aimed at analyzing the structure and performance statistics of all jobs running on the HPC system, as well as the behavior of different hardware and software components of the supercomputer that can influence the job flow execution. What is the average performance (in terms of the CPU/GPU load, memory usage intensity, etc) of all jobs running on the supercomputer? Which supercomputer partition is least efficiently used and why? In what subject areas the frequency of large jobs is higher and what is their efficiency? Unfortunately, administrators and management of supercomputer centers usually cannot answer most of these questions, perceiving the supercomputer functioning as a "black box". In such situation, the system performance may be easily lost due to the inefficient use of the computational resources caused by the lack of information of the supercomputer job flow behavior.

To solve this problem, the development of an approach to analyzing the HPC system workload has been started. For this purpose, we collect and integrate a variety of input data on the supercomputer functioning and performance, such as job performance data; information on users, projects, scientific areas; data on used packages, compilers, libraries and file system usage data to develop a unified way of studying different aspects of the supercomputer behavior. This work is done by developing a TASC software suite (described in [17]), a software solution created by the authors of this paper to help any HPC user detect performance issues in their jobs, find root causes and determine possible ways to eliminate them. TASC consists of 3 subsystems. The first one automatically performs primary analysis of all jobs to detect possible performance issues and promptly notify users about them. The second subsystem is aimed at the detailed analysis of a particular program, using the information about possible root causes of efficiency degradation collected during primary analysis. The third one, being the goal of the current research described in this paper, is devoted to help system administrators and HPC center management study efficiency of the overall supercomputer functioning.

The main contribution of this paper is the development and implementation of new methods for collection, integration and analysis of the supercomputer workload data to help administrators and management, as well as HPC center users, study its behavior. TASC software is planned to be made open-source, with its source code available via Github in the near future.

The rest of the paper is organized as follows. Section 2 describes current HPC facility we are working with, as well as the works related to the topic of this paper. In Sect. 3 we describe the process of collecting and integrating different input data and show the resulting technological solution. The analysis methods are analyzed and the real-life examples of such performed analysis are presented in Sect. 4. Section 5 provides conclusions and our plans for the near future.

2   Background and Related Work

At the Supercomputer center of Moscow State University, we have 2 main systems: Lomonosov (6200 nodes, 1.7 PFlop/s peak) and Lomonosov-2 (1700 nodes, 5 PFlop/s peak), being #3 and #1 systems in Russia according to the current Top50 list [1] published in September 2019. The overall number of active users is ∼1000, the daily amount of finished jobs ranges around 500–1000. Scientists from all major scientific areas like medicine, physics, chemistry, biology, etc work at our Supercomputer center, leading to a great variety in job size, duration, behavior and execution peculiarities, as well as a big set of different application packages, libraries and compilers being used. This means that the job flow behavior can be very diverse and hard to manually analyze, causing the need for an automated solution for the supercomputer workload studies.

The idea of analyzing the supercomputer workload is not new, so there are a number of research papers devoted to that. Some of them are aimed at deep and insightful analysis of many aspects of using a particular HPC center or system, for example the report on Blue Waters usage [8] or NERSC systems [4]. The authors of these papers do not only study the resource utilization, but also focus on the system performance analysis. Other works are dedicated to less thorough but wider analysis, such as [6,7], which are devoted to the resource utilization and pattern detection in the usage of several different HPC centers in the USA. Another work [18] combines mentioned ideas and provides a deep workload analysis of a variety of different HPC systems. This report helps to answer such questions as: Do jobs tend to use a larger number of cores over time? Are there any significant differences in the memory usage between specific scientific areas? Are there any trends in resource consumption in different scientific areas?

We have found only a few similar works performed in Russia, and they are not as detailed and holistic as the aforementioned ones. One of the most interesting among them is the paper describing a monitoring system for the general utilization analysis of a particular supercomputer [15]. Another related paper is devoted to the research and comparison of several supercomputers performance based on different real-life applications [11].

It should be also noted that there is a significant amount of papers devoted to the study of a particular aspect of HPC system performance like efficiency of resource utilization performed by resource manager, analysis of network activity or GPU usage, but such types of research do not focus on the analysis of workload itself and therefore are not considered here.

All mentioned studies are interesting examples of performed workload analysis, but they can't be directly applied within our HPC center. Firstly, these studies do not cover all aspects of supercomputer behavior we are interested in that can significantly influence its performance. For example, none of these papers address the questions of analyzing performance issues in applications, compiler usage or errors in hardware components. And in our research, we want to cover such aspects as well. Secondly, these related works are focused exclusively on several selected HPC systems and are not intended to be ported.

In such situation, it was decided that a new holistic approach should be developed. But it should be noted that the aforementioned papers helped us to develop our vision on supercomputer workload analysis described in this paper.

3   Collecting and Integrating Data on Supercomputer Workload

3.1   Input Data Sources

Our workload analysis is currently implemented on Lomonosov-2 supercomputer. Prior to this research, we were already collecting primary data on the jobs running on the supercomputer, used in other subsystems of TASC software suite. This primary job data contain information from Slurm resource manager about each job launch (queue/start/finish time, partition name, user account name, number of used nodes, launch command, etc) and general characteristics of each job performance, such as average values (describing CPU load, number of cache misses per second, amount of bytes and packages sent/received per second, GPU load, etc) for the entire duration of the program, collected by the monitoring system. Another source of primary data is a set of derived metrics calculated after the job execution. Using such metrics, for example, the network locality of the job is calculated: how compact it is "located" in the communication network in terms of the number of switches involved. Other derived metrics are used to obtain the name of the used application package (if any), size of MPI packets and memory locality characteristics.

Such data are stored in MongoDB database in a job-centric way, i.e. for each job there is a set of MongoDB documents that describe its properties, contain the results of the job performance analysis, other conclusions or metrics, etc. It was decided to use MongoDB instead of more traditional relational database solutions like PostgreSQL, since storing data in a relational model and working with data through SQL queries is not efficient for our case due to largely heterogeneous data.

Within this research, we need to significantly expand the amount of collected data. To perform a holistic analysis of the supercomputer workload, we want to add and integrate a variety of different input data:

1. more "intellectual" data from other analysis tools,
2. data on projects, organizations, subject areas,
3. information about used compilers, linked libraries,

4. data on file system performance,
5. supercomputer components health data.

Within this research, we implemented collection and integration of data from all these sources, except for the last type. This type of data includes information about failed jobs and root causes of their failures, temperature data, statistics on network or ECC errors, etc. This data is planned to be imported from the Octotron system [19] developed at our Research Computing Center of Lomonosov State University (RCC MSU).

Further, we will describe what specific data are collected and how they are implemented for each source of data.

The first type of data implies information on job performance collected from external job analysis tools. This includes, for example, data on abnormally inefficient applications obtained by anomaly detection tool [16]. Using this machine learning techniques implemented in the tool, we can test each running job for being abnormal, suspicious or normal in terms of resource usage activity. Another important source of such data is a list of performance issues detected for each job using TASC software suite. Examples of such issues are bad memory access locality, inefficient size of MPI packets or low utilization of all available resources. Such data are added to our database in a straightforward manner, i.e. for each job, several integer values need to be stored and updated every time new information arrives during this job execution.

Data on projects, organizations and subject areas as well as their relation with the executed jobs are needed because they allow not only evaluate the efficiency of individual jobs and users, but also assess the activity of more global entities. For example, with these data it is possible to find out the most active scientific areas in terms of node-hours or network usage, the projects using the CPU in the least efficient manner or the users causing the most active use of LAMMPS application package. Such data are especially important since our Supercomputer center provides project-based resource allocation, i.e. computational resources are allocated to projects (each project is a virtual entity devoted to solve a particular scientific task and carried out by a group of users). The source of these data in our case is the Octoshell management system [12] which serves as a single entry point for all users and administrators at our Supercomputer center [13].

To get more insights on the applications running on our supercomputer, we also collect information on the compilers and linked libraries used for each job. For getting such information, XALT tool is used [3]. This tool replaces standard ld and mpirun commands, allowing to obtain necessary data on link and launch steps. Adding these data to the aforementioned database enables us to evaluate the supercomputer job flow behavior from a new perspective. For example, we can find out the most popular compilers in different subject areas or compare the efficiency of the same applications built using different compilers.

Maintaining high and stable I/O performance is typically one of the most challenging tasks for system administrators. In this case, there is no surprise that we are interested in collecting data on the file system performance.

On Lomonosov2, we use Lustre file system without any local disks on the compute nodes; in this case, such data can be obtained in different ways: 1) analyzing network activity (since all I/O operations are performed using Infiniband communication network); 2) reading Lustre/proc files on compute nodes; 3) collecting Lustre information from its servers. We have implemented the first two options using our DiMMon monitoring system [20], collecting such data as the amount of data read/written per second or the number of opened/closed as well as created/deleted files.

3.2 Collected Data Management

Integrated, all these sources of data made up MongoDB database with a quite complex architecture. So, a number of modifications and optimizations had to be implemented in order to make the resulting solution to be easy-to-use and productive. One of the issues were the queries getting too complicated to write. To solve this issue, we tried to combine all collections into one virtual collection using the View database object, but it resulted in slowing down the execution of all queries, since a significant amount of unused data had to be loaded and then discarded. Instead of it, we decided to use MongoDB Aggregation Pipeline framework. Queries that use Aggregation Pipeline can perform necessary transformations on the data, for example, addition, grouping, and fetching data from other collections. This provides a convenient way for data processing on the database side. Using this mechanism, we developed a set of templates to extract and aggregate necessary data for different types of analysis. For example, such template is used to calculate average job activity within an organization or to group all package launches by the number of nodes used.

We have as well performed another optimization to make queries easier to write. Initial way of storing data on projects, organizations and scientific areas in Octoshell system is quite complex, with a lot of entities and relations between them. In order to simplify the date reading process, the data were combined into one collection. This led to data redundancy, but the overall volume of this information is small (megabytes of data), so the query execution speed didn't change.

The implemented database works on a separate server directly connected to the supercomputer. The database contains information about all job launches (data from Slurm resource manager as well as information on projects, organizations and subject areas) executed since the beginning of Lomonosov-2 functioning. Information about used compilers and libraries have been collected using XALT for the last 2 years. Performance data, as well as data from TASC primary analysis and anomaly detection tool, have been collected since the end of last year. The module for collecting file system performance is the most recent one, so this type of data is available only since the end of November 2019. The overall database size is about 1 GB of data. Although the size of database is not big, the complexity of its architecture makes query execution rather time consuming. For example, the query of top application packages by interconnect usage within this year takes about 20 s to complete. But since the proposed

– Statistics on application package usage. It is important to collect the information on how efficiently and actively different packages (like Gromacs, VASP, NAMD, etc) are used. Some of the important questions to be answered here are: what packages use GPUs in the most efficient way? With what intensity do the applications using GROMACS package load the Infiniband communication network? Are there any packages that have been used much less within the last year? For this purpose, the following information should be provided (represented both as statistics for a chosen time period and dynamics of change over time):
  • comparison of package popularity (changes in the number of unique users and launches as well as spent node-hours);
  • frequency the performance issues found by TASC primary analysis (see [17]), among the jobs that use certain packages;
  • comparison of average performance characteristics (CPU or GPU load, network usage intensity, etc.) between packages.
– Statistics on job behavior types. There are many job execution peculiarities that make it possible to determine specific program features concerning its performance or behavior. Using such peculiarities, we can detect, for example, the applications with lower memory access locality, programs well suited for supercomputer execution, suspiciously inefficient GPU-based jobs, abnormally inefficient applications etc. To analyze such information, it is proposed to use the same data presentation formats as described for package analysis.
– Comparison of large and small jobs. The comparison of large jobs that use a significant proportion of the available computational resources with all other jobs is of particular interest. In this case, it is proposed to divide jobs based on a number of processes (more and less than N processes). It can be done to evaluate the difference in behavior and efficiency of the large jobs, e.g. to determine what projects execute large jobs with greater frequency and efficiency. At this point, it is proposed to analyze the following data:
  • the ratio of large and small jobs in terms of the number of launches, node-hours, integral values of dynamic characteristics, amount of performance issues detected by TASC primary analysis;
  • distribution of the shares of large jobs by used application packages, projects, scientific areas or by behavior types.
– Statistics based on the results of the primary analysis. Here, the most complete information should be provided on the performance issues discovered in primary analysis practice. In this case, it is proposed to analyze the same data presentation formats as in the case of package analysis. This information can answer, for example, which scientific area performs the largest percentage of jobs with detected critical performance issues (indicating incorrect job launches)? It is also worth showing the feedback data from users, i.e. information about the detected issues confirmed (or rejected) by users.

The main goal of the other slice devoted to work activity analysis is to show information about the activity and overall performance of the main entities using the supercomputer, let those be individual users, projects or organizations. Information provided in this slice should cover, for example, the following: users of which scientific area use Gaussian application package most actively, what is the distribution of file system usage between projects within last month, what organization is the most resource consuming in the astronomy scientific area, what compilers are the most popular in physics area projects.

One of the possible ways to provide such information is to make a general table with each row corresponding to a particular user (project, organization) and each column showing the value of a metric needed to evaluate. Among such metrics, there are: number of job launches, amount of node-hours consumed, share of jobs of a particular behavior type, amount of performance issues detected by TASC primary analysis, average values of performance characteristics. Sorting by a chosen column allow to assess top users (or other entities) as well as distribution based on the selected metric. It should be noted that analysis performed within this slice is quite similar to the analysis of package usage, so another way of this data study is to provide the same presentation formats as described for package analysis.

In order to provide access to the results of using described methods on real data, a web interface is developed. This interface is built using Redash [2] — business intelligence tool that eases the process of connecting and querying the database as well as visualizing the collected data. Within this interface, it is planned to provide a number of dashboards with a static set of graphs, one (or several) for each slice described in the previous subsection. Each dashboard will have a number of filters and options to change the considered period of time, choose a package or a scientific area of interest or to select the desired performance characteristics to configure the graphs for current needs.

4.2   Real-Life Examples of Workload Analysis

This subsection will provide several real-life examples of analysis we can already perform and that will be available within the two aforementioned slices. All these results were achieved using Lomonosov-2 supercomputer.

Modern large-scale supercomputers allow executing very large jobs, in terms of the number of nodes used, but how often are large jobs executed in practice? This is the question the HPC center management wants to answer. Figure 1 shows the distribution of the share of the node-hours spent on various subject areas, collected on Lomonosov-2 for the first half of the year 2019. From the Figure, it can be concluded that the average job size can vary greatly for different areas. For example, in "Fundamentals of engineering" there are hardly any small jobs executed: the share of jobs using 1–3 nodes is just 0.5%, while the share of large jobs covers more than 20% of all node-hours. This is typically considered as a desirable distribution of supercomputer jobs, since there are other, usually more appropriate solutions for running small jobs like clouds or separate servers.

On the contrary, the area of "Mathematics, computer science, mechanics" shows ~20% of small jobs with 1–3 nodes and no jobs using 64+ nodes. It is clear that this behavior can be caused by various reasons, but for system administrators it means that there may be some obstacles that prevent users from running larger jobs (applications not scaling well, users do not know how to run packages on more nodes, etc). So, the administrators need to do a further investigation.

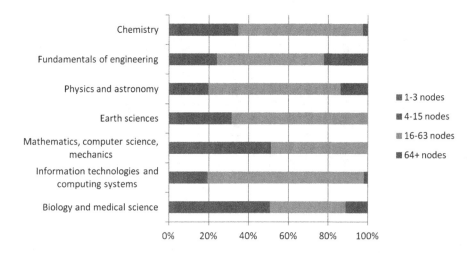

Fig. 1. Distribution of the share of node-hours spent on various subject areas

Talking about HPC center workload analysis, another important topic to study is the application package usage efficiency. Efficiency can be interpreted in different ways, with one of the possible approaches being to analyze the data provided by our anomaly detection tool. This tool performs online classification of all jobs running on Lomonosov-2 supercomputer into abnormal (job runs incorrectly or hung), normal (no performance problems detected) or suspicious (job seems to have significant performance problems). Figure 2 compares the ratio of node-hours spent on jobs from different classes, for each package. The data were collected on Lomonosov-2 supercomputer during several months in the end of the year 2018. The package marked as "Unknown" represents jobs that were not identified as package launches (usually these are proprietary programs developed by users themselves). The purple line shows the absolute amount of node-hours spent on each package, in the log scale.

It can be noticed that the most popular packages (in the left part of the Figure) have lower share of abnormal and suspicious jobs, which means that they show quite high efficiency. One the best results is obtained by the LAMMPS package with almost no abnormal launches with this package (red part is hardly visible), while the share of normal jobs (blue part) is very high, exceeding 90%. For CABARET, the situation is quite the opposite, with the most of the node-hours spent on suspicious jobs and a large share of abnormal jobs. This undesirable behavior of CABARET may be caused by its peculiarities that prevent a more efficient use of this package, but in any case it is another reason for system administrators to pay close attention to the use of this package.

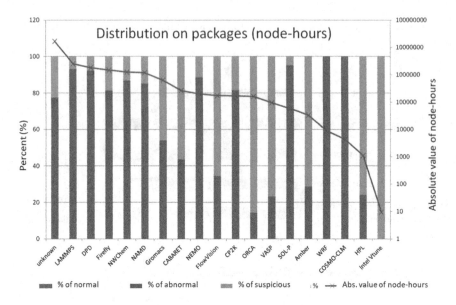

Fig. 2. Ratio of node-hours spent on jobs from different classes determined by anomaly detection tool, for each application package (Color figure online)

Since Lomonosov-2 supercomputer has a lot of GPUs installed, 1–2 on each compute node, it is vital to us to evaluate the efficiency of their use. An example of the graph enabling us to do so, developed with Redash-based web interface, is shown in Fig. 3. Here we can see the normalized distribution of resource consumption based on the GPU usage between scientific areas, only for jobs that utilize GPU in any way. For each area, we show the percentage of node-hours consumed this year by Lomonosov-2 jobs with the average GPU load within a certain range. It can be seen that in different areas the distribution varies greatly. For example, Biology and medical science jobs usually utilize GPUs very efficiently, with only 10% of the node-hours spent on jobs with less than 10% GPU load, while almost half of node-hours falls on the jobs with the average GPU load exceeding 50%. A slightly worse situation is seen for "Chemistry" or "Physics and astronomy", but generally it is preferable. The worst case is with "Other" area (projects marked as uncategorized by users), since all the jobs here showed a very low GPU load; but it should be noted that jobs from this area consume a very small fraction of the total node-hours. Analyzing such information for areas or application packages help us understand the efficiency of GPU resources use within the considered period of time.

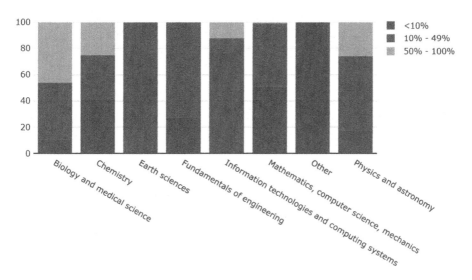

Fig. 3. Normalized distribution of resource consumption based on GPU usage between scientific areas. Includes only the jobs using GPU in any way

Talking about workload analysis, it is not only the supercomputer workload at a specific point in time or within a time period that needs to be analyzed; the changes that occur over time also need to be considered. Figure 4 shows the changes in the amount of node-hours spent on Lomonosov-2 jobs with different execution time since the beginning of this year. Integer values of axis X represent month serial number, each line is related to jobs of a certain execution time range, with duration under 1 h ("<1h"), from 1 h to 8 h ("1h–8h"), from 8 h to 1 day ("8h–1d"), from 1 day to 3 days ("1d–3d"), and over 3 days ("3d+"). We see that the longest jobs consume the biggest amount of supercomputer time, which is typically a good sign for an HPC center since its management is usually more interested in running resource consuming applications and not small jobs that could be run on a server or in a cloud. This situation changed only in October and November, when the amount of node-hours spent on jobs running for over 3 days significantly decreased, while the node-hour value for jobs running for 1–3 days increased. The reason is simple: the job time limit on Lomonosov-2 was changed from 7 days to 3 days in October. The rest of the jobs consumed noticeably less node-hours, showing that small jobs occupy only a small share of the Lomonosov-2 supercomputer resources.

4.3   Adjacent View of Proposed Analysis Methods

The proposed analysis levels and examples above have been originally developed for the needs of management and administration of a large HPC center. At the same time, efficient functioning of any supercomputer greatly depends on appropriate user activity coming from efficient coding for the target platform, reasonable batch queuing and so on.

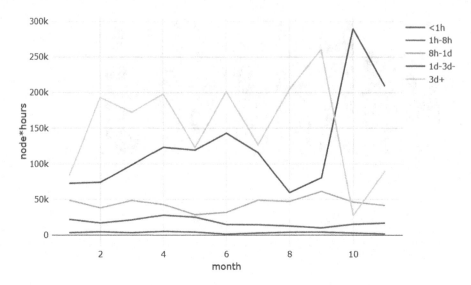

Fig. 4. Timeline of the amount of node-hours spent on jobs with different execution time

The techniques described above can contribute a lot to the work of regular users, with one of the key questions being to reduce the amount of available information to a still sufficient dataset, represented in a clear way to provide end users just as much data on their jobs and activities as it is needed to avoid confusion caused by the exhaustive amount of data. Some methods of tackling these questions have been proposed recently [12]. In this, a set of complementary multilevel pie charts representing various job distributions by finish states and consumed resources as an extension to regular table data presentation is implemented. Another example is using timeline statistics to analyze the queuing dynamics and workload uniformity.

Another question concerns the ways of modifying the proposed levels of analysis when analyzing smaller machines with specific workload. For example, a lab supercomputer aimed at chemistry domain tasks would run mostly jobs with specific algorithmic kernels, induced by the task domain. This severely influences the scalability of applications, their run duration etc.

Of course, all these factors make an impact on the workflow of the end users, so we believe that system holders should work such cases out. The influence made by such workload patterns is a special subject for further research.

5   Conclusions and Future Work

This paper describes our approach to the holistic workload analysis of large-scale supercomputers. Within this approach, we combine data from multiple sources describing HPC system behavior, which enables us to study various aspects of supercomputer functioning. Based on our longstanding experience of maintaining and ensuring efficiency of large-scale HPC center, we have developed a set of 7 slices, each of which devoted to perform insightful analysis of a particular aspect. Within this paper, we mostly focus on two slices aimed at analysis of supercomputer job flow structure and study of work activity. Several real-life examples showing types of analysis available within these slices are described.

Our main plans for future are as follows. We will add collection and integration of supercomputer components health data to correlate job performance issues with failures occurring within the supercomputer. Another scope of study is the software implementation of other slices described in this paper. Finally, we plan to make TASC software open-source and publicly available.

Acknowledgments. The results described in this paper, except for examples shown in Sect. 4.2, were achieved at Lomonosov Moscow State University with the financial support of the Russian Science Foundation (agreement No. 17-71-20114). The results described in Sect. 4.2 were obtained with the financial support of the grant from the Russian Federation President's Fund (MK 2330.2019.9). The research is carried out using the equipment of shared research facilities of HPC computing resources at Lomonosov Moscow State University.

References

1. Current rating of the 50 most powerful supercomputers in CIS. http://top50.supercomputers.ru/?page=rating
2. Redash homepage. https://redash.io/
3. Agrawal, K., Fahey, M.R., McLay, R., James, D.: User environment tracking and problem detection with XALT. In: 2014 First International Workshop on HPC User Support Tools, pp. 32–40. IEEE (2014). https://doi.org/10.1109/HUST.2014.6
4. Brian, A., et al.: 2014 NERSC workload analysis. Technical report (2015)
5. Das, A., Mueller, F., Hargrove, P., Roman, E., Baden, S.: Doomsday: predicting which node will fail when on supercomputers. In: Proceedings of the International Conference for High Performance Computing, Networking, Storage, and Analysis, p. 9. IEEE Press (2018)
6. Hart, D.L.: Measuring TeraGrid: workload characterization for a high-performance computing federation. Int. J. High Performance Comput. Appl. **25**(4), 451–465 (2011)
7. Hart, D.L.: Longitudinal user and usage patterns in the XSEDE user community. In: Proceedings of the 1st Conference of the Extreme Science and Engineering Discovery Environment: Bridging from the eXtreme to the Campus and Beyond, p. 53. ACM (2012)
8. Jones, M.D., et al.: Workload Analysis of Blue Waters (2017)
9. Martino, C.D., Kalbarczyk, Z., Iyer, R.K., Baccanico, F., Fullop, J., Kramer, W.: Lessons learned from the analysis of system failures at Petascale: the case of blue waters. In: 2014 44th Annual IEEE/IFIP International Conference on Dependable Systems and Networks, pp. 610–621. IEEE (2014). https://doi.org/10.1109/DSN.2014.62
10. Meuer, H., Strohmaier, E., Dongarra, J., Simon, H.D.: Top500 supercomputer sites. In: Proceedings of SC, pp. 10–16 (2001)

11. Moskovskii, A.A., Perminov, M.P., Sokolinskii, L.B., Cherepennikov, V.V., Shamakina, A.V.: Research performance family supercomputers SKIF Aurora on industrial problems (in Russian). Vestnik Yuzhno-Ural'skogo Universiteta. Seriya Matematicheskoe Modelirovanie i Programmirovanie **6**, 66–78 (2010)
12. Nikitenko, D., Zhumatiy, S., Paokin, A., Voevodin, V., Voevodin, V.: Evolution of the octoshell HPC center management system. In: Sokolinsky, L., Zymbler, M. (eds.) PCT 2019. CCIS, vol. 1063, pp. 19–33. Springer, Cham (2019). https://doi.org/10.1007/978-3-030-28163-2_2
13. Nikitenko, D.A., Voevodin, V.V., Zhumatiy, S.A.: Driving a petascale HPC center with octoshell management system. Lobachevskii J. Math. **40**(11), 1817–1830 (2019). https://doi.org/10.1134/S1995080219110192
14. Oliner, A., Rudolph, L., Sahoo, R., Moreira, J., Gupta, M.: Probabilistic QoS guarantees for supercomputing systems. In: 2005 International Conference on Dependable Systems and Networks (DSN 2005), pp. 634–643. IEEE (2005). https://doi.org/10.1109/DSN.2005.80
15. Safonov, A., Kostenetskiy, P., Borodulin, K., Melehin, F.: SUSU supercomputer system monitoring system (in Russian). Russian Supercomputing Days, pp. 662–666 (2015)
16. Shaykhislamov, D., Voevodin, V.: An approach for dynamic detection of inefficient supercomputer applications. Procedia Comput. Sci. **136**, 35–43 (2018)
17. Shvets, P., Voevodin, V., Zhumatiy, S.: HPC software for massive analysis of the parallel efficiency of applications. In: Sokolinsky, L., Zymbler, M. (eds.) PCT 2019. CCIS, vol. 1063, pp. 3–18. Springer, Cham (2019). https://doi.org/10.1007/978-3-030-28163-2_1
18. Simakov, N.A., et al.: A Workload Analysis of NSF's Innovative HPC Resources Using XDMoD, p. 93 (2018)
19. Sobolev, S.I., et al.: Evaluation of the Octotron system on the Lomonosov-2 supercomputer. In: Parallel Computational Technologies (PCT) 2018: Proceedings of International Scientific Conference (2–6 April 2018, Rostov-on-Don), pp. 176–184 (2018)
20. Stefanov, K., Voevodin, V., Zhumatiy, S., Voevodin, V.: Dynamically reconfigurable distributed modular monitoring system for supercomputers (DiMMon). Procedia Comput. Sci. **66**, 625–634 (2015). https://doi.org/10.1016/j.procs.2015.11.071

Parallelization of Algorithms in Set@l Language of Architecture-Independent Programming

Ilya I. Levin[1] , Alexey I. Dordopulo[2] , Ivan V. Pisarenko[2(✉)] ,
and Andrey K. Melnikov[3]

[1] Southern Federal University, Academy for Engineering and Technology,
Institute of Computer Technologies and Information Security, Taganrog, Russia
`iilevin@sfedu.ru`
[2] Supercomputers and Neurocomputers Research Center, Taganrog, Russia
`{dordopulo,pisarenko}@superevm.ru`
[3] InformInvestGroup CJSC, Moscow, Russia
`ak@iigroup.ru`

Abstract. Traditional programming languages for parallel computer systems do not efficiently separate the algorithm description from the details of its hardware implementation. As a result, a significant code modification is required to port the same algorithm from one computational architecture to another. To simplify and speed up the porting procedure, an architecture-independent Set@l programming language based on the aspect-oriented programming paradigm and set-theoretical code view is proposed to be used. In contrast to conventional parallel programming tools, Set@l defines the information structure of a problem as sets, subsets of different types and relations between them. Various aspects of implementation such as processing method, parallelization, optimization and others transform an architecture-independent source code according to the certain architecture and configuration of the computer system. Set@l offers new opportunities for efficient architecture- and resource-independent parallel programming. It ensures porting of parallel applications without the source code modification, transitions between different algorithm implementations with regard to the computer system's features or user's preferences, and the indefinite description of calculations. This paper deals with the essential techniques and specificities of algorithm parallelization in the Set@l aspect-oriented programs. The demonstrated examples are adopted from programs in the Set@l language. The lower-upper decomposition code demonstrates the conversion of set types by parallelism during the architectural adaptation, and the Jacobi algorithm code introduces indefinite collections (classes and semisets) for the description of effective computing structure for reconfigurable computer systems. In addition, we consider several instances of memory use for the organization of parallel calculations.

Keywords: Software porting · Architecture-independent parallel programming · Aspect-oriented approach · Set@l programming language · Algorithm parallelization

© Springer Nature Switzerland AG 2020
L. Sokolinsky and M. Zymbler (Eds.): PCT 2020, CCIS 1263, pp. 31–45, 2020.
https://doi.org/10.1007/978-3-030-55326-5_3

1   Introduction

The design of heterogeneous, hybrid and reconfigurable computer systems, containing processors as well as other types of computing devices [1], is an urgent scope of studies in the field of supercomputer engineering. The variety of architectures used in hybrid computer systems and lack of efficient methods and tools for architecture-independent parallel programming considerably complicate the process of software porting. In the conventional programming languages, the computational problem solution algorithm and features of its implementation are described by indivisible code fragments. As a result, the change of the parallelizing method caused by porting a program between computer systems with different architectures or configurations requires the development of a new code. Traditional approaches to the interarchitecture porting problem have significant disadvantages: they are based on the specialized translation algorithms (e.g. the Pyfagor language of functional programming [2]) or fix the procedural parallelization model (e.g. the OpenCL (Open Computing Language) standard [3]).

The high-level COLAMO (Common-Oriented Language for Architecture of Multi-Objects) programming language [4] solves the majority of problems related to the programming of FPGA-based reconfigurable computer systems. In COLAMO, algorithm parallelization is described implicitly by the declaration of access types for arrays and indexation of their elements. However, COLAMO is aimed at the structural and procedural organization of calculations, and porting a parallel application in COLAMO to computer systems with different architectures seems to be very problematic.

To solve the aforementioned problem, we propose an architecture-independent Set@l programming language that develops the essential ideas of COLAMO and the set-theory-based programming language SETL (SET Language) [5]. Set@l is based on a paradigm of aspect-oriented programming (AOP) [6] and describes an algorithm and its implementation features as separate program modules. A program in the Set@l language represents the information structure of a computational problem as sets, subsets and relations between them. The decomposition and typing of collections define different variants of parallelization and other aspects of the algorithm implementation. In contrast to other set-theory-based programming languages (e.g. SETL), Set@l classifies sets by various criteria and operates indefinite collections according to the alternative set theory (AST) of P. Vopenka [7].

This paper focuses on the features and techniques of algorithm parallelization in Set@l using classification of collections by parallelism and definiteness.

2   Description of Algorithm Parallelization in Set@l

For the implicit declaration of algorithm parallelization, the collections classified according to the parallelism of their elements during processing are introduced into the Set@l programming language. The basic parallelism types of collections and format of their description in Set@l are given in Table 1.

Table 1. Key types of collections classified by parallelism and formats of their description in Set@l

Type of collection	Processing type	Symbolic notation	Format of description
Tuple	Sequential	$[1, 2, \ldots, p]$	`seq(1...p)`
Pipeline tuple	Pipeline	$\langle 1, 2, \ldots, p \rangle$	`pipe(1...p)`
Set	Parallel-independent	$\{1, 2, \ldots, p\}$	`par(1...p)`
Set of processing by iteration	Parallel-dependent	$\overrightarrow{\{1, 2, \ldots, p\}}$	`conc(1...p)`
Implicit collection	Type is defined in other aspect	$[[1, 2, \ldots, p]]$	`imp(1...p)`

If the method for parallelization of the collection elements is clearly defined, the following types by parallelism are used: "tuple" (`seq` – sequential processing), "pipeline tuple" (`pipe` – pipeline processing), "set" (`par` – parallel-independent processing) and "set of processing by iterations" (`conc` – parallel-dependent processing).

However, in some aspects it is impossible to specify the exact type of collections by parallelism because any information about the architecture of computer system is not available. In this case, a special type `imp` (implicit or undefined) is applied. The typing of `imp` collections is defined in the architectural aspect by means of the following syntax structure:

```
type(<name of collection>)='<type of collection>';
```

Each collection of the Set@l program has the only parallelism attribute, but it can be changed when passing from one abstraction level to another.

Consider the information graph G of a computational problem that includes six operational vertices V. In the architecture-independent Set@l source code, the graph is declared as an implicit collection because at this abstraction level we do not know the architecture and configuration of a computer system and, consequently, the parallelism of vertices:

$$G = [[V_1, V_2, V_3, V_4, V_5, V_6]]. \tag{1}$$

The aspects of the Set@l program describe the parallelization method through the partition of graph G and specification of parallelism types. Several examples of G decomposition and typing for the reconfigurable architecture of a computer system are shown in Fig. 1. If G is a parallel-dependent collection without subsets, there are six hardwarily implemented vertices V in the computing structure (see Fig. 1-a). When G is divided into two subsets, and each subset is a parallel-dependent collection with three operational vertices V, we have three hardwarily realized vertices V, and data flow $\langle D_1, D_2 \rangle$ is organized (see Fig. 1-b). If G is a pipeline collection without subsets, only one operational vertex V is hardwarily implemented, and data flow $\langle D_1, D_2, D_3, D_4, D_5, D_6 \rangle$ is organized (see Fig. 1-c).

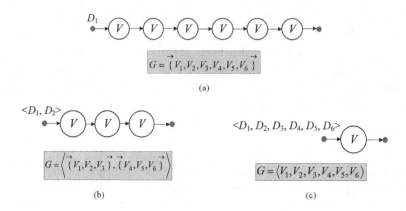

(a)

(b)   (c)

Fig. 1. Application of collection types by parallelism and set partition for the description of algorithm parallelization in the case of the reconfigurable architecture: parallel dependent implementation (a); pipeline implementation with three parallel-dependent hardware vertices (b); pipeline implementation with one hardware vertex (c)

The proposed classification of collections by parallelism provides the architectural independence of the source code in Set@l and makes it possible to describe various parallelizing methods for an algorithm as an entire program. To switch between implementations, the user activates the corresponding processing method, architectural and configuration aspects, while the source code of an algorithm remains unchanged.

3   Program of Lower-Upper Decomposition in Set@l

Lower-upper (LU) decomposition is a well-known computational algorithm for the solution of linear equation systems, which represents a $n \times n$ matrix as the product of a lower and an upper triangular matrix.

The source code of the LU-decomposition program in the Set@l language is shown in Fig. 2. It declares the problem's information graph G and describes its operational vertices in terms of sets, attributes and relations between them.

The previously represented Set@l program of the Gaussian elimination [8] used the partition and typing of the sets of row (I), column (J) and iteration (K) numbers for the specification of the algorithm parallelism. By contrast, the aforementioned code operates with special collection G that includes operational vertices of the information graph.

This approach allows considering the membership relations between sets and their elements in the traditional mathematical sense.

```
int(n);                // size of matrix;
set(I,J,K);            // sets of row, column and iteration numbers;
I=1...n;   J=1...n;   K=1...n-1;
set(a);                // matrix;
{set,graph,imp}(G);    // information graph;

attribute LU_it(k,s1,s2):              // attribute of LU iteration;
  operand(element(k),set(s1),set(s2)); // operand types;
  (forall i in s1 and j in s2 | i<=k or (i>k and j<k)):
    a(k+1,i,j)=a(k,i,j);               // transit elements;
  end(forall);
  (forall i in s1 | i>k):              // recalculated elements;
    a(k+1,i,k)=a(k,i,k)/a(k,k,k);
    (forall j in s2 | j>k):
      a(k+1,i,j)=a(k,i,j)-a(k+1,i,k)*a(k,k,j);
    end(forall);
  end(forall);
end(LU_it);

(forall k in K):                       // graph description;
  G(k)=LU_it(k,I,J);                   // attribute assignment;
end(forall);
```

Fig. 2. The source code of the LU-decomposition program in Set@l

Within the source code, the parallelism type and partition of set G are unknown. Therefore, the special parallelism type imp (implicit) is applied.

The aspect of processing method for the LU-decomposition program in the Set@l language is given in Fig. 3. This component of the program determines the partition of the initial information graph G into subgraphs described by subsets with unknown type of parallelism.

```
// Decomposition by rows:
(forall k in K):
  R(k,i)=(G(k,i,j)|j in J);          // i-th row at k-th iteration;
  RB(k,p)=imp(R(k,i)|i in ((p-1)*s+1...p*s)); //p-th row subset;
  GR(k)=imp(RB(k,p)|p in (1...N));           // row-built subgraph;
end(forall);

// Decomposition by columns:
(forall k in K):
  C(k,j)=(G(k,i,j)|i in I);     // j-th column at k-th iteration;
  CB(k,q)=imp(C(k,j)|j in ((q-1)*c+1...q*c)); // column subset;
  GC(k)=imp(CB(k,q)|q in (1...M));      // column-built subgraph;
end(forall);

// Decomposition by iterations:
It(k)=(G(k,i,j)|i in I and j in J);          // k-th iteration;
IB(l)=imp(It(k)|k in ((l-1)*ni+1...l*ni));  // iteration subset;
G=imp(IB(l)|l in (1...T));          // iteration-built graph;
```

Fig. 3. The aspect of processing method for the LU-decomposition program in Set@l

The aspect of processing method defines the partition of collection G into the row (RB(k,p)), column (CB(k,q)) and iteration (IB(l)) blocks with implicit type of parallelism. Each block consists of rows R(k,i), columns C(k,j) or iterations It(k). These partitions allow describing the processing methods by rows, columns, cells and iterations [8] applied for the parallelizing of the linear algebra algorithms. The values of decomposition parameters s, N, c, M, ni and T depend on the computer system architecture and are declared in the configuration aspect.

The aspect of architecture for the LU-decomposition program in Set@l specifies the parallelism types of the basic set G and its subsets GR(k), GC(k), RB(k,p), CB(k,q) and IB(l). The code of the architectural aspect for the reconfigurable architecture of computer system is given in Fig. 4.

```
// Rows:
s=n;                         // number of rows in subset RB(k,p);
N=1;                         // number of row subsets;
type(RB(k,p))='pipe';        // pipeline processing of rows;
type(GR(k))='pipe';

// Columns:
c=n;                         // number of columns in subset CB(k,q);
M=1;                         // number of colum subsets;
type(CB(k,q))='pipe';        // pipeline processing of columns;
type(GC(k))='pipe';

// Iterations:
ni=floor(R/R0);              // number of iterations in subset IB(l);
T=(n-1)/ni;                  // number of iteration subsets;
type(IB(l))='conc';          // parallel-dependent processing of
                             // iterations in each subset;
type(G)='pipe';              // G typing;
```

Fig. 4. The aspect of architecture for the LU-decomposition program in Set@l designed for the reconfigurable architecture of computer system

According to the code represented above, the algorithm of LU-decomposition is parallelized by iterations in the case of implementation on a reconfigurable computer system. In the architectural aspect, the characteristics of set partitions are calculated using the parameters of configuration (R and R0) and the size of processing matrix (n).

If the algorithm of LU-decomposition is realized on a multiprocessor computer system, the architectural aspect is given as in Fig. 5. In contrast to a reconfigurable computer system, the implementation of the LU-decomposition algorithm on a multiprocessor supercomputer assumes the parallelization by cells. Analogously to the previous version of the architectural aspect, the decomposition of sets depends on the configuration parameters (q1 and q2) and the matrix size (n).

The aspect of computer system's configuration substitutes the specific values of configuration parameters R (available computing resource) and R0 (resource of basic subgraph) for the reconfigurable architecture or q1 and q2 (numbers of processors by rows and by columns) for the multiprocessor architecture into other modules of the program and completes the forming of graph set G.

Owing to the parallelism typing of collections in the Set@l programming language, it is possible to describe different variants of the algorithm parallelizing in a single aspect-oriented program.To switch the method of implementation during the inter-architecture porting, one should change the architectural aspect of a program, but the source coderemains unchanged.

```
// Rows:
s=q1;                     // number of rows in subset RB(k,p);
N=n/s;                    // number of row subsets;
type(RB(k,p))='par';      // parallel-independent processing of rows
                          // in each subset;
type(GR(k))='seq';        // sequential processing of subsets;

// Columns:
c=q2;                     // number of columns in subset CB(k,q);
M=n/c;                    // number of column subsets;
type(CB(k,q))='par';      // parallel-independent processing of
                          // columns in each subset;
type(GC(k))='seq';        // sequential processing of subsets;

// Iterations:
ni=n-1;                   // number of iterations in subset BI(k);
T=1;                      // number of iteration subsets;
type(IB(l))='seq';        // sequential processing of iterations;
type(G)='seq';            // G typing;
```

Fig. 5. The aspect of architecture for the LU-decomposition program in Set@l designed for the reconfigurable architecture of computer system

4   Indefinite Collections in Set@l

If the aspects do not modify an algorithm during its architectural adaptation, the computational problem can be solved within the Cantor–Bolzano set theory [9]. However, the functionality of the aspects is not limited to the algorithm parallelization. In some cases, it is reasonable to modify an algorithm according to the architectural features of the computer system used for calculations. Then some collections are indefinite and are not classified as sets; in that case, it is impossible to describe them using the concepts of the traditional set theory.

The architecture-independent Set@l programming language for high-performance computer systems describes various implementations of an algorithm in a unified aspect-oriented program. For this purpose, the classification of collections by the definiteness of their elements is introduced into Set@l (see Table 2) [10].

In the Set@l programming language, indefinite collections are described by special mathematical objects (classes and semisets).The class and semiset concepts were formulated by P. Vopenka within the AST [7].

The type "set" (**set**) describes a sharply defined and definite collection of certain objects. For a set, we always exactly know if one or another object belongs

Table 2. Collections' classification by definiteness of their elements in Set@l

Type of collection	Description	Symbolic notation	Keyword
Set	Sharply defined collection of elements	{ }	set
Semiset	Indefinite collection of elements	{ ? ? } {? i}	sm
Class	Collection with the type that is not defined unambiguously	? i	cls

to it or not. In Set@l, a set can be specified using the direct enumeration of its elements or by means of the relation calculus. Relational structures have the following format:

```
<name of set>=<type>(<variable>|<predicate>);
```

Sets of numbers forming a numerical sequence with the fixed step size are defined as follows:

```
<name>=<type>(<1st element>,<2nd element>, ..., <last element>);
```

If the step size equals to 1, only the first and the last elements of a numerical sequence are mentioned:

```
<name of set>=<type>(<1st element>...<last element>);
```

In fact, plenty of naturally organized collections are not sets, because their elements are not clearly defined. The AST analyses the phenomenon of indefinite collections with the help of special mathematical objects (classes and semisets).

The type "semiset" (sm) indicates a collection whose indefiniteness is a fundamental characteristic and can not be eliminated by the aspects of a program. The relation of inclusion usually connects a semiset with some sharply defined set. Therefore, to specify a semiset as an object of the Set@l program, it is necessary to form a suitable superset and declare the appropriate relation of inclusion. If the algorithm for the solution of the system of linear algebraic equations (SLAE) by the Jacobi method is implemented with one verification of the termination condition after several computational iterations, the collection of iterations represents the example of a semiset.

A class (cls) is the most common and multipurpose type of collections in the Set@l programming language. If the type and structure of a collection are not sharply defined on the current level of abstraction, it is declared as a class and is used in code analogously to standard sets. Owing to the extension of the class definition in the aspects of a program, it is possible to specialize the type and partition of the collection during translation. The application of classes provides the unification of objects' names in all units of an aspect-oriented program in Set@l. The indefiniteness of collections by parallelism denoted by `imp` attribute (see Table 2) can also be described with the help of classes. To specify a collection as a class in the Set@l program, one has to assign `cls` attribute to this collection and give its possible attributes using the following syntax construction:

```
cls(<name of class>);
typing(<name of class>):'<type 1>' or '<type 2>';
```

The collections of rows, columns and iterations I, J, and K (see Fig. 1) are the examples of classes because in the source code they are considered as objects with unknown types and decompositions.

Using classes, sets and semisets, one can describe various modifications methods for an algorithm in a unified aspect-oriented Set@l program.

5   Program of Jacobi Method in Set@l

The basic parallelization techniques for the computer-aided solution of linear equation systems by the Jacobi iterative method are shown in Fig. 6.

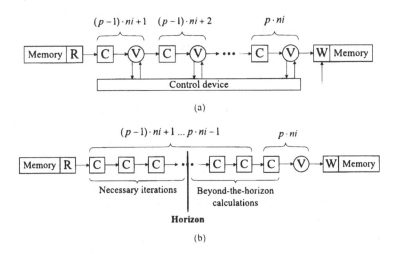

Fig. 6. Parallel implementations of the Jacobi algorithm for the solution of linear equation systems: with verification during each iteration (a) and with one verification after several computational iterations (b)

In the case shown in Fig. 6-a, each iteration of the Jacobi algorithm contains the following operations: calculation of the column-vector of unknown variables (block C); the verification of the termination condition (block V) given by $err(k) \leq \delta$, where err is the residual; k is the number of iteration; δ is the fixed value of tolerance. If the condition is true, the control device transfers data via the untapped blocks C and V and saves the result into a specially allocated distributed memory area. In practice, the considered approach is efficient, but not for all computational architectures. Each verification block V performs the resource-intensive and time-consuming operation of $err(k)$ calculation. The hardware resource required for the implementation of V block is comparable with C, and time costs are equivalent too.

For better hardware usage efficiency and higher operation speed, it is reasonable to modify the Jacobi algorithm in case of the reconfigurable architecture. This modification assumes single verification of the termination condition in a cadr (see Fig. 6-b) with ni iterations. In the case being discussed, a cadr is a set of hardwarily implemented operations. These operations are united into an indivisible computing structure, which performs the functional transformation of input data flows into out-put ones [11]. If the condition is fulfilled before the operation of verification (in iterations with numbers from $(p-1) \cdot ni + 1$ to $p \cdot ni - 1)$,), further iterations will not affect the calculation results. At the same time, the algorithm modification reduces the operation time: hardware resource freed from V blocks can be used for the placement of additional C blocks. The quantity of C blocks in the cadr is defined by ni parameter. It is worth noting that both considered variants of the calculations are suitable for multiprocessor computer systems as well as for reconfigurable ones.

In the case of verification during each iteration (see Fig. 6-a), collection K of the algorithm iteration numbers is a typical set because its first, intermediate and last elements are sharply defined. The number of the last iteration I_m is unknown in advance, but it is explicitly determined by the termination condition $err(k) \leq \delta$.

If there is one verification in each cadr (see Fig. 6-b), the Cantor-Bolzano set theory provides the description of only one special case K^* corresponding to the fulfillment of the termination condition at the iteration with $T \cdot ni$ number, where T is the number of the last iteration block. Otherwise, collections K and K^* describe different mathematical objects: set K^* contains not only necessary but also excessive iterations, and it is impossible to specify the exact location of fulfillment point for termination condition $err \leq \delta$. In terms of the AST, the collection of iterations K is a semiset, i.e. a class that has the subset relation $K \subseteq K^*$ with set K^*. The iteration at which the termination condition $err \leq \delta$ becomes true represents a horizon. Due to the implementation features of the algorithm, it is impossible to point out the horizon position precisely. Depending on various factors (e.g., initial approximation or matrix properties), the horizon can move and form different variants of collection K. In the case of the condition fulfillment at iteration $T \cdot ni$, the horizon transforms to a sharp boundary, and semiset K becomes a definite set of operations, which coincides with set K^*. In general, the set difference of K^* and K corresponds to the semiset of special iterations. During these iterations, the condition $err \leq \delta$ is true, but calculations are not terminated because it is impossible to check the condition. The aforementioned semiset describes beyond-the-horizon calculations, which are not necessary from the mathematical point of view, but do not lead to the degradation of results. These calculations occur due to the features of the considered approach to the Jacobi algorithm implementation.

The source code of the Jacobi program in the Set@l language declares the attributes of calculation and verification operations C and V and utilizes them for the description of full information graph G, which is a class within the source code shown in Fig. 7.

Collection K of iteration numbers and graph G can be sets (Fig. 6-a) as well as semisets (Fig. 6-b). It depends on the approach to the algorithm implementation, which is described in the aspect of the processing method. When a source code of the program is developed, the implementation details are still unknown. Therefore, the types of collections K and G cannot be unambiguously determined, and they are marked as classes. Within the source code, only one variant of implementation with verification at each iteration is considered since it corresponds to the mathematical sense of the Jacobi algorithm.

```
set(I,J);                    // sets of row and column numbers;
cls(K);                      // class of iteration numbers;
{cls,graph}(G);              // information graph;
set(a,b);                    // matrix and vector-column;

attribute C(k):              // calculation attribute;
   operand(element(k));      // operand type;
   k=1 -> x(k,*)=x_init(*);  // initial approximation;
   (forall i in I):
      x(k+1,i)=(b(i)-sum(a(i,j)*x(k,j)|j in J and
                  j!=i))/a(i,i);   // recalculation;
   end(forall);
end(C);

attribute V(k):              // verification attribute;
   operand(element(k));      // operand type;
   err(k)=max(abs(x(k+1,i)-x(k,i))|i in I);
   err(k)<=delta  ->  x_res(*)=x(k+1,*);
end(V);

(forall k in K):             // graph definition;
   G(k)={conc(C,V)}(k);      // subgraph description;
end(forall);
```

Fig. 7. The source code of the Jacobi program in the Set@l language

The fragment of the processing method aspect that describes two variants of the Jacobi algorithm implementation is given in Fig. 8.

At first, possible types of collection K (a clearly defined set or an in-definite semiset) are deckared. Then, set K^* is specified, and its partition is suitable for both reconfigurable (ni>1) and multiprocessor (ni=1) computer systems. If computer system has the multiprocessor architecture, aspects choose the implementation variant with verification during each iteration (see Fig. 6-a). In this case, K is a set, and its elements are described using reference set K^*. For the reconfigurable architecture, it is reasonable to verify the termination condition after several computational iterations (see Fig. 6-b). In this case, K is a semiset,

```
typing(K)='set' or 'sm';   // possible types of K;
K*={set,imp}(IT(k)|IT(k)={set,imp}((k-1)*ni+1 ... k*ni)
                        and (k=1 or err((k-1)*ni)>delta) );
```
```
// Verification during each iteration:
type(K)='set';
K=(k|k in K* and (k=1 or err(k-1)>delta));
```
```
// One verification after ni computational iterations:
type(K)='sm';
K=sub(K*);  // subset relation;
G(k)=(P|mod(k,ni)=0 -> P={conc(C,V)}(k) and
        mod(k,ni)!=0 -> P=C(k)); // subgraph modification;
```

Fig. 8. The fragment of the processing method aspect that describes two variants of the Jacobi algorithm implementation

and its subset relation with set K is declared and the subgraph of iteration G(k) is modified according to its number.

Owing to indefinite collections in the Set@l programming language, one can change the implementation method of an algorithm by means of aspect linking without the modification of the architecture-independent source code.

6   Memory Usage in Set@l Programs

The usage of computer system's memory for the organization of high-performance parallel calculations is an important detail of the algorithm implementation. In traditional tools for parallel programming, the model of memory usage is fixed, and special syntax elements describe the loading and saving of data. As a result, the porting of parallel programs requires the thorough analysis and revision of a source code. Set@l makes it possible to create special aspects, which are independent from a source code and control the loading and saving of intermediate data and results during parallel calculations. The common architectures of computer systems utilizes several basic techniques of the memory usage. If the parallel implementation of an algorithm demands a special approach, one can develop and link special aspects of memory to a source code. Otherwise, some standard techniques are applied, and it is not necessary to develop additional aspects.

The memory aspect code that illustrates the essential principles of memory usage description in Set@l is shown in Fig. 9. We introduce the simplified attributes of data reading (R) and saving (W) for the Jacobi algorithm and include the elements with appropriate R and W attributes into information graph set G. To consider the reconfigurable parallel architecture, the memory aspect includes a special attribute cadr that describes the hardwarily implemented computing structure. The reading of A elements and saving of intermediate results in set B are pipeline operations with pipe type of parallelism. Collection D of a cadr has conc type because all included operations are performed parallel-dependently. In Fig. 9, reconfigurable computer systems with single-port distributed memory are considered. The resulting graph GG has the following structure:

$$GG = \langle \overrightarrow{\{R^{(A)}, G_1, W^{(B)}\}}, \overrightarrow{\{R^{(A)}, G_2, W^{(B)}\}}, \overrightarrow{\{R^{(A)}, G_3, W^{(B)}\}}, \ldots \rangle. \quad (2)$$

Here, we assume that the initial graph G is partitioned into cadr subsets. The memory aspect given above declare special memory sets A and B (type memset) with n elements (size(n)), which correspond to storage areas of distributed memory in reconfigurable computer systems.

Analogously to the aforementioned single-port memory case, the memory aspect for reconfigurable computer systems with dual-port memory type can be developed:

```
{memset,float,size(2*n)}(A);              // memory set;
GG=pipe(cadr(A,G(f),A)|f in Cn);          // graph;
```

```
attribute R(A):    // reading attribute;
   operand(set(A));
   x(k,*)=A(*);
end(R);
attribute W(A):    // saving attribute;
   operand(set(A));
   A(*)=x(k+1,*);
end(W);
attribute cadr(A,F,B):
   operand(set(A),element(F),set(B));
   set(D);
   D=conc({pipe,R}(A),F,{pipe,W}(B));
end(cadr);
Cn=((k/ni)|k in K and mod(k,ni)=0);    // cadr numbers;
{set,graph}(GG);                 // modified information graph;
{memset,float,size(n)}(A,B);          // memory sets;
GG=pipe(z(f)|f in Cn and
                (mod(f,2)=1 -> z(f)=cadr(A,G(f),B)) and
                (mod(f,2)=0 -> z(f)=cadr(B,G(f),A)));
```

Fig. 9. The memory aspect code in Set@l designed for reconfigurable computer systems with single-port distributed memory

The structure of GG is the same as in (2), but $W^{(B)}$ is replaced by $W^{(A)}$ due to the presence of two ports in memory blocks.

For the multiprocessor architecture, attribute cadr is not necessary, and the memory aspect is given below:

```
{memset,float,size(2*n)}(A);              // memory set;
GG=seq(seq(par,R(A),G(f),par,W(A))|f in K);
```

Here, the collection of iteration numbers K is used for the forming of GG, and the graph set has the following structure:

$$GG = \left[\left[R^{(A)}, G_1, W^{(A)}\right], \left[R^{(A)}, G_2, W^{(A)}\right], \left[R^{(A)}, G_3, W^{(A)}\right], \ldots\right]. \quad (3)$$

Thus, using special attributes, set types and partitions, one can describe different variants of memory usage as independent aspects of a unified aspect-oriented program in the Set@l language.

7   Conclusions

The Set@l programming language applies the AOP paradigm, AST and relation calculus in order to represent the algorithm for the problem solution as the architecture-independent source code and to separate it from the description of implementation features, which are defined in aspects by means of the partition and classification of sets. Using the aspects of processing method, architecture and configuration, one can provide the efficient porting of parallel applications between computer systems with different architectures and adapt programs to various modifications of hardware resources.

Application of the Set@l language gives fundamentally new possibilities for efficient porting of software to various architectures of computer systems, including heterogeneous and reconfigurable ones, without the change of a source code. This paper summarizes the development of theoretical principles of algorithm parallelizing in Set@l: the distinctive features of the language are thoroughly considered, and the fundamental syntax elements and code examples are given. Currently, we are working at the development of the translator prototype for the Set@l programming language.

Acknowledgment. The reported study was funded by the Russian Foundation for Basic Research, project number 20-07-00545.

References

1. Mittal, S., Vetter, J.: A survey of CPU-GPU heterogeneous computing techniques. ACM Comput. Surv. **47**(4), 69 (2015). https://doi.org/10.1145/2788396
2. Legalov, A.I.: Functional language for creation of architecture-independent parallel programs. Comput. Technol. **10**(1), 71–89 (2005). (in Russian)
3. OpenCL: The open standard for parallel programming of heterogeneous systems. https://www.khronos.org/opencl/
4. Dordopulo, A.I., Levin, I.I., Kalyaev, I.A., Gudkov, V.A., Gulenok, A.A.: Programming of hybrid computer systems in the programming language COLAMO. Izvestiya SFedU. Eng. Sci. **11**, 39–54 (2016) (in Russian). 10.18522/2311-3103-2016-11-3954
5. Dewar, R.: SETL and the evolution of programming. In: Davis M., Schonberg E. (eds.) From Linear Operators to Computational Biology, Springer, London (2013). https://doi.org/10.1007/978-1-4471-4282-9_4

6. Dessi, M.: Spring 2.5 Aspect-Oriented Programming. Packt Publishing Ltd., Birmingham (2009)
7. Vopenka, P.: Introduction to Mathematics in Alternative Set Theory. Alfa, Bratislava (1989). (in Czech)
8. Levin, I.I., Dordopulo, A.I., Pisarenko, I.V., Melnikov, A.K.: Approach to architecture-independent programming of computer systems in aspect-oriented Set@l language. Izvestiya SFedU. Eng. Sci. **3**, 46–58 (2018) (in Russian). 10.23683/2311-3103-2018-3-46-58
9. Haggarty, R.: Discrete Mathematics for Computing. Pearson Education, Harlow (2002)
10. Levin, I.I., Dordopulo, A.I., Pisarenko, I.V., Melnikov, A.K.: Description of Jacobi algorithm for solution of linear equation system in architecture-independent Set@l programming language. Izvestiya SFedU. Eng. Sci. **5**, 34–48 (2018) (in Russian). 10.23683/2311-3103-2018-5-34-48
11. Kalyaev, I.A., Levin, I.I., Semernikov, E.A., Shmoilov, V.I.: Reconfigurable Multipipeline Computing Structures. Nova Science Publishers, New York (2012)

Solving the Problem of Detecting Similar Supercomputer Applications Using Machine Learning Methods

Denis Shaikhislamov$^{(\boxtimes)}$ and Vadim Voevodin

Research Computing Center of Lomonosov Moscow State University,
Leninskie Gory, 1, bld. 4, Moscow, Russia
sdenis1995@gmail.com, vadim@parallel.ru

Abstract. Top HPC centers provide administrators with a huge amount of data on the running applications collected by the monitoring system, resource manager and other system software. Using this data, it is possible to study and optimize different aspects of supercomputer performance. One of the possible ways to do this is related to finding similar applications. The solution to this issue would be beneficial for the administrators by enabling them, for example, to predict the application execution time, but also for the users, helping the cluster jobs for further analysis (e.g. anomaly detection). In this article, the authors considered both static analysis methods based on binary application files analysis and dynamic data analysis methods using autoencoders and DTW variations for detection of similar supercomputer applications. All proposed methods have been evaluated on the Lomonosov-2 supercomputer.

Keywords: Supercomputers · Time-series analysis · Similar applications detection · DTW · Doc2Vec

1 Introduction

Modern supercomputers provide a lot of useful information about jobs executed on them: the names of used application packages, job performance data collected by the monitoring system, detailed job launch information collected by the resource manager, etc. Analyzing this data, it is possible to identify the features and patterns among the applications that would help both users and administrators to obtain various statistics about the running applications. One of the tasks the Lomonosov-2 supercomputer administrators would like to solve is to identify similar applications, since it will not only provide information about the users working on similar problems, but also help solving more important tasks, such as predicting the execution time of a recently launched application to improve job scheduling.

The task of detecting similar applications can be solved in many ways, such as by source code analysis. In this article we focus on solutions based on statistical

© Springer Nature Switzerland AG 2020
L. Sokolinsky and M. Zymbler (Eds.): PCT 2020, CCIS 1263, pp. 46–57, 2020.
https://doi.org/10.1007/978-3-030-55326-5_4

and machine learning methods. Methods for identifying similar applications can be divided into static analysis that does not require to run the application, and dynamic analysis which implies analysis of the dynamics of application execution. It is worth noting that such solutions are very relevant for large supercomputer systems due to the large amount of data impossible to be processed manually.

In this article we propose two different methods for detecting similar applications on a supercomputer, analyzing the supercomputer job flow from different angles. Depending on the necessary granularity of detection, one or another method can be selected.

All the experiments described in this paper were conducted on Lomonosov-2 [17] supercomputer with ~5 Petaflops of peak performance, ~1700 nodes, located in Lomonosov Moscow State University.

The paper is organized as follows. Section 2 describes the background and reviews the works related to the topic of this paper. Section 3 describes the static analysis method and its results. The details on the selected method for dynamic analysis and the experiment results are shown in Sect. 4. Section 5 contains the conclusion and plans for future work.

2   Background and Related Work

The objective of this study is to find a suitable way to detect similar applications for further use in real-time supercomputer job flow analysis. To do this, the concept of similar applications needs to be defined. We distinguish two definitions: for static and dynamic analyses.

Static analysis implies analyzing an application without running it, for example, by analyzing the source code/executable files of the application. We found only one work [18] that employs machine learning methods in static analysis for detecting similar applications. In this article, the authors compare applications to predict their further energy consumption using historical data. The analysis is based on the so-called symbols extracted from executable files. Since the authors showed promising results, it was decided to adapt this method for our needs as described in detail in Sect. 3 as a part of the static method for detecting similar applications. Thus, in static analysis, two applications are considered similar if they have no or little difference in the symbols of their executable files.

Dynamic analysis involves analyzing the dynamics of the applications' execution, hereinafter referred to as dynamic characteristics, acquired by running the applications on a particular system. Dynamic characteristics data is provided in real time by the monitoring system. For each application, such characteristics as utilization of CPU, GPU, Infiniband network, number of level 1 cache misses per second, etc. are collected. In dynamic analysis, applications are considered similar if both the behavior of the dynamic characteristics and their absolute values are similar to each other. An example of applications similar in behavior is shown in Fig. 1. It shows the dynamic characteristics of two launched applications represented by different line types. It can be seen that the overall behavior is very similar to each other, though there are big local differences.

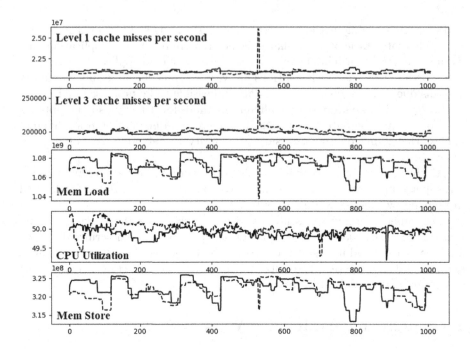

Fig. 1. Values of dynamic characteristics during execution of two applications similar in behavior, one shown with solid line, the other — with dashed line

We found a lot more works in the field of dynamic analysis. In [4] and [9], the authors detect similar phases of application execution using clustering methods. IPC (Instructions Per Cycle) and Instructions retired are collected during execution of the applications to be grouped by CPU bursts referring to the sequential computations between MPI communications. In this work authors show that these two characteristics only are sufficient to divide the application phases into clusters, but this approach is not applicable to our problem: our monitoring system groups the data not by CPU bursts, but by 2 min intervals. Consequently, the core of the approach is not applicable to our system.

The authors could not find any other works directly related to detecting similar applications using the analysis of dynamic characteristics. Therefore, the search for the method for detecting similar applications comes down to the methods of multidimensional time series comparison. After searching for existing methods, the following groups were identified.

2.1   Dynamic Time Warping for Time Series Comparison

Dynamic Time Warping (DTW, [2]) is one of the most popular and effective methods for calculating the distance between time series. The algorithm finds the optimal match for the points in time series that have similar shapes but are not time-aligned. That is, for the selected distance function f between the points of the time series, the algorithm for the time series X and Y produces such a sequence of pairs (x_i, y_j) that the sum of $f(x_i, y_j)$ is minimal. When comparing time series, the standard distance function is Euclidean distance. As the complexity of the naive DTW algorithm is quadratic $(O(NM))$, there have been various options proposed to accelerate its work: FastDTW [15], Pruned-DTW [16], SparseDTW [1] etc. that introduce a set of restrictions to significantly reduce the number of necessary operations.

One of the examples of the DTW use is [8]. In this article, the authors use a DTW modification to calculate the distance between multidimensional time series. They divide the initial time series into intervals, transforming each into one point, significantly reducing the length of the time series. This was done to further reduce the DTW operating time, and this approach may be of interest in the future when the selected dynamic analysis method of will be employed on a supercomputer in real time.

DTW is a fairly old and simple but effective method. However, DTW has its drawbacks: the frequently used Euclidean distance is very sensitive to scaling and shifts. In this regard, other distance functions that solve such problems are also considered. For example, in [6], the authors propose to compare approximate derivatives instead of the points themselves. In [10], the authors propose analyzing the time series not as a sequence of points, but as a sequence of vectors (point-to-point), and then using cosine similarity as a distance function between the vectors. This makes the aforementioned algorithms resistant to shifts along the Y axis, and these methods can also be used for multidimensional time series, since the proposed distance functions are less sensitive to multidimensionality than Euclidean distance. But these methods have drawbacks, such as sensitivity to short-term oscillations, making it necessary to "smooth out" the time series before applying any of the above methods.

2.2   Structural Analysis

Another option is structural analysis. These methods are based on a set of so-called primitives, or basic figures which can be used to describe parts of the time series (for example, constant, line, sine, etc.), and then the time series is converted to a sequence of primitives to be further analyzed. Detailed information on this method is given in [11]. The analysis can be carried out both with the formal grammar theory and with statistical analysis methods using the acquired sequence of symbols/primitives as a feature vector. These methods usually do not consider the absolute values of the series, which is not suitable for us. For this reason, this approach is replaced with such methods of time series analysis as DTW and neural networks, since they show good results for being simpler both in terms of implementation and computational complexity.

2.3   Neural Networks in Time Series Similarity Detection

There are several works on detecting similarity of time series using neural networks. In [13] the authors suggest using a Siamese neural network, which provides the probability of how similar the input time series can be. The method is based on the principle of two recurrent networks with the same parameters (weights, biases, etc.), receiving time series as an input for further comparison. The main task of the recurrent networks is to detect features and patterns in the time series that are important for further comparison, which are then fed as an input to the feedforward neural network that will decide whether these time series are similar or not.

Neural networks can also be used for dimensionality reduction, after which the traditional methods of time series comparison are applied to the resulting series. Such neural networks are called autoencoders. Architecture of such networks include a so-called bottleneck layer, with a less number of neurons than the number of dimensions in the input vector. The neural network itself is intended to recreate the original signal at its output, which allows for unsupervised learning, and during training the network will learn how to detect common features and patterns in the time series. There are two parts of the network: the encoder, i.e. apart of the network with layers up to bottleneck that encodes the input signal into a signal of lesser dimensions, and the decoder, i.e. layers of the neural network after the bottleneck, intended to decode the input signal using encoded data of a lower dimension. The encoder can be used as a method of dimensionality reduction. An example of using an encoder can be seen in [5]. After reducing the dimension of the time series by the encoder, the result is used as an input for another neural network, which calculates the correspondence matrix of the elements of time series. This information is used to calculate the distance between time series.

All aforementioned methods that use neural networks to compare time series are theoretically suitable for solving our problem, but they are very difficult use due to the lack of a large number of labeled applications available for the training.

Another example of using an encoder for time series comparison is shown in [3]. In this article, the authors used recurrent encoders to obtain a more compact representation of a multi-dimensional time series, and then used the Stratified Locality Sensitive Hashing to obtain a fixed-length vector (hash). Then, they used these hashes for comparison and classification using the 1-Nearest-Neighbor (1NN) algorithm. The method showed good results in predicting acute hypotensive episodes, but the authors of the paper did not show how they compared the resulting hashes for time series, which is a very important part of this algorithm.

3   Static Analysis

The first type of analysis selected to test was static analysis. It involves only static information, such as source code, or, in our case, executable files of the application. Obviously, applications can behave differently depending on their launch parameters, so this type of analysis can only be used to detect groups of applications similar in content. In terms of one user, the method will allow us to determine a number of different classes of problems solved by the given user.

In terms of different users, this type of analysis can show, for example, which users work on the same problems or use the same applications packages/libraries, which is also interesting information about the supercomputer operation.

Since the method described in [18] is theoretically suitable to solve our problem, we decided to implement and evaluate it in our conditions. At first, it was necessary to extract information from the executable file for subsequent analysis. To do so, the standard Linux OS utility nm was used to extract symbols from the executable files. It was further proposed to use the Paragraph Vector model [7] to convert sets of symbols of the executable files into a fixed-length vectors that can be later compared with each other. The authors propose to use distributed bag of words version of Paragraph Vector (PV-DBOW), where the input data is unordered, since the symbols extracted from the executable files are not ordered. The PV-DBOW algorithm is very similar to the skip-gram model of word2vec: if in a skip-gram a neural network learns predicting source context words given a target word, then in PV-DBOW a network learns predicting source context words given a document (set of words). During training, the model trains not both weights of the neural network, but also the weights related to words and documents that can be used to describe a corresponding word or a document. During prediction, it is possible to input a new document, and then all the neural network parameters, except for the weights related to the documents, would be fixed, and the network would be trained to obtain the weights specific to this new document.

In our work, we used the gensim [14] package for Python, which has a doc2vec model applicable in DBOW mode. To train the PV-DBOW model, we used application symbols from over 7,000 unique executable files launched in 2019.

After obtaining a fixed-length vector representation of the document, and in our case, an executable file, it was proposed to use cosine similarity to compare applications. This metric provided the cosine of the angle between the vectors, thereby showing how similar they were.

To test the resulting model operability, it was decided to use the following approach:

- manually build clusters based on similarity between the applications;
- use the developed static analysis method to obtain an estimate of the distance between each pair of applications;
- cluster applications using the estimated distance;
- compare manual and obtained clustering.

To manually build clusters for verification, it was decided to use the following assumption. The paths to executable files often include the names of both solved problems and their versions; which is why we consider applications similar if their path names are the same, but the versions are different. Thus, 132 applications were selected and manually divided into 17 different groups. With the automatic clustering method, a hierarchical agglomerative clustering method from sklearn package for Python was selected [12]. The method is based on the fact that initially all elements belong to different clusters, after which, depending on the distance between the elements, some clusters merge into one. That is why there was no need for this clustering method to specify the number of clusters, as it varied depending on the distance between the elements. To compare the resulting clusters, MoJoFM was selected [19]. It is the measure of how close two clustering configurations are to each other, 1 being exact match and 0 — exact opposite to each other. The formula for calculating MoJoFM is shown in Formula (1). $mno(A, B)$ shows how many operations of cluster merging and moving an element from one cluster to another are required to obtain cluster configuration B (reference) from A (automatic clustering).

$$MoJoFM(A, B) = 1 - \frac{mno(A, B)}{max(mno(\forall A, B))} \tag{1}$$

After clustering the selected 132 applications and calculating MoJoFM, we got a score of 0.96, which means that the resulting clusters were very close to manual clustering. It is worth noting that almost all of the jobs were clustered correctly, except for four clusters, each of which divided into two separate clusters. This behavior can be explained as an error in the vector description of the application, as well as an error in the assumption during manual clustering.

Despite the fact that manual clustering was only done for verification purposes, this type of verification is only suitable for static analysis, and it was also based on the strong assumption that applications are similar if their paths are similar. We wanted to get more accurate reference clustering from the user working on the Lomonosov-2 supercomputer. We asked one of the users to perform manual clustering for usto test both static and dynamic methods of detecting similar applications. But it should be noted that the number of jobs in this manual clustering was significantly lower, with only 52 labeled applications and only 38 executable files possible to access. Besides application classes, the user marked out the subclasses that describe different launch options within a specific class. For focusing on the application classes only, the static methods results were good, with MoJoFM score being 0.97. Considering subclasses is a challenge for static analysis: MoJoFM score is 0.65, as static analysis does not have information about launch parameters of the applications, and therefore it has no information about the kind of subtask the application is trying to solve.

In conclusion, the static analysis method described in [18] is applicable to detecting similar applications detection on the Lomonosov-2 supercomputer and can be used to isolate different groups of applications both within a single user and within several users. However, it cannot account for more subtle differences within individual application classes.

4 Dynamic Analysis

4.1 Description of the Proposed Method

The next type of analysis is dynamic analysis, which compensates the drawback of the method described in the previous section: despite showing great results in detecting different classes of applications, it cannot isolate subclasses of applications defined by different launch parameters/solved sub-tasks. Dynamic analysis methods can use dynamic characteristics that describe its behavior to extract the features, specific to different subclasses. These features can be used to differentiate one subclass from another. In our case, each application is described by 16 dynamic characteristics, i.e. 16 time series, which include utilization of GPU, CPU, Infiniband network, loadavg, the number of level 1 cache misses per second etc. The selected method of dynamic analysis consists of data preprocessing, dimensionality reduction and comparison of the obtained time series.

All dynamic characteristics were noticed to follow normal distribution; therefore, in our case, preprocessing consists of standardization of all dynamic characteristics.

For dimensionality reduction, we decided to test both the autoencoder and the well-known Principal Component Analysis method, which is very often applied in this case. New features obtained after dimensionality reduction also need to be standardized for further comparison, since they can have different amplitudes and the features can contribute to the distance between points in a very uneven manner.

To compare time series, it was decided to use DTW which shows good results in this area. It is worth noting that in the articles which involve DTW as a time series comparison method, the length of the time series is often fixed. In our case, time series length may vary, which is a problem: the distance returned by the DTW algorithm depends on the length of the compared time series, making it impossible to compare the $DTW(A, C)$ and $DTW(A, C)$ values for series A, B and C if their lengths are different. To solve this problem, we use the Formula 2 to calculate the distance between time series:

$$dist(A, B) = \frac{DTW(A, B)}{len(optimal\, match\, sequence)}, \tag{2}$$

which shows the average distance between the points of the obtained time series after calculating the optimal match between them.

DTW has several modifications that solve the problem when there are shifts in the data on Y axis iand it is necessary to get the correct match between the points. But in the DTW variations described in Sect. 2, the algorithms do not consider absolute values, which is unacceptable in our case. That is why it was decided to use the original DTW version. Due to the quadratic complexity of DTW, in our experiments the accelerated version, FastDTW, from the fastdtw package for Python with a linear complexity was used.

Many metrics can be used as a distance function, but we decided to look into the Euclidean distance and cosine similarity and compare them with each other. The Euclidean distance has the advantage of considering the absolute values of the time series points, while cosine similarity shows good results when comparing large-dimensional data, as, for example, shown in static analysis.

4.2   The Experiments and Results

Since static analysis enables you to quickly and accurately determine the classes of applications, the dynamic analysis methods are only required to solve the problem of identifying subclasses of applications that differ in behavior due to, for example, different launch parameters. One way of testing the method is to take the data manually labeled by the user as described in Sect. 3 and use it to test the static analysis method. Out of the 52 applications labelled by the user, we collected dynamic characteristics data of about only 42: the remaining applications were discarded due to their short execution time, making it impossible to collect sufficient amount of data from the monitoring system on the dynamics of their work. Testing will be carried out in the same way as for static analysis: DTW provides the estimated distance between time series, which enables us to conduct hierarchical clustering and to compare its result with manual clustering and evaluate it with the MoJoFM score. This test method will be used to identify the best combination of possible dynamic analysis methods.

Table 1. MoJoFM accuracy scores for different types of dynamic analysis

	Autoencoder		PCA		Without reduction
	3 feat	5 feat	3 feat	5 feat	
DTW with cosine similarity	0.8	0.66	0.71	0.69	0.77
DTW with Euclidean distance	**0.85**	0.83	0.78	0.78	0.78

Table 1 shows the test results of all possible combinations of the considered methods. You may notice that, if only the distance functions are analyzed, cosine similarity shows worse results than the Euclidean distance, regardless of the dimensionality reduction method applied. That is why the Euclidean distance was chosen. In case of dimensionality reduction methods, the autoencoder shows the results much better than the PCA or if no dimensionality reduction whatsoever is applied. It is worth noting that the results of DTW+PCA and DTW without dimensionality reduction do not differ much; that is why it was decided to use an autoencoder. In case of the autoencoder, the results of dimensionality reduction to 3 features do not differ much from reduction to 5 features, but is still better, so in the future this option will be used. So, in the end, the autoencoder with the dimensionality reduction to 3 features and the DTW that uses Euclidean distance were selected.

During the tests for selecting the best method combination, we also selected the hierarchical clustering parameters such as, in particular, the distance between the time series, after which they do not belong to one cluster. This led us to detecting the exact value d, which in practice can be used to identify applications belonging to the same subclass.

When analyzing the results of clustering based on the selected configuration, one subclass of manual clustering was found to be that divided into 3 smaller clusters by means of hierarchical clustering. According to the selected method, it means that within this subgroup of applications there are 3 different application types that differ in behavior. In our case, such a result is acceptable, since it is often possible to find applications with different behavior within one subclass of applications. However, to verify this fact, we manually analyzed the behavior of applications from the three clusters and found that even though the user assigned the applications to one subclass, their behavior was significantly different. Therefore, in this case, our method worked correctly. Besides, two clusters were found to have been merged in the clustering process; that is what we would like to avoid. Jobs from these two clusters belong to the same applications class, but were started with different launch parameters and solve different tasks. Consequently, the chosen dynamic characteristics analysis method makes it possible to distinguish subclasses of applications according to the dynamics of their work, but there are some problematic subclasses with behavior that may be found slightly different according to this method.

The next step of testing was the analysis of new previously unseen jobs. For this purpose, the jobs of the same user who previously provided us with the labelled data were analyzed. In the testing procedure, DTW was used in the same way as in the majority of time series classification methods, or as 1NN method, that is, after calculating the distance between time series the new application is classified as the application nearest to it. This method is more similar to the scenario in which this method is intended to be used in real time due to its simplicity and speed. However, some changes to this method have been introduced: after finding the application with the smallest distance, the distance value was checked. If the distance was less than d, the new job was labelled as a new subclass. In the future, new applications will be compared not only to existing subclasses, but also to the applications of that new subclass. This will break the dependence on currently known subclasses, making it possible to detect applications that solve new tasks. To test the operability of this approach, ~40 new applications of the selected user were picked, compared with the existing labelled applications, classified and, if necessary, new subclasses were identified. The results were sent to the user for verification. After receiving the results, similar tests were carried out. As new labelled data can be perceived as manual clustering, the MoJoFM score was calculated; the result was equal to 0.82, which is a good result. After a detailed comparison of automatic and manual clustering, type-II errors were discovered: three jobs were classified incorrectly even within the application class. It was also found that the subclass divided into three clusters in the previous experiment, was now divided by an automatic clustering algorithm into three clusters with different behavior. But as mentioned above, these differences are explained by the fact that the user labelled the applications by the subtasks solved by the jobs, without considering the difference in behavior that it may yield.

5   Conclusions and Future Work

In this article, we present two different methods for comparing applications running on a supercomputer complex. Based on static analysis of applications, the first method shows good results in identifying similar application classes (MoJoFM score ~0.97), but has difficulties in detecting subclasses due to the lack of data on the dynamics of the application, since subclasses are identical for static analysis methods but differ in their behavior.

In case of dynamic analysis, a method for detecting similar applications based on dynamic characteristic analysis was proposed and the best configuration for our system was selected. The results of the experiments show good results in detecting applications similar in behavior (MoJoFM score $\sim$0.85), but the method still has room for improvement, because there were cases of type-II error that had to be avoided. It might be done through more precise preprocessing of the dynamic characteristics and will be addressed in future work.

The next step in our work is to use the proposed methods for online clustering of both the users' applications and the entire supercomputer job flow of the Lomonosov-2 supercomputer. Also, one of the directions of future work is the application of the developed methods to solve other tasks of interest, such as, for example, predicting the execution time of applications.

Acknowledgments. The research was carried out with the financial support of the grant from the Russian Federation President's Fund (MK-2330.2019.9).

References

1. Al-Naymat, G., Chawla, S., Taheri, J.: Sparsedtw: A novel approach to speed up dynamic time warping. In: Conferences in Research and Practice in Information Technology Series, vol. 101 (2012)
2. Berndt, D.J., Clifford, J.: Using dynamic time warping to find patterns in time series. In: Proceedings of the 3rd International Conference on Knowledge Discovery and Data Mining, AAAIWS 1994, pp. 359–370. AAAI Press (1994)
3. Dhamala, J., Azuh, E., Al-Dujaili, A., Rubin, J., O'Reilly, U.: Multivariate time-series similarity assessment via unsupervised representation learning and stratified locality sensitive hashing: Application to early acute hypotensive episode detection. CoRR abs/1811.06106 (2018)
4. Gonzalez, J., Gimenez, J., Labarta, J.: Automatic detection of parallel applications computation phases. pp. 1–11 (2009). https://doi.org/10.1109/IPDPS.2009.5161027
5. Grabocka, J., Schmidt-Thieme, L.: Neuralwarp: Time-series similarity with warping networks. CoRR abs/1812.08306 (2018)
6. Keogh, E., Pazzani, M.: Derivative dynamic time warping. First SIAM International Conference on Data Mining, vol. 1 (2002). https://doi.org/10.1137/1.9781611972719.1
7. Le, Q.V., Mikolov, T.: Distributed representations of sentences and documents. CoRR abs/1405.4053 (2014)
8. Li, Z.X., Li, K.W., Wu, H.S.: Similarity measure for multivariate time series based on dynamic time warping. In: Proceedings of the 2016 International Conference on Intelligent Information Processing, ICIIP 2016, pp. 15:1–15:5. ACM, New York (2016). https://doi.org/10.1145/3028842.3028857
9. Llort, G., Servat, H., Gonázlez, J., Giménez, J., Labarta, J.: On the usefulness of object tracking techniques in performance analysis. In: SC 2013: Proceedings of the International Conference on High Performance Computing, Networking, Storage and Analysis, pp. 1–11 (2013). https://doi.org/10.1145/2503210.2503267

10. Nakamura, T., Taki, K., Nomiya, H., Seki, K., Uehara, K.: A shape-based similarity measure for time series data with ensemble learning. Pattern Anal. Appl. **16**(4), 535–548 (2012). https://doi.org/10.1007/s10044-011-0262-6
11. Olszewski, R.T.: Generalized feature extraction for structural pattern recognition in time-series data. Ph.D. thesis, Carnegie Mellon University, Pittsburgh, PA (2001)
12. Pedregosa, F., et al.: Scikit-learn: machine learning in Python. J. Mach. Learn. Res. **12**, 2825–2830 (2011)
13. Pei, W., Tax, D.M.J., van der Maaten, L.: Modeling time series similarity with siamese recurrent networks. CoRR abs/1603.04713 (2016)
14. Řehůřek, R., Sojka, P.: Software framework for topic modelling with large corpora. In: Proceedings of the LREC 2010 Workshop on New Challenges for NLP Frameworks, pp. 45–50. ELRA, Valletta, Malta (2010)
15. Salvador, S., Chan, P.: Toward accurate dynamic time warping in linear time and space. Intell. Data Anal. **11**(5), 561–580 (2007)
16. Silva, D.F., Giusti, R., Keogh, E., Batista, G.E.A.P.A.: Speeding up similarity search under dynamic time warping by pruning unpromising alignments. Data Min. Knowl. Disc. **32**(4), 988–1016 (2018). https://doi.org/10.1007/s10618-018-0557-y
17. Voevodin, V.V., et al: Supercomputer Lomonosov-2: large scale, deep monitoring and fine analytics for the user community. Supercomput. Front. Innov. **6**(2), 4–11 (2019). 10.14529/jsfi190201
18. Yamamoto, K., Tsujita, Y., Uno, A.: Classifying jobs and predicting applications in HPC systems. In: Yokota, , R., Weiland, M., Keyes, D., Trinitis, C. (eds.) High Performance Computing, pp. 81–99. Springer, Cham (2018). https://doi.org/10.1007/978-3-319-92040-5_5
19. Zhihua, W., Tzerpos, V.: An effectiveness measure for software clustering algorithms. In: Proceedings. 12th IEEE International Workshop on Program Comprehension, vol. 2004, pp. 194–203 (2004). https://doi.org/10.1109/WPC.2004.1311061

Integrated Computational Environment for Grid Generation Parallel Technologies

Valery Il'in[1,2(✉)]

[1] Institute of Computational Mathematics and Mathematical Geophysics SB RAS,
6, Ac. Lavrentieva Avenue, Novosibirsk 630090, Russia
`ilin@sscc.ru`
[2] Novosibirsk State University, 1, Pirogova Street, Novosibirsk 630090, Russia

Abstract. This paper is devoted to the conception and general structure of the integrated computational environment for constructing multi-dimensional large grids (with 10^{10} nodes and more) for high-performance solutions of interdisciplinary direct and inverse mathematical modelling problems in computational domains with complicated geometrical boundaries and contrast material properties. This includes direct and inverse statements which are described by the system of differential and/or integral equations. The constructed computational grid domain consists of subdomains featuring a grid, which may be of different types (structured or non-structured); discretization at the internal boundaries can be consistent or non-consistent. The methodology of such quasi-structured meshes makes it possible to use various algorithms and codes in the subdomains, as well as different data structure formats and their conversion. The proposed technologies include grid quality control, the generation of dynamic grids adapted to singularities of input geometric data of structures and multigrid approaches with local refinements, taking into account information about the solution to be obtained. The balanced grid domain decomposition, based on hybrid programming at the heterogeneous clusters with distributed and hierarchical shared memory, supports scalable parallelization. In addition, the paper outlines the technological requirements to provide a successful long-life cycle for the proposed computational environment. In a sense, the considered development presents a stable software ecosystem (integrated grid generator DELAUNAY) for supercomputing modelling in the epoch of big data and artificial intellect.

Keywords: Multi-dimensional boundary value problems · Grid computational domain · Adaptive quasi-structured grids · Grid generation methods · Data structures · Scalable parallelization

V. Il'in—This work was supported by Russian Foundation of Basic Research grant 18-01-00295.

1   Introduction

Grid generation is an important stage of mathematical modelling of real processes and phenomena in various applications. In the ideology of the Integrated Computational Environment (ICE, [1]) for mathematical modelling, the discretization of the original problem to be solved follows after the geometrical and functional modelling and precedes the approximation and algebraic solution steps. The mesh quality defines the efficiency of the numerical methods for differential and/or integral equation systems, or corresponding variation statements. There are many papers and books on mesh constructing algorithms (see [1,9], for example). Also, many conferences, special journal issues and internet sites deal with these problems [10,11] and numerous program implementations for grid generation, commercial or free available packages, such as Netgen [18] and Gmesh [12]. Also, in gas-oil applications, PEBI (Perpendicular Bisection [20]) and Corner Point [21] are popular discretization approaches.

Grid problems include many mathematical and technological issues. Numerical approaches to the discretization of computational domains are based on the conformal and non-conformal mapping [2], variational principles [4], differential geometry methods [1], self-organizing neural networks for unsupervised learning [6], and various empirical algorithms [3,4,7,8]. Grid quality estimation and an approach to control, improve and optimize the mesh are very important questions. Also, there are many open problems in this respect, and a formulaic definition of the optimal grid has not been established yet. In a sense, the problem to construct a good mesh can be more difficult than to solve the original boundary value problem as a means of simple discretization.

From the practical point of view, the ultimate success of the grid generation consists in the mesh data structure (MDS). It must support different advanced discretization approaches: local mesh refinement based on a posteriori or/and a priori analysis of the solution to be sought, multi-scale and super-element algorithms, multigrid methodologies, as well as the domain decomposition methods (DDM), which are intended for providing the scalable parallelism and high-performance computing. Here, it is important to remark that at the further modeling stages, the parallel computations by means of synchronous solving of the auxiliary problems in subdomains with the DDM technology require to organize a distributed data structure in advance, i.e. at the step of grid generation.

We will mainly consider stationary adaptive quasi-structured grids. Their first characteristic feature means that the vertices of a computational domain coincide with the grid nodes. Also, in this case, we assume that the edges and faces of the computational domain coincide, or "almost coincide" with the grid faces, respectively. The term *quasi-structured* defines the grids which consist of the grid subdomains that each of them presents a structured or a non-structured grid. In the structured grid, for each mesh node, the number of the neighboring nodes can be computed by means of a simple formula. For each mesh point of the non-structured grid, a set of the neighboring nodes can be defined by enumeration only. Also, each grid subdomain can consist of different types of mesh elements.

We will further describe the conception of the integrated computational environment for generating a wide class of quasi-structured grids. In general, the properties of grids are defined by their MDS. Let us consider the numerical solution of the multidimensional and multi-disciplinary initial boundary value problems (IBVPs) in the computational domain with complicated contrast material properties and geometry with piecewise smooth multi-connected boundaries. We suppose that the grid generation is performed once, in advance to the general computational process. Of course, in some problems, it is necessary to construct dynamic meshes, but such computing tasks require further research.

In the inverse problems, input data are presented in the parametrized form, and we must find the solution which would provide the minimum of the prescribed objective functional, under some additional constraints. In these cases, the optimization methods include solving a set of the direct problems which use, probably, different grids. Such cases take place, for example, in the actual shape optimization problems which are connected with the geometrical and topological modifications of a computational domain.

The numerical solution of IBVPs is based on the approximation principles: finite difference, finite volume, finite element, discontinuous Galerkin methods of various order of accuracy as well as collocation and the least squares approaches. So, the grid generation stage of a mathematical simulation has an intermediate place between the geometrical modeling and approximation steps. Also, from the practical point of view, it is important to introduce a concordance between MDS and CAD products (CAE, CAM, PLM, etc.) which have an essential market and popular data formats.

In general, the grid issues include a large set of mathematical problems, algorithms, computational and program technologies which would hardly be unified in the single software product. So, we consider the concept of the integrated instrumental surrounding for the new generation of the grid constructing tool which presents not the group project but community project. The projects of FOAM, DUNE (Distributed Unified Numerical Environment), INMOST (the development of Marchuk Institute of Numerical Mathematics, RAS) and BSM [3, 22–24, 29] are some examples of such an integrated approach.

The considered numerical methods and technologies are implemented in the framework of the integrated grid generator DELAUNAY which is a separate part of BSM [22] and interacts with other subsystems responsible for the corresponding modeling stages just via data structures. In particular, DELAUNAY is constructed with the native CAD system HERBARIUM [25]. More exactly, the input data for grid constructing are presented by the geometrical and functional data structures (GDS and FDS) which are formed by the subsystem VORONOI responsible for constructing a continuous model of the original problem and CAD system usage. In other words, DELAUNAY presents a library of grid generators, which not only contains its own algorithms but can also efficiently use external program products. The grid construction (MDS, together with GDS and FDS) results in a discretized model for the BSM approximation stage, i.e. the CHEBYSHEV subsystem [24].

The content of this paper is as follows. Section 2 reviews a formal statement of the continuous direct and inverse problems to be solved. Section 3 describes grid objects, structures and their specifications, as well as main mesh operations which are necessary to be implemented in the numerical methods. Section 4 discusses the conception, main components and technical requirements for the integrated grid generation environment.

2   Formal Statement of the Interdisciplinary Direct and Inverse Problems

Since our end goal is to discretize direct and inverse IBVPs, we need to consider, from the formal point of view, specifications of the problems to be solved.

Let there be N_Ω subdomains Ω_k in the closed computational domain $\bar{\Omega}$ with the boundary Γ

$$\bar{\Omega} = \Omega \bigcup \Gamma = \bigcup_{k=1}^{N_\Omega} \bar{\Omega}_k, \tag{1}$$

and it is necessary to find the vector solution $\mathbf{u} = \{u_\mu, \quad \mu = 1, ..., \bar{m}\}$ to satisfy the differential equation

$$L\mathbf{u}(\mathbf{x}, t) = \mathbf{f}(\mathbf{x}, t), \ \mathbf{x} \in \Omega, \tag{2}$$

as well as the boundary and initial conditions

$$l\mathbf{u} = \mathbf{g}(\mathbf{x}, t), \ \mathbf{x} \in \Gamma, \ \mathbf{u}(\mathbf{x}, 0) = \mathbf{u}^0(\mathbf{x}), \tag{3}$$

where $\mathbf{x} = (x_1, ..., x_d)$ and t are the spatial and temporal independent variables, respectively. Here Ω is an open one-connected bounded domain in the d-dimensional euclidean space ($d = 1, 2, 3$), L and l are the operators of the original equation and the boundary condition. For example, operator L can have the form

$$L = A\frac{\partial}{\partial t} + \nabla B \nabla + C \nabla + D. \tag{4}$$

Here A, B, C, D are the matrices, which entries depend on the time t and Cartesian or some other type of the coordinate $\mathbf{x}$. In the nonlinear case, the matrix and the right-hand side $\mathbf{f}$ in (2) can also depend on the unknown solution $\mathbf{u}$. In the interdisciplinary problem, the unknown functions u_μ present the fields of different physical nature: velocities, pressure, temperature, densities, etc. The boundary Γ of the computational domain is presented by two parts, on each one of them the boundary conditions of the first (Dirichlet), the second (Neumann) or the third type (Robin) are given:

$$\mathbf{u} = \mathbf{g}_D, \ \mathbf{x} \in \Gamma_D, \quad D_N \mathbf{u} + A_N \nabla_n \mathbf{u} = \mathbf{g}_N, \ \mathbf{x} \in \Gamma_N, \tag{5}$$

where ∇_n denotes the external normal derivative on Γ_N, and A_N, D_N are, in general, the matrix of coefficients, in general. Let us remark that at the joint boundary of the contacted subdomains, some internal (interface) conditions can be given. The initial boundary value problem (IBVP) is called the interdisciplinary (or multi-physical) for the case $\bar{m} > 1$ in the sense that every scalar function u_μ corresponding to a different physical subdomain and different equations in system (1) describes various phenomena. Equation (2) can be have a different form in different subdomains Ω_k. Usually it means that we have different material properties in various computational subdomains.

Equations (1)–(5) describe the direct IBVPs. But in real cases the ultimate goal of the research consists in solving inverse problems, which means, for example, the identification of the model parameters, the optimization of some processes, etc. The universal optimization approach to solving the inverse problems is formulated as minimization of the objective functional

$$\Phi_0(\mathbf{u}(\mathbf{x}, \ t, \ \mathbf{p}_{opt})) = \min_{\mathbf{p}} \Phi_0(\mathbf{u}(\mathbf{x}, \ t, \ \mathbf{p})), \tag{6}$$

which depends on the solution $\mathbf{u}$ and on some vector parameter $\mathbf{p}$ which is included in the input data of the direct problem. The constrained optimization is carried out under the linear conditions

$$p_k^{min} \le p_k \le p_k^{max}, \ k = 1, ..., m_1, \tag{7}$$

and/or under the functional inequalities

$$\Phi_l \mathbf{u}(\mathbf{x}, \ t, \ \mathbf{p})) \le \delta_l, \ l = 1, ..., m_2. \tag{8}$$

Formally, the direct problem can be considered as the state equation and can be written down as follows:

$$L\mathbf{u}(\mathbf{p}) = \mathbf{f}, \ \mathbf{p} = \{p_k\} \in \mathcal{R}^m, \ m = m_1 + m_2. \tag{9}$$

There are two main kinds of optimization problems. The first one consists in the local minimization. This means that we look for a single minimum of the objective function in the vicinity of the initial guess $\mathbf{p}^0 = (p_1^0, \ ..., \ p_m^0)$. The second problem is more complicated and presents the global minimization, i. e. the search for all extremal points of $\Phi_0(\mathbf{p})$. The usual way to find the solution of the inverse problem consists in solving a set of direct problems with different parameter values.

We suppose that the subdomains Ω_k are one-connected and have no intersections. Its boundary can be presented in the form

$$\Gamma_k = \Gamma_k^{(e)} \bigcup \Gamma_k^{(h)} = \bigcup_{k' \in \omega_k} \Gamma_{k,k'}, \Gamma_{k,k'} = \bar{\Omega}_k \bigcap \bar{\Omega}_{k'} = \Gamma_{k,k'} \ k, k' = 1, ..., N_\Omega.$$

$$\tag{10}$$

Here ω_k is a set of subdomain numbers which are neighboring to ω_k, and $\Gamma_k^{(e)}$ and $\Gamma_k^{(h)}$ are the surface segments, which belong to external and internal parts of Γ respectively. Formally, we can write $\Gamma_k^{(e)} = \Gamma_{k,0}$, i.e. the index "0" denotes the number of external subdomains. Each fragment Γ_{k,k^0} presents the joint boundary of the contacted subdomains, without any boundary condition in the statement of the original IBVP.

In general, the computational domain Ω consists of the following geometric objects: subdomains or macro-volume, Ω_k, $k = k_1, ..., N_\Omega$, the vertices $V_p = (x_p, y_p, z_p)$, $p = 1, ..., N_V$, the surface segments (macro-faces) F_q, $q = 1, ..., N_f$, and macro-edges (curvilinear fragments) E_l, $l = 1, ..., N_f$. For the analytical description of the geometric primitives the global and local coordinate systems are defined (Cartesian, cylindrical or spherical). In many cases, it is convenient to describe the line or surface segments in the parametric form:

$$x = x(\tau), y = y(\tau), \ or \ x = x(\tau^{(1)}, \tau^{(2)}), y = y(\tau^{(1)}, \ \tau^{(2)}), z = z(\tau^{(1)}, \tau^{(2)}).$$
(11)

For particular situations, there are a lot of different approaches, to describe geometric objects. An advanced simulation system should have various possibilities for efficient representations and the converters of the input formats.

The connections between geometric objects are described by means of incident matrix with bit entries:

$$M_{VE} = \{m_{i,j}^{VE} : i = 1, ..., N_V; \ j = 1, ..., N_E\} - (\text{vertices} - \text{edges}),$$
$$M_{EF} = \{m_{i,j}^{EF} : i = 1, ..., N_E; \ j = 1, ..., N_F\} - (\text{edges} - \text{faces}),$$
$$M_{F\Omega} = \{m_{i,j}^{F\Omega} : i = 1, ..., N_F; \ j = 1, ..., N_\Omega\} - (\text{faces} - \text{subdomains}).$$

In total, configuration information on the computational domains is included in geometric data structure (GDS). The functional description of the IBVPs (the system of equations, to be solved. the boundary conditions, and their coefficients) is presented in the functional data structure (FDS), see [6] for details. Generally speaking, this input (for DELAUNAY) information presents the continuous model of the problem to be solved. It is formed by the VORONOI system, which is responsible for the geometric and functional modeling in BSM.

Let us remark, that some of the input data can be given in parametrized form, in order to solve the inverse, or the set of direct problems. In such cases the additional information to organize the computational experiments should be given.

3   Grid Objects, Operations and Data Structure

In this section, we define the main specifications of the grids, the operations with the mesh objects. as well as principles of the constructing the mesh data structure.

3.1   Main Grid Notions and Specifications

The discretization of the computational domain Ω consists in the construction of the grid computational domain Ω^h, which includes the following mesh objects, similar to the "macro-objects" of Ω:

grid subdomain $\bar{\Omega}_k^h \colon \bar{\Omega}^h = \bigcup_k \bar{\Omega}_k^h, \ \bar{\Omega}_k^h = \Omega_k^h \bigcup \Gamma_k, \ k = 1, ..., N_\Omega^h, \ \Omega_k^h \approx \Omega_k,$

grid subdomain boundaries: $\Gamma_{k,k'}^h = \bar{\Omega}_{k'}^h \bigcap \bar{\Omega}_k^h, \ \Gamma_k^h = \bigcup_k \Gamma_{k,0}^h, \ \Gamma_k^h \approx$
$\Gamma_k, k, k' = 1, ..., N_\Gamma^h,$

grid nodes: $V_p^h = V_{e,e'}^h \equiv \bar{E}_e^h \bigcap \bar{E}_{e'}^h = (x_p, y_p, z_p), \ V_{l,l',l''} \equiv$
$\bar{F}_l^h \bigcup \bar{F}_{l'}^h \bigcup \bar{F}_{l''}^h, p = 1, ..., N_V^h,$

grid edges: $E_s^h = E_{l,l'}^h \equiv \bar{F}_l^h \bigcap \bar{F}_{l'}^h, \quad s = 1, ..., N_E^h,$

grid faces: $F_l^h = F_l^h = F_{m,m'}^h \equiv \bar{T}_m^h \bigcap \bar{T}_{m'}^h, \quad l = 1, ..., N_F^h,$

grid (finite) volumes, or elements: $T_m^h, m = 1, ..., N_T^h, \quad \bar{\Omega}^h = \bigcup_m \bar{T}_m^h,$

where $N_\Omega^h, N_\Gamma^h, N_V^h, N_E^h, N_F^h, N_T^h$ means the total numbers of the grid subdomains, boundaries, nodes, edges, faces, and volume elements respectively.

The grid objects have the following topological connections. The mesh domain and subdomains are the unions of the corresponding elements. The mesh faces are the joint boundaries of the contacted finite elements. The grid edges present the intersection of the neighboring faces. The nodes are the joint points of the intersected edges or faces. Similar to macro-object, we can define rectangular matrice of the different types, which characterize the connections between the grid objects. For example, a node-edge matrix is $M_{V,E}^h = \{m_{i,j}^{V,E}, i = 1, ..., N_N^V, j = 1, ..., N_E^h\}$ where each bit entry $m_{i,j}^{V,E} = 1$ if the j-th grid edge is connected to the i-th grid node, and $m_{i,j}^{V,E} = 0$ otherwise. In a similar way, we can define a node-volume matrix $M_{V,T}^h = \{m_{i,j}^{V,T}, i = 1, ..., N_V^h, j = 1, ..., N_T^h\}$, an edge-force matrix $M_{E,F}^h = \{m_{i,j}^{V,T}, i = 1, ..., N_E^h, j = 1, ..., N_F^h\}$, as well as other types of the matrix: $M_{S,T}^h, M_{V,F}^h, \quad M_{F,E}^h = (M_{E,F}^h)^T$, etc. All types of grid objects have their own global and local (on subdomains) numbering, as well as the functions for their remembering. The geometric and topological specification of the object can be formed via the MDS. It also includes necessary connections with the functional data structure for defining the type of an equation and the values of its coefficients in each element T_m^h, as well as for describing the boundary conditions at the grid faces F_l^h.

3.2   Grid Operations and Data Structure

Advanced numerical methods of mathematical modelling are based on the complicated matrix transformations, which are connected with the graph structure of a grid. The main consuming step of the numerical simulation is to solve a system of linear algebraic equations (SLAEs) which is repeated many times if the original problem is inverse or non-stationary and non-linear. The modern large SLAE can be characterized as a sparse matrix of order 10^9 or higher and the medium number of non-zero entries which are about 100 in each row.

The arithmetic operations must be done in the standard double precision format (64 bits, or 8 bytes).

The high performance solution in such cases is provided by means of scalable parallelizm based on the domain decomposition approaches. The advanced numerical tools include the two-level multi-preconditioned iterative methods in the Krylov subspaces. At the upper level, the distributed version of the additive block Swarz algorithm is implemented with the help of the MPI (Message Passing Interface between cluster nodes) functions. The lower level consists of the synchronous solution to the auxiliary SLAEs in subdomains using the multi-thread computing (Open MP technologies) and vectorization of the operations by AVX instruments. Also, some auxiliary algebraic subsystems can be solved by means of special algorithms on the super-fast graphic accelerators (GP GPU or Intel Phi).

For the real mathematical problems, which are solved on a non-structured grid, an important circumstance for implementation of the algebraic methods is that the matrices are presented in the compressed sparse formats, such as CSR, because the allocation of the non-zero matrix entries in the rows corresponds to the connection of the respective grid nodes to their neighbors. In this case, only non-zero matrix entries and references between them are saved. It is obvious that such forced technologies make access to matrix values in memory too slow and expensive.

The considered approach defines the matrix portrait which is isomorphic, in a sense, to the graphic structure of a non-structured grid. So, the algebraic data structure (ADS) which is the base for iterative algorithms is fairly similar to the MDS, and the CSR format can be the base for the grid data structure because each matrix row corresponds to a mesh node or another primitive.

In some cases, the situation is more complicated. If we solve an interdisciplinary problem which is described by a system of partial differential and/or integral equations, then after discretization in each grid node, several unknown variables can be defined, and we have to generalize the CSR format to the block one (BCSR).

Because of a big volume of the grid data structure, the domain decomposition techniques must be implemented at the mesh generation stage. It should include a description of the grid computational subdomains and a description of the MDS informational arrays in the memory of the corresponding cluster node. This means that the corresponding approximation stage will be done efficiently in parallel. From the algorithmic point of view, the decomposition is performed in two steps. The first one consists in disassembling the mesh computational domain into parts without intersections. At the second stage, the grid subdomains are extended to a necessary number of mesh layers. By the end of this stage, we must obtain the distributed MDS for decomposition with parameterized overlapping of the subdomains. At the following modeling steps, it will be the base for the parallel implementation of the total computational process.

Also, at this step, the multigrid data structure is performed if needed. Theoretically, this approach provides the optimal in order algebraic solvers. The complementation technology requires forming a hierarchical data structure for a set of embedded grids. Computational schemes in these cases are rich in various numerical operations: pre-smoothing and post-smoothing, restriction and prolongation, course grid correction, etc. Automatic construction of such algorithms is a highly intellectual problem, and the grid transformation support is very important in such a challenging task.

In recent decades, the advanced approaches have appeared for solving multi-scale problems based on super-element technologies and constructing the immersion type meshes, which are aimed at the application of a special type of non-polynomial basis functions. In such cases, a special kind of MDS is required to take into account the geometric details of the under-cell scale.

The above brief review of the grid construction problems reflects many mathematical issues. In particular, there are valuable computational geometry tasks, see [26, 27]. Automatic construction of mesh algorithms and their mapping on the computer architecture present a challenging intellectual problem of great practical importance. In fact, we need to create a mathematical knowledge base with expandable sets of computational methods and technologies. A prototype of such a base is presented in [28].

4   ICE Structure and Technical Features

We will consider the ICE for the grid generation as a DELAUNAY subsystem of the Basic System of Modeling (BSM [15]). In fact, it presents a library for grid generation and transformation algorithms, as well as a set of system instruments for data processing and for supporting the external and internal communications. Because of the rich functionality and a large volume of the ICE code, the following technical requirements must be provided in order to ensure an efficient long life cycle of the proposed applied program product for the grid generation.

- Flexible extendability of the IBVPs to be solved by means of the BSM as well as a manifold of applicable numerical methods and technologies for the grid generation without program limitations on the degrees of freedom (d.o.f.) and the number of computer nodes, cores and other hardware.
- Adaptation to the evolution of computer architectures and platforms; automatic mapping of the algorithm structures onto hardware configuration.
- Compatibility of the flexible data structures with the conventional formats to provide the efficient re-use of the external products for grid generation, which presents great intellectual potential.
- High-performance of the developed software, scalable parallelism based on the hybrid programming tools, minimization of expensive communications, and code optimization on the heterogeneous multi-processor supercomputers with distributed and hierarchical shared memory.
- Multi-language interaction and consistency of various program components, enabling working contacts for different groups of developers, as well as creating friendly interfaces for the end users of different professional background. The considered integrated program environment must have valuable system

support, aimed at the maintenance, collective operations, information security and further high productive development. The corresponding intelligent instruments and big data support must constitute the infrastructure to support the following system procedures.

– Automatic verification and validation of the codes, as well as testing and comparative experimental analysis of the algorithms.
– Generating the multi-variant program configurations for a specific application by assembling the functional modules.
– Data structures control and transformations to provide the internal component compatibility and re-using of the external products.
– Deep learning of the grid generation issues: creating a knowledge database on mesh constructing methods, grid quality analysis, automatic selection of the available algorithms based on cognitive technologies.

5   Conclusion

The conception, main components and data structure of the integrated computational environment for constructing a wide class of multi-dimensional quasi-structured grids have been considered. The objective of the development is to provide high-productive program tools for the modern supercomputers with a long life cycle aimed at efficient support of the mathematical modelling in various applications.

References

1. Il'in, V.P.: Mathematical Modeling, Part I: Continuous and Discrete Models. SBRAS Publ, Novosibirsk (2017). (in Russian)
2. Godunov, S.K., Romenski, E.L., Chamakov, G.A.: Grid generate ioncomplicated domains by qusuconformal mapping. Trudy IM SBRAS Novosibirsk **18**, 75–84 (1990). (in Russian)
3. Terekhov, K., Vassilevski, Y.: Mesh modification and adaption within INMOST programming platform. In: Proceedings of the 9th International Conference NUM-GRID 2018, Moscow, LNCSE, vol. 131, pp. 243–255 (2018)
4. Garanzha, V.A.: Variational principles in grid generation and geometric modelling. Numer. Lin. Alg. **11**, 535–563 (2003)
5. Bronina, T.N., Gasilova, I.A., Ushakova, O.V.: Algorihms fort hree- dimensional structured grid generation. Zh. Vychisl. Mat. Met. Fiz. **43**, 875–883 (2003). (in Russian)
6. Fritzke, B.: Growing cell structuring - a self-organizing networks for unsupervised learning. Neural Netw. **7**(9), 1441–1460 (1994)
7. Ivanenko, S.A.: On the existence of equation for description of the classes of non-singular curvilinear coordinates on arbitrary domain. Zh. Vgchisl. Mat. Phys. **42**, 47–52 (2002). (in Russian)
8. Liseikin, V.D.: Grid Generation Methods. SC. Springer, Cham (2017). https://doi.org/10.1007/978-3-319-57846-0
9. Il'in, V.P.: DELAUNAY: technological media for grid generation. Sib. J. Industr. Appl. Math. **16**, 83–97 (2013). (in Russian)

10. Funken, S.A., Schmidt, A.: Ameshref: a Matlab-toolbox for adaptive mesh refinement in two dimensions. In: Proceedings of the 9th International Conference NUMGRID 2018, Moscow, LNCSE, vol. 131, pp. 269–279 (2018)

11. Li, S.: Mesh curving refinement based on cubic bezier surface for high-order discontinuous Galerkin methods. In: Proceedings of the 9th International Conference NUMGRID 2018, Moscow, LNCSE, vol. 131, pp. 205–216 (2018)

12. Ushakova, O.V. (ed.): Advances in Grid Generations. Nova Sci. Publ, New York (2007)

13. Zint, D., Grosso, R., Aizinger, V., Kostler, H.: Generation of block structured grids on complex domains for high performance simulation. In: Proceedings of the 9th International Conference NUMGRID 2018, Moscow, LNCSE, vol. 131, pp. 87–99 (2018)

14. Khademi, A., Korotov, S., Vatne, J.E.: On equivalence of maximum angle conditions for tetrahedral finite element meshes. Performance Simulation. In: Proceedings of the 9th International Conference NUMGRID 2018, Moscow, LNCSE, vol. 131, pp. 101–108 (2018)

15. Gartner, K., Kamenski, L.: Why do we need voronoi cells and delaunay Meshes? In: Proceedings of the 9th International Conference NUMGRID 2018, Moscow, LNCSE, vol. 131, pp. 45–60 (2018)

16. International Meshing Roundtable: www.imr.sandia.gov/18imr

17. Internat. J. for Numer. Math. in Eng., 58(2), special issue "Trends in Unstructured Mesh Generation" (2003)

18. Schoberl, J.: Netgen-an advancing front 2D/3D-mesh generator based on abstract rules. Comput. Visualizat. Sci. **1**, 41–52 (1997)

19. Geuzain, C., Remacle, J.-F.: Gmsh: a 3-D finite element mesh generator with built-in pre- and post-processing facilities. Int. J. Num. Methods Eng. **79**, 1309–1331 (2009)

20. Meng, X., Duan, Z., Yang, Q., Liang, X.: Local PEBI grid generation method for reverse faults. Comput. Geosci. **110**, 73–80 (2018)

21. Ponting, D.K.: Corner point geometry in reservoir simulation. In: Proceedings of the 1st European Conference on Mathematics in Oil Recovery, Cambridge, pp. 45–65 (1989)

22. DUNE.URL: http://www.dune-project.org. Accessed 15 Jan 2016

23. Il'in, V.P., Gladkih, V.S.: Basic system of modeling (BSM): the conception, architecture and methodology. In: Proceedings of the International Conference on Modern Problems of Mathematical Modeling, Image Processing and Parallel Computing. (MPMMIP & PC-2017) DSTU Publ. Rostov-Don, pp. 151–158 (2017). (in Russian)

24. Ilin, V.P.: The Conception, Requirements and Structure of the Integrated Computational Environment. In: Voevodin, Vladimir, Sobolev, Sergey (eds.) RuSCDays 2018. CCIS, vol. 965, pp. 653–665. Springer, Cham (2019). https://doi.org/10.1007/978-3-030-05807-4_56

25. HERBARIUM. http://tflex.ru/about/publications/detail/index.php?ID=3846

26. Cottrell, J., Hughes, T., Bazilevs, Y.: "Isogeometric Analysis", Towards Integration of CAD and FEA. Wiley, Singapore (2009)

27. Delfour, M., Zolesio, J.-P.: Shape and Geometries. Metrics, Analysis, Differential Calculus, and Optimization, SIAM Publication, Philadelphia (2011)

28. ALGOWIKI. https://algowiki-project.org

29. Ilin, Valery: On an Integrated Computational Environment for Numerical Algebra. In: Sokolinsky, Leonid, Zymbler, Mikhail (eds.) PCT 2019. CCIS, vol. 1063, pp. 91–106. Springer, Cham (2019). https://doi.org/10.1007/978-3-030-28163-2_7

Parallel Numerical Algorithms

Optimal Control of Chemical Reactions with the Parallel Multi-memetic Algorithm

Maxim Sakharov[1]($\boxtimes$), Kamila Koledina[2,3], Irek Gubaydullin[2,3], and Anatoly Karpenko[1]

[1] Bauman MSTU, Moscow, Russia
max.sfn90@gmail.com
[2] Institute of Petrochemistry and Catalysis of RAS, Ufa, Russia
[3] Ufa State Petroleum Technological University, Ufa, Russia

Abstract. This paper deals with the parallel multi-memetic algorithm based on the Mind Evolutionary Computation (*MEC*) technique for solving optimal control problems of various chemical reactions. The article describes the algorithm outline along with its parallel software implementation, which was utilized to obtain the optimal control for two chemical reaction models. The first model describes the thermally-stimulated luminescence of polyarylenephthalides; the second one - the catalytic hydroalumination of olefins. Both processes are of significant practical importance. In this work, the optimal control problem was reduced to a non-linear high-dimensional global minimization problem and was solved with the proposed algorithm. The numerical experiment results are presented in the paper.

Keywords: Optimal control · Memetic algorithms · Global optimization · Chemical reactions

1 Introduction

Nowadays real-world global optimization problems in various fields are computationally expensive due to a large number of optimized parameters and the non-trivial landscape of an objective function. Solving such problems within a reasonable time requires parallel computing systems. In addition, to solve the optimization problems efficiently, optimization algorithms shall take into account a parallel system architecture [1]. This implies developing specialized parallel optimization methods.

This work deals with grid systems made of heterogeneous personal computers (desktop grids), which are widely used for scientific computations [2]. Their

Irek Gubaydullin—This research was performed due to the Russian Science Foundation grant (project No. 19–71–00006) and RFBR according to the research projects No. 18–07–00341.

popularity is caused by a relatively low cost and simple scaling. On the other hand, the desktop grids require intermediate software to organize task scheduling and communication between the computing nodes.

Many real-world global optimization problems are frequently solved using various population-based algorithms [3]. One of their main advantages is a high probability to localize solutions that are close to the global optimum (sub-optimal solutions). Such solutions are often sufficient in practice. However, the efficiency of population-based algorithms heavily depends on the numeric values of their free parameters, which should be selected based on the characteristics of a problem in hand. However, it is not always feasible to tune the algorithm to every optimization problem. This especially applies to computationally expensive problems, when the number of evaluations is crucial as one evaluation can take dozens of minutes. This is why modern optimization techniques often utilize a problem preliminary analysis and pre-processing. This includes initial data analysis, dimensionality reduction of a search domain, landscape analysis of the objective function, etc. [4].

A two-level adaptation technique for the population-based algorithms was proposed in [5]. It was designed to extract information from an objective function prior to the optimization process at the first level and provide adaptation capabilities for fine-tuning at the second level. The first level is based on the proposed landscape analysis (LA) method, which utilizes a Lebesgue integrals concept and allows grouping objective functions into three categories. Each category suggests using specific values of the free parameters included in the basic algorithm. At the second level, it was proposed to utilize multi-memetic hybridization of the basic algorithm with suitable local search techniques.

When solving an optimization problem on parallel computing systems in general, and on desktop grids in particular, one of the main difficulties is the optimal mapping problem [6] – how to distribute sub-problem groups over the processors. A static load balancing method for desktop grids was proposed in [7] to cope with this problem. It minimizes the number of interactions between the computing nodes. The static load balancing is based on results of the LA procedure and helps to allocate more computational resources to the promising search sub-areas of the domain. Such tight integration between the algorithm and the load balancing provides the algorithm consistency with the computing system architecture.

In this work, the Simple Mind Evolutionary Computation (Simple MEC, SMEC) algorithm [8] was selected for investigation because it is highly suitable for parallel computations, especially for desktop grids. This paper presents a modified parallel MEC algorithm with an incorporated static load balancing method. An outline of the algorithm is described in this work along with its parallel software implementation, which was utilized to obtain optimal control for models of two chemical reactions. The first model describes the thermally-stimulated luminescence of polyarylenephthalides; the second one - the catalytic hydroalumination of olefins. Both processes are of significant practical importance. In this work, the optimal control problem was reduced to a non-linear high-dimensional global min-imization problem and was solved with the proposed algorithm.

2 Problem Formulation

The paper considers a deterministic global constrained minimization problem

$$\min_{X \in D \subset R^n} \Phi(X) = \Phi(X^*) = \Phi^*, \tag{1}$$

Here $\Phi(X)$ is the scalar objective function; $\Phi(X^*) = \Phi^*$ is the required minimal value; $X^* = (x_1, x_2, ..., x_n)$ is the n-dimensional vector of variables; R^n is the n-dimensional arithmetical space; D is the constrained search domain.

Initial values of vector X are generated within a domain D_0, which is defined as follows

$$D_0 = \left\{ X | x^{min} \leq x_i \leq x^{max}, i \in [1:n] \right\} \subset R^n. \tag{2}$$

The death penalty technique was utilized to handle constraints [1], meaning that if any vector X is located outside the search domain D it is removed from the population and replaced with a new one.

3 Parallel Multi-memetic Global Optimization Algorithm

The original *SMEC* algorithm [9] is inspired by human society and simulates some aspects of human behaviour. An individual s is considered as an intelligent agent, which operates in group S made of analogous individuals. During evolution, every individual is affected by other individuals within the group. In order to achieve a high position within the group, an individual has to learn from the most successful individuals in this group, while the groups should follow the same logic to stay alive in the intergroup competition.

The modified parallel multi-memetic *MEC* algorithm (*M3MEC–P*), in turn, starts with the proposed *LA* procedure, which is based on the concept of Lebesgue integral [5] and divides the range space of the objective function into levels based on values $\Phi(X)$. This stage can be described as follows.

1. Generate N quasi-random n-dimensional vectors within domain D_0. In this work LP_τ sequence was used to generate the quasi-random numbers since it provides high-quality coverage of the domain.
2. For every X_r, $r \in [1: N]$ calculate the corresponding values of the objective function Φ_r and sort those vectors in ascending order of values Φ_r, $r \in [r: N]$.
3. Equally divide a set of vectors $(X_1, X_2, \ldots, X_N)$ into $|K|$ sub–domains, so that sub–domain K_1 would contain the lowest values of $\Phi(X)$.
4. For every sub–domain K_l, $l \in [1: |K|]$ calculate its diameter d_l – a maximum Euclidean distance between any two individuals within this sub–domain (Fig. 1).
5. Build a linear approximation for the dependency of diameter $d(l)$ on the sub–domain number, using the least squares method.

6. Put the objective function $\Phi(X)$ into one of three categories [10] based on the calculated values (Table 1). Each of the three categories represents a certain topology of an objective function $\Phi(X)$.

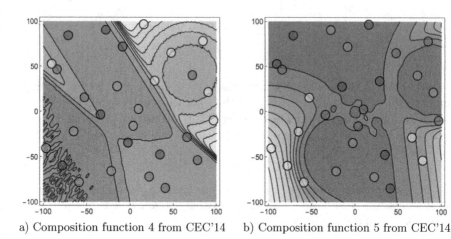

a) Composition function 4 from CEC'14   b) Composition function 5 from CEC'14

Fig. 1. Results of the landscape analysis for a few benchmark functions ($n = 2$)

There are three possible cases for approximated dependency $d(l)$: d can be an increasing function of l; d can decrease as l grows; $d(l)$ can be neither decreasing nor increasing. Within the scope of this work it is assumed that the latter scenario takes place when a slope angle of the approximated line is within $\pm 5°$. Each case represents a certain set of the numeric values of *M3MEC-P*'s free parameters suggested based on the numeric studies [10].

Table 1. The classification of objective functions based on the *LA* results.

$d(l)$ increases	$d(l)$ neither increases nor decreases	$d(l)$ decreases
Nested sub-domains with the dense first domain (category I)	Non-intersected sub-domains of the same size (category II)	Distributed sub-domains with potential minima (category III)

The *M3MEC–P* algorithm also modifies the local search stage of *SMEC* in order to include multi-memetic optimization. The meme is selected by a simple random hyper-heuristic [11,12]. Once the most suitable meme is selected for a specific sub-population, it is applied randomly to half of its individuals for $\lambda_{ls} = 10$ iterations. The dissimilation stage of *SMEC*, which governs a global search, was not modified in *M3MEC–P*.

In this work, three local search methods were utilized, namely, Nelder–Mead method [13], Hooke–Jeeves method [14] and Monte–Carlo method [14]. Only zero-order methods were used to deal with the problems where the objective function derivative is not available explicitly and its approximation is computationally expensive.

The local search and dissimilation stages are performed in parallel independently for each sub-population. To map those sub-populations on the available computing nodes, the static load balancing was proposed by the authors [7] specifically for desktop grids.

We have modified the initialization stage described above so that at step 2, apart from calculating values of the objective function Φ_r, the time required for those calculations t_r would also be measured. The proposed adaptive load balancing method can be described as follows.

1. For each sub-population K_l, $l \in [1: |K|]$, we analyse all time measurements t_r for the corresponding vectors X_r, $r \in [1: N/|K|]$ whether there are outliers or not.
2. All found outliers t^* are excluded from the sub-populations. A new sub-population is composed of those outliers and it can be investigated by the user's request after the computational process is over.
3. All available computing nodes are sorted by their computation power; then the first sub-population K_1 is sent to the first node.
4. The individuals of other sub-populations are re–distributed between neighbouring sub–populations starting from K_2 so that the average calculation time would be approximately the same for every sub-population. Balanced sub-populations K_l, $l \in [2: |K|]$ are then mapped onto the computational nodes.

The modified local search and dissimilation stages are launched on each computing node with the specific values of free parameters in accordance with the results of the landscape analysis. Each computing node utilizes the computational process stagnation as a termination criterion, while the algorithm in general works in a synchronous mode so that the final result is calculated when all computing nodes have completed their tasks.

4 Optimal Control Problems

The Modified Multi-Memetic *MEC* algorithm (*M3MEC*) and the utilized memes were implemented by the authors in Wolfram Mathematica. The software implementation has a modular structure, which helps to easily modify the algorithm and extend it with additional assisting methods and memes. Mathematica also offers some tools to manage computational nodes in the desktop grids. The proposed parallel algorithm and its software implementation were used to solve two computationally expensive optimization problems: controlling the thermally-stimulated luminescence of polyarylenephthalides (*PAP*) and controlling the reaction of catalytic hydroalumination of olefins.

4.1 Thermally-Stimulated Luminescence of Polyarylenephthalides

Nowadays, organic polymer materials are widely used in the field of optoelec-
tronics. The polyarylenephthalides (PAP) are high-molecular compositions that
belong to a class of unconjugated cardo polymers. The PAP exhibit good optical
and electrophysical characteristics along with the thermally-stimulated luminescence. Determining the origins of PAP luminescent states is of both fundamental
and practical importance [15]. In the Ufa Federal Research Centre of the Russian
Academy of Science, the experimental and theoretical studies of the thermally-
stimulated luminescence are conducted under the supervision of professor Khur-
san S. L. and senior research fellow Antipin V. A. [15].

Physical experiments [15,16] suggest that there are at least two types of
stable reactive species, which are produced in PAP. These species have different
activation energy levels or, in other words, trap states. Table 2 shows the reaction
network and the activation parameter values [16].

The model studied in this work includes the following process: recombination
of stable ion-radicals (Y_1, Y_2); repopulation of ion radical trap states; luminescence or, in other words, deactivation of excited state Y_3 and emission of Y_4 light
quantum.

Table 2. The kinetic model of the PAP thermally-stimulated luminescence.

Reaction network	lnk^0, sec^{-1}	E, J/mol
1)$2Y_1 \rightarrow Y_3$	5.32	69944
2)$2Y_2 \rightarrow Y_3$	12.2	101360
3)$Y_3 \rightarrow 2Y_2$	18.4	21610
4)$Y_3 \rightarrow Y_4$	9.95	69944

Table 2 contains kinetic parameters for every stage of the reaction network:
lnk^0 is a logarithm of the pre-exponential factor in an Arrhenius equation, sec^{-1},
E is the stage activation energy, J/mol. Here Y_1, Y_2 represent the initial stable
species of various nature; Y_3 is a certain excited state where Y_1 and Y_2 are headed
to; Y_4 – quants of light. $T(t)$ is the reaction temperature. The luminescence
intensity $I(t)$ is proportional to the recombination rate $I(t) = K \cdot Y_3(t)$ [17].
The $I(t)$ is measured in relative units. Initial concentrations of the species in the
reaction are equal to $Y_1(0) = Y_1{}^0$; $Y_2(0) = Y_2{}^0$; $Y_3(0) = Y_4(0) = 0$.

The kinetic model takes into account how the temperature affects the luminescence. It is required to determine the law of temperature variation over time
$T(t)$, which would guarantee the desired law of intensity variation for the PAP
thermally-stimulated luminescence $I(t)$. The physical experimental setup for
chemical reaction imposes restrictions on the minimal and maximal values of
the temperature 298K$\leq T(t) \leq$ 460K.

The optimal control problem was transformed in this work to a global optimization problem in the following manner. Integration interval $[0; 2000]$ is discretized so that the length of one section $[t_i; t_{(i+1)}]$ meets the restrictions

imposed by the experimental setup on the velocity of the change in temperature $T(t)$. The values of $T(t_i)$, are the components of vector $X = (x_0, ..., x(n-1))$. The piecewise linear function was selected to approximate function $T(t)$. The following objective function was proposed in this study:

$$J(T(t)) = \int_0^{2000} |I_{ref}(t) - I(T(t))|^2 dt \rightarrow \min_{T(t)}. \qquad (3)$$

The global minimization problem 3 was solved using the proposed *M3MEC-P* algorithm and its software implementation. The dimension of the problem $n = 201$. The following values of the algorithm free parameters were utilized: the number of groups $\gamma = 40$; the number of individuals in each group $|S| = 20$; the stagnation iteration number $\lambda_{stop} = 100$; and the tolerance used for identifying stagnation was equal to $\varepsilon = 10^{-5}$. All computations were performed using a desktop grid made of eight personal computers that did not communicate with each other. The number of sub-populations $|K| = 8$ was selected to be equal to the number of computing nodes. To increase the probability of localizing global optima, the multi-start method with 25 launches was used.

The first set of experiments was devoted to studying dynamics of the kinetic model at constant temperature values in the reaction within the range of $T = 150\,°C..215\,°C$, with a 10-degree step (Fig. 2). The obtained results demonstrate that at any constant temperature the luminescence intensity $I(t)$ decreases over time. Furthermore, the higher the temperature is, the faster the luminescence intensity decreases.

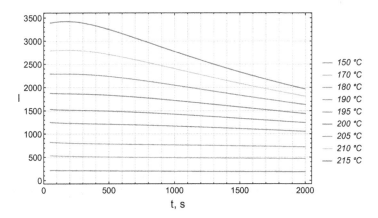

Fig. 2. The *PAP* luminescence intensity at various constant temperature values $T \in [150..215]\,°C$.

The second set of experiments was devoted to maintaining the constant value of the luminescence intensity $I(t)$. The results obtained for the target value $I_{ref}(t) = 300$ are displayed in Fig. 3. They suggest that in order to maintain the constant value of $I(t)$, the reaction temperature has to increase similar to the linear law of variation (approximately by $1\,^{\circ}C$ every 300 s). This implies the restriction on the reaction time as it is impossible to maintain the constant growth of the temperature in the experimental setup.

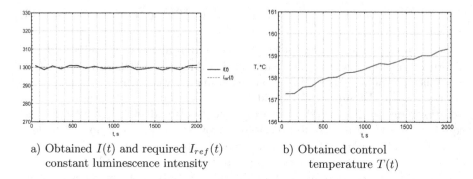

a) Obtained $I(t)$ and required $I_{ref}(t)$
constant luminescence intensity

b) Obtained control
temperature $T(t)$

Fig. 3. The optimal control for maintaining a constant value of PAP luminescence intensity $I(t)$

The third set of experiments was conducted to determine the law of temperature $T(t)$ variation that would provide the required pulse changes in the luminescence intensity $I(t)$. Figure 4 presents the obtained results. The optimal control trajectory repeats the required law of $I(t)$ variation with the addition of a linear increasing trend.

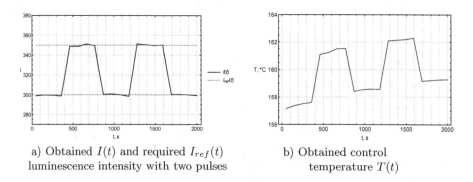

a) Obtained $I(t)$ and required $I_{ref}(t)$
luminescence intensity with two pulses

b) Obtained control
temperature $T(t)$

Fig. 4. The optimal control for providing pulse changes in the luminescence intensity $I(t)$

Figure 5 displays the results of the numerical experiments that were conducted to determine the law of reaction temperature $T(t)$ variation that would provide a constant rate of the luminescence changes, while Fig. 6 demonstrates the results that provide constant acceleration of the luminescence changes.

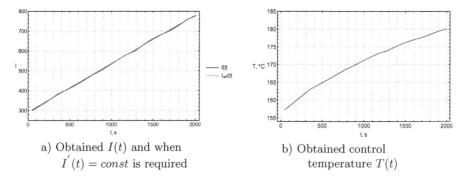

a) Obtained $I(t)$ and when $I'(t) = const$ is required

b) Obtained control temperature $T(t)$

Fig. 5. The optimal control for providing constant change rate of the luminescence intensity $I(t)$

4.2   Catalytic Hydroalumination of Olefins

The reaction of catalytic hydroalumination of olefins belongs to the metal complex catalysis reactions and produces cyclic and acyclic organoaluminium compounds with the given structure. The reaction is important in terms of practice [18]. In [19–21] the kinetic model of hydroalumination of olefins is presented with diisobutylaluminum chloride $ClAlBu_2^i$ ($DIBACH$), catalysed by Cp_2ZrCl_2 ($Bu = C_4H_9$, $Cp = C_5H_5$) (Table 3).

Table 3. Kinetic model of hydroalumination of olefins with $ClAlBu_2^i$ catalyzed by Cp_2ZrCl_2.

Reaction network	lnk^0, sec^{-1}	E, J/mol
1)$Y_{15} + Y_9 \rightleftarrows Y_{18} + Y_{11}$	46.7; -13.1	31.0; 1.1
2)$Y_{18} + Y_9 \rightarrow Y_{10} + Y_{13}$	20.15	12.4
3)$Y_{10} + Y_9 \rightarrow Y_2 + Y_{13} + Y_{11}$	24.28	10.6
4)$2Y_2 \rightleftarrows Y_1$	16.62; 9.37	7.02; 4.50
5)$Y_2 + Y_3 \rightarrow Y_4 + Y_5$	33.74	16.2
6)$Y_1 + Y_5 \rightarrow Y_8 + Y_2$	12.7	6.07
7)$Y_2 + Y_5 \rightarrow Y_8$	18.53	7.00
8)$Y_8 + Y_3 \rightarrow Y_4 + 2Y_5$	42.32	26.0
9)$Y_4 + Y_5 \rightarrow Y_7 + Y_6$	19.52	7.04
10)$Y_1 + Y_9 \rightarrow Y_8 + Y_{10}$	22.67	12.3
11)$Y_7 + Y_5 \rightarrow Y_2$	24.21	14.2
12)$Y_7 + Y_9 \rightarrow Y_{10}$	45.42	26.7
13)$Y_6 + Y_{11} \rightleftarrows Y_{19} + Y_9$	36.43; 16.4	18.2; 19.0
14)$Y_{15} + Y_5 \rightleftarrows Y_{10}$	9.39; 12.67	13.6; 10.1
15)$Y_{10} + Y_5 \rightleftarrows Y_2 + Y_9$	25.71; 4.97	11.1; 8.00

Here $Y_1 = [Cp_2ZrH_2 \cdot ClAlBu_2]_2$, $Y_2 = [Cp_2ZrH_2 \cdot ClAlBu_2]$, $Y_3 = CH_2CHR$, $Y_4 = Cp_2ZrCl(CH_2CH_2R)$, $Y_5 = HAlBu_2 - DIBACH$, $Y_6 = Bu_2Al(CH_2CH_2R)$, $Y_7 = Cp_2ZrHCl$, $Y_8 = [Cp_2ZrH_2 \cdot HAlBu_2 \cdot ClAlBu_2]$, $Y_9 = ClAlBu_2 - DIBACH$, $Y_{10} = [Cp_2ZrHCl \cdot ClAlBu_2]$, $Y_{11} = Cl_2AlBu$, $Y_{12} = [Cp_2ZrHBu \cdot ClAlBu_2]$, $Y_{13} = C_4H_8$, $Y_{14} = AlBu_3$, $Y_{15} = Cp_2ZrCl_2$, $Y_{16} = [Cp_2ZrH_2 \cdot HAlBu_2 \cdot 2(ClAlBu_2)]$, $Y_{17} = [Cp_2ZrH_2 \cdot HAlBu_2 \cdot ClAlBu_2]$, $Y_{18} = Cp_2ZrClBu$, $Y_{19} = ClBuAl(CH_2CH_2R)$, $Y_{20} = Cp_2ZrHBu \cdot ClAlBu_2$, $R = C_5H_{11}, C_6H_{13}, C_7H_{15}, C_8H_{17}, Bu = C_4H_9, Cp = C_5H_5$.

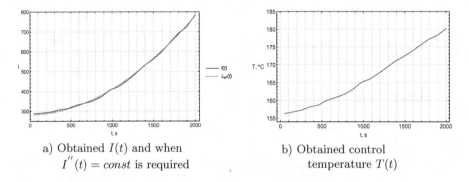

a) Obtained $I(t)$ and when
$I''(t) = const$ is required

b) Obtained control
temperature $T(t)$

Fig. 6. The optimal control for providing the constant acceleration of changes in the luminescence intensity $I(t)$

The reaction products are the organoaluminum compounds of a higher order: Bu_2AlR (Y_6) and $ClBuAl(CH_2CH_2R)$ (Y_{19}). The reaction temperature in physical experiments is set to be constant. However, the kinetic model allows one to take the temperature changes into account through the Arrhenius equations.

For the chemical reaction, it is required to maximize the output volume of Y_6 at $t = 250$ min. The physical experimental setup for the chemical reaction imposes restrictions on the minimal and maximal temperature values $0\,°C \leq T(t) \leq 30\,°C$.

The optimal control problem was transformed in this work to a global optimization problem in the same manner as the previous one. The integration interval $[0; 250]$ is discretized. Since *M3MEC-P* was developed for solving minimization problems, the following objective function was proposed:

$$J(T(t)) = -Y_6(250) \to \min_{T(t)}. \tag{4}$$

Global minimization problem 4 was also solved using the proposed *M3MEC–P* algorithm and its software implementation with same values of the algorithm free parameters, excluding the tolerance used to identify the stagnation; for this study the tolerance was equal to $\varepsilon = 10^{-7}$. The dimension of the problem $n = 251$. All computations were performed using the same desktop grid system. The multi-start method was utilized with 20 launches.

The first set of experiments was devoted to studying dynamics of the kinetic model at constant temperature values of the reaction within the range of $T = 1\,°C..30\,°C$ (Fig. 7). The maximum output concentration of Y_6 equals 0.954 mmol/l and is achieved approximately at $t = 74$ min ($T = 30\,°C$) while at $t = 250$ the maximum value is 0.789 mmol/l and corresponds to $T = 23\,°C$. The obtained results demonstrate that the higher the temperature produces a greater maximum value but the reaction time decreases.

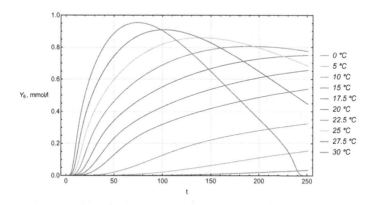

Fig. 7. The output $Y_6(t)$ at various constant temperature values $T \in [0...30]\,°C$

The second set of experiments was devoted to maximizing the output of Y_6 at $t = 250$ with the constraint that the temperature change cannot be greater than $1\,°C$ per 5. The results presented in Fig. 8 display that the obtained maximum ($Y_6(250) = 0.944$ mmol/l) exceeds the value previously found at the constant temperature.

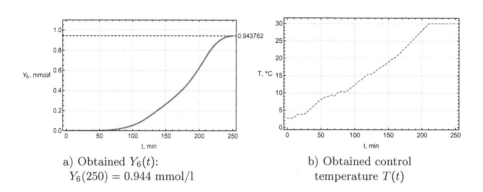

a) Obtained $Y_6(t)$:
   $Y_6(250) = 0.944$ mmol/l

b) Obtained control
   temperature $T(t)$

Fig. 8. The optimal control when maximizing $Y_6(250)$ with constraints on temperature changes: $1\,°C$ per 5 min

The third set of experiments was conducted under slightly relaxed constraints. The temperature change could not be greater than $5\,°C$ per $5\,min$. As a result, the maximum value of $Y_6(250)$ was increased even further - up to $0.952\,mmol/l$ (Fig. 9).

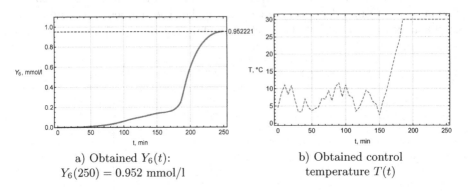

a) Obtained $Y_6(t)$:
$Y_6(250) = 0.952\ mmol/l$

b) Obtained control
temperature $T(t)$

Fig. 9. The optimal control when maximizing $Y_6(250)$ with the constraints on temperature changes: $5\,°C$ per $5\,min$

However, both the second and the third sets of experiments did not allow obtaining a maximum value that would exceed the maximum received at the constant temperature of 30 $(Y_6(74) = 0.954\ mmol/l)$. To overcome this issue, the next set of experiments was conducted without any constraints on the temperature change. As a result, a new maximum value was obtained $(Y_6(250) = 1.0\ mmol/l)$. Figure 10 demonstrates that the control temperature fluctuates heavily; this is why it is difficult to implement this control temperature in practice.

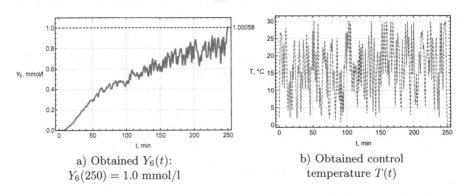

a) Obtained $Y_6(t)$:
$Y_6(250) = 1.0\ mmol/l$

b) Obtained control
temperature $T(t)$

Fig. 10. The optimal control when maximizing $Y_6(250)$ without constraints on temperature changes

Such fluctuations of the obtained control temperature $T(t)$ can be caused by the fact that reactions of metal composite catalysis are multistage ones and contain many intermediate complexes (Table 3). This is why local temperature $T(t)$ fluctuations affect concentrations of the intermediate substances (both stable and non-stable), which define the output of a target product.

5   Conclusions

This paper presents the new modified global optimization algorithm based on the parallel population and designed for desktop grids. In addition, the article describes its software implementation. The *M3MEC–P* algorithm is based on the adaptation strategy and the landscape analysis procedure originally proposed by the authors and incorporated into the traditional *SMEC* algorithm.

The algorithm is capable of adapting to various objective functions using both static and dynamic adaptation. The static adaptation was implemented through a landscape analysis, while the dynamic adaptation was made possible by utilizing several memes. The proposed landscape analysis is based on the Lebesgue integral concept and allows one to group objective functions into six categories. Each category suggests using a specific set of values for the algorithm free parameters. The proposed algorithm and its software implementation were proved to be efficient in solving a real-world computationally expensive global optimization problems: controlling the thermally-stimulated luminescence of polyaryleneph-thalides (*PAP*) and controlling the reaction of catalytic hydroalumination of olefins.

Further research will be devoted to multi-criteria control problems for the specified chemical reactions.

References

1. Karpenko, A.P.: Modern algorithms of search engine optimization. Nature-inspired optimization algorithms. Bauman MSTU Publication, Moscow, p. 446 (2014)
2. Sakharov, M.K., Karpenko, A.P., Velisevich, Y.I.: Multi-memetic mind evolutionary computation algorithm for loosely coupled systems of desktop computers. In: Science and Education of the Bauman MSTU, vol. 10, pp. 438–452 (2015). https://doi.org/10.7463/1015.0814435
3. Sakharov, M.K.: New adaptive multi-memetic global optimization algorithm for loosely coupled systems. In: Herald of the Bauman Moscow State Technical University, Series Instrument Engineering, no. 5, pp. 95–114 (2019). https://doi.org/10.18698/0236-3933-2019-5-95-114, (in Russia)
4. Mersmann, O. et al.: Exploratory landscape analysis. In: Proceedings of the 13th Annual Conference on Genetic and Evolutionary Computation. ACM, pp. 829–836 (2011). https://doi.org/10.1145/2001576.2001690
5. Sakharov, M., Karpenko, A.: Multi-memetic mind evolutionary computation algorithm based on the landscape analysis. In: Fagan, D., et al. (eds.) TPNC 2018. LNCS, vol. 11324, pp. 238–249. Springer, Cham (2018). https://doi.org/10.1007/978-3-030-04070-3_19
6. Voevodin, V.V., Voevodin, Vl. V.: Parallel Computations. SPb.: BHV-Peterburg, p. 608 (2004)

7. Sakharov, M.K., Karpenko, A.P.: Adaptive load balancing in the modified mind evolutionary computation algorithm. Supercomput. Front. Innov. **5**(4), 5–14 (2018). https://doi.org/10.14529/jsfi180401

8. Jie, J., Zeng, J.: Improved mind evolutionary computation for optimizations. In: Proceedings of the 5th World Congress on Intelligent Control and Automation, Hangzhou, China, pp. 2200–2204 (2004). https://doi.org/10.1109/WCICA.2004.1341978

9. Chengyi, S., Yan, S., Wanzhen, W.: A survey of MEC: 1998–2001. In: 2002 IEEE International Conference on Systems, Man and Cybernetics IEEE SMC2002, Hammamet, Tunisia. October 6–9. Institute of Electrical and Electronics Engineers Inc., vol. 6, pp. 445–453 (2002). https://doi.org/10.1109/ICSMC.2002.1175629

10. Sakharov, M., Karpenko, A.: Performance investigation of mind evolutionary computation algorithm and some of its modifications. In: Abraham, A., Kovalev, S., Tarassov, V., Snasel, V. (eds.) Proceedings of the First International Scientific Conference Intelligent Information Technologies for Industry (IITI 2016). AISC, vol. 450, pp. 475–486. Springer, Cham (2016). https://doi.org/10.1007/978-3-319-33609-1_43

11. Ong, Y.S., Lim, M.H., Zhu, N., Wong, K. W.: Classification of adaptive memetic algorithms: a comparative study. In: IEEE Transactions on Systems, Man, and Cybernetics, Part B: Cybernetics, pp. 141–152 (2006)

12. Karpenko, A.P., Sakharov, M.K.: New adaptive multi-memetic global optimization algorithm. In: Herald of the Bauman Moscow State Technical University, Series Natural Science, no. 2, pp. 17–31 (2019). https://doi.org/10.18698/1812-3368-2019-2-17-31

13. Nelder, J.A., Meade, R.: A simplex method for function minimization. Comput. J. **7**, 308–313 (1965)

14. Karpenko, A.P.: Optimization Methods (Introductory Course). http://bigor.bmstu.ru/

15. Antipin, V.A., Mamykin, D.A., Kazakov, V.P.: Recombination luminescence of poly(arylenephthalide) films induced by visible light. High Energy Chemistry **45**(4), 352–359 (2011)

16. Akhmetshina, L.R., Mambetova, Z.I., Ovchinnikov, M.Yu.: Mathematical modelling of thermoluminescence kinetics of polyarylenephthalides. In: V International Scientific Conference on Mathematical Modelling of Processes and Systems, pp. 79–83 (2016)

17. Sakharov, M., Karpenko, A.: Parallel multi-memetic global optimization algorithm for optimal control of polyarylenephthalide's thermally-stimulated luminescence. In: Le Thi, H.A., Le, H.M., Pham Dinh, T. (eds.) WCGO 2019. AISC, vol. 991, pp. 191–201. Springer, Cham (2020). https://doi.org/10.1007/978-3-030-21803-4_20

18. Parphenova, L.V., Pechatkina, S.V., Khalilov, L.M., Dzhemilev, U.M.: Study of hydroalumination of olefins catalysed with Cp2ZrCl2. In: Izv. RAS, Series Chemistry, vol. 2, pp. 311–322 (2005)

19. Gubaydullin, I., Koledina, K., Sayfullina, L.: Mathematical modelling of induction period of the olefins hydroalumination reaction by diisobutylaluminiumchloride catalysed with Cp2ZrCl2. Eng. J. **18**(1), 13–24 (2014)

20. Koledina, K.F., Gubaidullin, I.M.: Kinetics and mechanism of olefin catalytic hydroalumination by organoaluminum compounds. Russian J. Phys. Chem. A **90**(5), 914–921 (2016)

21. Nurislamova, L.F., Gubaydullin, I.M., Koledina, K.F., Safin, R.R.: Kinetic model of the catalytic hydroalumination of olefins with organoaluminum compounds. Reaction Kinetics, Mech. Catalysis **117**(1), 1–14 (2016)

On the Implementation of Taylor Models on Multiple Graphics Processing Units for Rigorous Computations

Nikolay M. Evstigneev and Oleg I. Ryabkov[(⊠)]

Federal Research Center "Computer Science and Control",
Institute for System Analysis, Russian Academy of Science, Moscow, Russia
evstigneevnm@yandex.ru, roi-techsup@yandex.ru

Abstract. Taylor Models present the polynomial generalization of the simple interval approach for rigorous computations of differential equations suggested by Martin Berz. These models are used to obtain better estimates for guaranteed enclosures of solutions. The authors of the recent papers describing the mathematical framework suggest some improvements of the models. These models are assumed to be applied to high dimensional systems of equations. Therefore, a parallel version of these models is required. In this paper improved Taylor models (TM) are implemented on the architecture of the general-purpose graphics processing units (GPU). Algebraic operations and algorithms are implemented with a particular optimization for the computational architecture, namely: addition, multiplication, convolution to two TMs, substitution of independent variable, integration with respect to a variable and boundary interval estimations. For this purpose interval arithmetic is implemented using intrinsic functions. The multiple GPU version is also implemented and its scalability is verified. In terms of acceleration various test examples are presented. The reduction operation is discovered to the bottleneck of the GPU performance. Using the GPU version for sufficiently large problem dimensions is suggested.

Keywords: Taylor model · GPU computations · Self-validated methods · Interval methods · Rigorous computations

1 Introduction

Validated numerics or rigorous computation is a numerical computation that includes mathematically strict error evaluation. The question of rigorous computations arises in many theoretical and applied problems. Practical problems include guaranteed enclosures and bounds in particle physics and astronomy problems [1,2], global optimization [3], data analysis [4], electrical engineering [5], chemical processes [6], ray tracing [7], uncertainty bounds in Kalman filters

This work is supported by RFBR grant no. 18-29-10008 mk.

L. Sokolinsky and M. Zymbler (Eds.): PCT 2020, CCIS 1263, pp. 85–99, 2020.
https://doi.org/10.1007/978-3-030-55326-5_7

[8], economics [9] etc. Theoretical problems are related to rigorous computations and verified proofs in graph theory, combinatorics [10], mathematical logic, the metric theory of continued fractions [11], hydrodynamics [12], dynamical systems [13] and other fields related with constructive algorithms. Interval arithmetic [14] is used in all these results to obtain guaranteed bounds of the computed values.

The authors are applying rigorous computations to prove the existence of periodic trajectories in systems of ordinary differential equations (ODEs). There are several methods for this problem; however, all of them involve the application of interval arithmetic [14,15] and deducing interval bounds of the ODEs solutions. These bounds are deduced with Picard iterations extended on the interval arithmetics. The simplest extension of interval Picard iterations is known as Moore's method [14]. Interval estimates are well known to be of little use for deducing the bounds with the Moore method from the computational point of view because of the so-called packing effect [16], whereby the interval estimate exponentially increases when moving along the trajectory. To overcome this drawback, a new object, the interval Taylor model, or simply the *Taylor model* (TM), was proposed in [2,17,18] by Martin Berz and co-authors and implemented in COSY software [19]. The Taylor Models are the polynomial generalization of simple interval approach for rigorous computations in differential equations. The COSY software demonstrates good bounds for the small size systems of ODEs. In the latest release, parallel computations are implemented for the solution of the ODE systems in the shared memory computational architecture. However, the TM calculation method requires $\mathcal{O}(m^{2n})$ computational operations, where m is the ODE system size and n is the degree of polynomials used in a TM. Such computational difficulty can become substantial for large m and n values. To overcome these drawbacks, the authors came with a different variant of Taylor Models in [20,21], where arithmetic operations, algorithms, estimates and rigorous theory are constructed both for the new TMs and computer-assisted proofs of periodic trajectories. These TMs were developed with the parallel implementation in mind. Such TMs now have $\mathcal{O}(m^n)$ computational complexity and can be extended to parallel implementation of all required operations. These operations include addition, multiplication, convolution to two TMs, independent variable substitution, and integration with respect to a variable and boundary interval estimations.

Therefore, the present research deals with the generalization of the authors' TMs for the computational architecture of the parallel multiple graphics processing units (GPUs). The extension of simple interval arithmetic on GPUs is implemented in NVIDIA CUDA Sample kit as well as in many publications related to ray tracing or computer graphics, e.g. [22,23]. However no research in the field of GPU implementation and applications to any extensions based on the interval arithmetics is known until now.

The paper is laid out as follows. Firstly, brief definitions and formal algorithms of TMs are presented. Then the implementation of these algorithms is shown and discussed. Performance measurements are followed with comparison to CPU version. Finally an example of a relatively big ODE system derived from the Bubnov–Galerkin approximation of the Navier-Stokes equations is considered.

All calculations are carried out on a personal multiple GPU cluster.

2 Definitions and Algorithms

The set of interval numbers is denoted by $\mathbf{R}$, a single valued interval is denoted by $[a,]$ for a value a, an interval $\mathbf{I}$ has upper $H(\mathbf{I})$ and lower $L(\mathbf{I})$ bounds, i.e. $\mathbf{I} = [L(\mathbf{I}), H(\mathbf{I})]$. The interval embedding operations are understood in the sense of interval arithmetics [14,15]. Following [20], one introduces an interval Taylor model, or a TM, of degree n of m variables $\vec{x} = (x_1, ..., x_m)^T$ and dimension q as a pair $<\vec{P}^n(\vec{x}), \vec{\mathbf{I}}>$. Here $\vec{P}$ is a vector polynomial of dimension q and degree no greater than n and $\vec{\mathbf{I}} = \mathbf{I}_1 \times ... \times \mathbf{I}_q \in \mathbf{R}_q$ is a vector interval, called the remainder. Component with number j of a vector TM of dimension q is defined as $\{TM\}_j = <\{\vec{P}^n(x_1, ..., x_m)\}_j, \{\vec{\mathbf{I}}\}_j>$, where $\{TM\}_j$ is the TM of dimension one and $j = 1, ..., q$. Any operation with a machine precision is denoted as $/ \cdot /$ requiring that the machine zero coincides with the exact zero. Interval median is denoted as $\bar{\mathbf{a}} = /0.5(L(\mathbf{a}) + H(\mathbf{a}))/$. Some additional sets are denoted as: $R_m(n)$ set of m tuples $(i_1, ..., i_m)$ of non-negative integers subject to inequality $i_1 + ... + i_m \leq n$, $S_m(n)$ the same set of m-tuples subject to equality $i_1 + ... + i_m = n$. A notation $TM(\vec{f})$ indicates that a vector function $\vec{f}$ satisfies a TM with the function defined on the domain $M = [-1,1]^m$. Three Taylor models TM, TM_1 and TM_2 are considered in the following algorithms for convenience. The coefficients of their polynomials are denoted by $\vec{a}_{n_1,...,n_m}$ for TM_1, $\vec{b}_{n_1,...,n_m}$ for TM_2, and $\vec{c}_{n_1,...,n_m}$ for TM, their interval parts (remainders) are denoted by $\vec{\mathbf{I}}^1$, $\vec{\mathbf{I}}^2$ and $\vec{\mathbf{I}}$ and the polynomial parts are denoted by $\vec{P}^{1,n}$, $\vec{P}^{2,n}$ and $\vec{P}^n$ respectively.

Now the basic operation algorithms are provided for performing necessary calculations. See [20,21] for rigorous definitions and theorems of these algorithms. The following algorithms are formulated in terms of mathematical operations as follows. Firstly, the name of the algorithm is given. Then the set of input parameters is provided, followed by the set of output parameters. Then the mathematical formulation of the algorithm is given without any implementation. A particular implementation is proved in the next section below.

Algorithm 1. *Sum of two TMs. Input(TM₁, TM₂). Output(TM).*

$$\vec{c}_{n_1,...,n_m} := /\vec{a}_{n_1,...,n_m} + \vec{b}_{n_1,...,n_m}/.$$

$$\vec{\mathbf{I}} := \vec{\mathbf{I}}_1 + \vec{\mathbf{I}}_2$$
$$+ \sum_{(n_1,...,n_m) \in R_m(n)} [-1,1] \cdot \left(\left([\vec{a}_{n_1,...,n_m},] + [\vec{b}_{n_1,...,n_m},] \right) - [\vec{c}_{n_1,...,n_m},] \right).$$

Algorithm 2. *Difference of two TMs. Input(TM₁, TM₂). Output(TM).*

$$\vec{c}_{n_1,...,n_m} := /\vec{a}_{n_1,...,n_m} - \vec{b}_{n_1,...,n_m}/.$$

$$\vec{\mathbf{I}} := \vec{\mathbf{I}}_1 + \vec{\mathbf{I}}_2$$

$$+ \sum_{(n_1,\ldots,n_m) \in R_m(n)} [-1,1] \cdot \left(\left([\vec{a}_{n_1,\ldots,n_m},] - [\vec{b}_{n_1,\ldots,n_m},] \right) - [\vec{c}_{n_1,\ldots,n_m},] \right).$$

Algorithm 3. *Sum of a TM and an interval. Input(TM$_1$, $\vec{\mathbf{y}}$).Output(TM).*

$$\vec{c}_{0,\ldots,0} := \Big/ \vec{a}_{0,\ldots,0} + \vec{\mathbf{y}} \Big/ ,$$

$$\vec{\mathbf{I}} := \vec{\mathbf{I}}_1 + ([\vec{a}_{0,\ldots,0},] + \vec{\mathbf{y}} - [\vec{c}_{0,\ldots,0},]).$$

$(n_1,\ldots,n_m) \neq (0,\ldots,0)$:

$$\vec{c}_{n_1,\ldots,n_m} := \vec{a}_{n_1,\ldots,n_m},$$

The addition and subtraction of TMs is commonly denoted as $TM_1 + TM_2$, $TM_1 - TM_2$ and $TM_1 + \vec{\mathbf{y}}$.

Algorithm 4. *Polynomial interval estimate. Input($\vec{P}^n$).Output($\vec{\mathbf{I}}$).*

$$\vec{\mathbf{I}} := \sum_{(n_1,\ldots,n_m) \in R_m(n)} [-1,1] \cdot [\vec{a}_{n_1,\ldots,n_m},].$$

Interval estimate of the polynomial is denoted as $\mathcal{B}(\vec{P}^{n,1})$.

Algorithm 5. *Product of two TMs. Input(TM$_1$, TM$_2$).Output(TM).*

$$\vec{c}_{n_1,\ldots,n_q} := \Big/ \sum_{j_1,\ldots,j_m;k_1,\ldots,k_m} \vec{a}_{j_1,\ldots,j_m} \cdot \vec{b}_{k_1,\ldots,k_m} \Big/ , \tag{1}$$

where indexes are: $(j_1,\ldots,j_m) \in R_m(n)$, $(k_1,\ldots,k_m) \in R_m(n)$, $j_1 + k_1 = n_1,\ldots,j_m + k_m = n_m$, $n_1 + \ldots + n_m \leq n$.

$$\vec{\mathbf{I}} := \vec{\mathbf{I}}_1\vec{\mathbf{I}}_2 + \vec{\mathbf{I}}_1\mathcal{B}(\vec{P}^{2,n}) + \vec{\mathbf{I}}_2\mathcal{B}(\vec{P}^{1,n}) + \vec{\mathbf{A}} + \vec{\mathbf{B}};$$

where:

$$\vec{\mathbf{A}} := \sum_{(n_1,\ldots,n_m) \in R_m(n)} [-1,1] \cdot$$

$$\cdot \left(\sum_{j_1,\ldots,j_m;k_1,\ldots,k_m} [\vec{a}_{j_1,\ldots,j_m},] \cdot [\vec{b}_{k_1,\ldots,k_m},] - [\vec{c}_{n_1,\ldots,n_m},] \right),$$

with the conditions listed above for the indices in (1) and

$$\vec{\mathbf{B}} := \sum_{\substack{0 \leq j,k \leq n \\ j+k > n}} \mathcal{B} \left(\sum_{(j_1,\ldots,j_m) \in S_m(n)} \vec{a}_{j_1,\ldots,j_m} x_1^{j_1}\ldots x_m^{j_m} \right) \cdot$$

$$\cdot \mathcal{B} \left(\sum_{(k_1,\ldots,k_m) \in S_m(n)} \vec{b}_{k_1,\ldots,k_m} x_1^{k_1}\ldots x_m^{k_m} \right).$$

The operation of product over interval vectors is their componentwise product. Here $\vec{\mathbf{A}}$ gives the interval estimate of the error in the computation of the coefficients of the product of polynomials, and $\vec{\mathbf{B}}$ is the interval estimate of the terms in the product of polynomials with the degree higher than n.

Algorithm 6. *Product of a TM and an interval. Input(TM$_1$, $\mathbf{u}$).Output(TM).*

$$\vec{c}_{n_1,\dots,n_m} := /\bar{u}\vec{a}_{n_1,\dots,n_m}/.$$

$$\vec{\mathbf{I}} := \mathbf{u}\vec{\mathbf{I}}_1 + \sum_{n_1,\dots,n_m \in R_m(n)} [-1,1]\left((\mathbf{u}[\vec{a}_{n_1,\dots,n_m},]) - [\vec{c}_{n_1,\dots,n_m},]\right).$$

The product operations are commonly denoted as TM_1TM_2 and $TM_1\mathbf{u}$.

Algorithm 7. *Independent variable substitution.   Input(TM$_1$,   k, C).Output(TM).*

$$\vec{c}_{n_1,\dots,n_m} := \begin{cases} /0/, n_k \neq 0; \\ /\displaystyle\sum_{j=0}^{l(n)} \vec{a}_{n_1,\dots,n_{k-1},j,n_{k+1},\dots,n_m} C^j / , n_k = 0. \end{cases}$$

$$\vec{\mathbf{I}} := \vec{\mathbf{I}}_1 + \sum_{(n_1,\dots,n_m) \in R_m(n)} [-1,1]\left(\vec{\mathbf{c}}_{n_1,\dots,n_m} - [\vec{c}_{n_1,\dots,n_m},]\right),$$

where:

$$\vec{\mathbf{c}}_{n_1,\dots,n_m} := \begin{cases} [0,], n_k \neq 0; \\ \displaystyle\sum_{j=0}^{l(n)} [\vec{a}_{n_1,\dots,n_{k-1}jn_{k+1},\dots,n_m},][C,]^j, n_k = 0. \end{cases}$$

with $l(n) = n - (n_1 + \dots + n_{k-1} + n_{k+1} + \dots + n_m)$.

The above algorithm substitutes the independent variable with number $k \in [1, m]$ of the TM_1 by the value of C; it can be designated as $TM_1|_{x_k=C}$.

Algorithm 8. *Integration of TM$_1$ over $[0, x_k]$. Input(TM$_1$, k).Output(TM).*

$$\vec{c}_{n_1,\dots n_k,\dots,n_m} := \begin{cases} /0/, n_k = 0; \\ /\dfrac{\vec{a}_{n_1,\dots n_k-1,\dots,n_m}}{n_k} / , n_k \neq 0. \end{cases}$$

$$\vec{\mathbf{I}} := [-1,1]\vec{\mathbf{I}}_1$$

$$+ \sum_{\substack{(n_1,\dots,n_m) \in R_m(n) \\ n_k \geq 1}} [-1,1]\left(\frac{[\vec{a}_{n_1,\dots n_k-1,\dots,n_m},]}{[n_k,]} - [\vec{c}_{n_1,\dots n_k,\dots,n_m},]\right)$$

$$+ \sum_{\substack{(n_1,\dots,n_m) \in S_m(n+1) \\ n_k \geq 1}} [-1,1]\frac{[\vec{a}_{n_1,\dots n_k-1,\dots,n_m},]}{[n_k,]}.$$

The above algorithm provides the integration with variable upper limit w.r.t. x_k and is denoted as $\int_0^{x_k} TM_1 dx_k$.

The following algorithm returns an interval containing the external boundary of a TM. Such operation is denoted as $\mathcal{EB}(TM_1)$.

Algorithm 9. *External boundary of* TM_1. *Input(*TM_1*).Output(*$\vec{\mathbf{I}}$*).*

$$\vec{\mathbf{I}} = \mathcal{B}(\vec{P}^{n,1}) + \vec{\mathbf{I}}^1.$$

The following algorithm verifies $TM_1 \subseteq TM_2$ and returns 1 if the inclusion is valid and 0 otherwise.

Algorithm 10. *Check TM inclusion. Input(*TM_1, TM_2*).Output(res).*

$$res := \begin{cases} 1, \textit{if } \{\mathcal{EB}\left(TM_1 - <\vec{P}^{2,n}, [\vec{0,}] >\right)\}_j \subseteq \mathbf{I}_j^2, \forall j = 1, ..., q, \\ 0, \textit{else.} \end{cases}$$

Algorithm 11. *Matrix of the TM product. Input(*TM_1, $\mathbf{W}$*).Output(TM).*

$$\{TM\}_l := \sum_{j=1}^q \mathbf{w}_{l,j}\{TM_1\}_j, \forall l = 1, ..., q.$$

The product of TM_1 with an interval matrix $\mathbf{W} \in \mathbf{R}^{q \times q}$ is denoted as $\mathbf{W}TM_1$.

The next algorithm provides the convolution of two TMs and is designated as $TM_1 * TM_2$. It is essentially built around TM multiplication and summation algorithms. Such operation is common in many nonlinear partial differential equations (PDEs) when approximated by spectral or finite element methods including nonlinear Schrodinger equation, Korteweg–de Vries equation, Navier-Stokes equation etc.

Algorithm 12. *Convolution of* TM_1 *and* TM_2 *weighted by interval matrix of weights* $\mathbf{W}$ *ordered by two general index matrices I, J. Input(*TM_1, TM_2, $\mathbf{W}$, I, J*). Output(TM).*

$$\{TM\}_j = \sum_{k=0}^{m_j} \mathbf{W}_{j,k}\{TM_1\}_{I_{j,k}}\{TM_2\}_{J_{j,k}}, \forall j \in [0,m].$$

Here m_j depends on whether the index is in or out of bounds. The index matrices represent the convolution rule. For example, for a 1D circular convolution one would use $I_{j,k} = k$ and $J_{j,k} = j - k$. The convolution is computationally extensive resulting in $\mathcal{O}(m^2)$ operations. A better approach to a circular convolution is to use the interval fast Fourier transform algorithm. Such algorithm is suggested to be applied to intervals in [24], though no generalization for the TMs is provided. This is the goal for future work.

The algorithms listed above are used in rigorous assertions of the periodic orbits of ODE systems. Other algorithms built around the provided ones are available in [20, 21] and are not considered here.

3 Parallel GPU Implementation

The implementation of the algorithms provided above is done in an object-oriented manner on CUDA C++ using templated classes with encapsulation of details that depend on the computational architecture. Implementation of CPU interval class is based on C++ BOOST library with optimization of setting one-sided rounding operations before a pure interval computational block. Implementation of GPU interval class is based on CUDA interval library header file available in CUDA Samples in the "Advanced" folder, where all basic operations are implemented with the correct rounding directions using intrinsics. Memory operation class encapsulates all necessary operations in a particular computational architecture. Classes of tensors (including vectors and matrices), polynomials, remainders and TMs are templated from classes of memory operations and intervals. The structure of a TM class is organized into two main groups: TM as a data storage (example shown in Listing 1.1) and TM operations storage where all operations are kept (example shown in Listing 1.2). Such structure is less common for C++ programs when the operators are usually overloaded. However, such organization may be used to avoid creation of temporal objects thus maximizing efficiency and minimizing memory consumption.

```
template<int Deg,class Int,class Memory>//Deg - polynomial degree,
    Int - interval class, Mem - memory manager
class taylor_model
{
private:
    int vars_n_, dim_; //number of independent polynomial variables
    polynoms_array_type polynoms_; //polynomial parts of TMs
    remainders_array_type remainders_; //remainder parts of TMs
    bnds_array_type bnds_; //polynomial boundary estimate
    deg_bnds_array_type deg_bnds_; //polynomial boundary estimate
        for deg>n
public:
...
}
```

Listing 1.1. TM storage class example

```
template<int Degree, class Interval, class Memory>
class taylor_model_operations
{
public:
    typedef taylor_model<Degree,Interval,Memory> taylor_model_type;
        //TM type of TM storage class
    void init_taylor_model(taylor_model_type& x)const; //inits TM
    void free_taylor_model(taylor_model_type& x)const; //frees TM
    ...
    //Sum of two TMs with x = TM_1, y = TM_2, z = TM:
    add(const taylor_model_type &x, const taylor_model_type &y,
        taylor_model_type &z);
    ...
}
```

Listing 1.2. TM operations class example with the sum of two TMs operation.

Summation of two TMs according to the Algorithm 1 is presented as CUDA implementation example in Listings 1.3, 1.4 and 1.5. In all implementations, the

polynomial parts and the remainder parts are separated into two device kernels
to decrease the register pressure, grouped in Listings 1.3 and 1.4.

```
template<int Deg,class Int,class Mem>//Deg - polynomial degree, Int
    - interval class, Mem - memory manager
__global__ void ker_add_remainders(
int tm_q, //TM dimension q
bnds_arr<Deg,Int,Mem> x_remainders, // TM₁ remainders
bnds_arr<Deg,Int,Mem> y_remainders, // TM₁ remainders
bnds_arr<Deg,Int,Mem> round_errors, // Rounding error intervals
bnds_arr<Deg,Int,Mem> z_remainders) // TM remainders
{
    typedef scalar<Deg,Int,Mem>     scalar_t;     // non interval
        scalar
    typedef interval<Deg,Int,Mem>   interval_t;   // interval scalar

    int i = blockIdx.x * blockDim.x + threadIdx.x;
    if (!(i < tm_q)) return; //running over all TMs

    interval_t  x_rem = x_remainders(i),
                y_rem = y_remainders(i),
                round_err = round_errors(i);

    interval_t  res = x_rem + y_rem + round_err;
    z_remainders(i) = res;
}
```

Listing 1.3. Cuda kernel for adding remainder parts in the summation algorithm.

```
template<int Deg,class Int,class Mem>
__global__ void ker_polynomial_add(int poly_len,
                                    poly_arr<Deg,Int,Mem> x,
                                    poly_arr<Deg,Int,Mem> y,
                                    poly_arr<Deg,Int,Mem> z,
                                    int_poly_arr<Deg,Int,Mem> z_err)
{
    typedef scalar<Deg,Int,Mem>     scalar_t;
    typedef interval<Deg,Int,Mem>   interval_t;

    int i = blockIdx.x * blockDim.x + threadIdx.x;
    if (!(i < x.size())) return; //running over all monomials of all
        polynomials

    int i_poly = i/poly_len,
        i_term = i%poly_len;

    scalar_t   res(0.f);
    interval_t exact_res(0.f);

    scalar_t    cum_1 = x(i_poly, i_term),
                cum_2 = y(i_poly, i_term);
    res = cum_1+cum_2;
    exact_res = interval_t(cum_1)+interval_t(cum_2);

    z(i_poly, i_term) = res;
    z_err(i_poly, i_term) = (exact_res - interval_t(res))*mon_bnd<
        interval_t>(i_term);
}
```

Listing 1.4. Cuda kernel for adding polynomial parts in the summation algorithm.

```
template<int Deg,class Int,class Mem>
void    taylor_model_operations<Deg,Int,Mem>::add(const
    taylor_model_type &x, const taylor_model_type &y,
    taylor_model_type &z)const //x=TM_1, y=TM_2, z=TM.
{
    int block_size = 256;  //setting number of blocks on a GPU

    detail::ker_polynomial_add<Deg,Int,Mem><<<((polylen()*dim()))/
        block_size)+1, block_size>>>(
        polylen(), x.polynoms(), y.polynoms(),
        z.polynoms(), interval_polynoms_array_); //TM.polynoms()
            returns polynomial part of a TM
        //polylen() - return number of monomials in a polynomial

    cub::DeviceReduce::ReduceByKey( //perform summation of accordant
        monomial coefficients for all vector elements.
        cub_reduce_by_key_temp1, cub_reduce_by_key_temp1_sz,
        taylor_model_index_array_.raw_ptr(), bnds_index_.raw_ptr(),
        interval_polynoms_array_.raw_ptr(), mul_errors_.raw_ptr(),
        cub_res_size.raw_ptr(), custom_plus_func(),
            taylor_model_index_array_.size());

    detail::ker_add_remainders<Deg,Int,Mem><<<(dim()/block_size)+1,
        block_size>>>(
        dim(), x.remainders(), y.remainders(), mul_errors_, z.
            remainders()); //add to the remainders
}
```

Listing 1.5. Sum of two TMs according to the Algorithm 1.

One can check that for almost all algorithms in the previous section for monomial coefficients estimation a sum operation is required. Thus, as one can see in Listing 1.5, a reduction operation must be executed in each algorithm. The reduction part takes about 20% of time in the CPU implementation, but the GPU implementation depends heavily on the reduction efficiency. For simple algorithms like addition and subtraction reduce by key operation takes about 80% of execution time. Three variants of reduction by key were tested, namely *thrust::reduce_by_key*, *cub::DeviceReduce::ReduceByKey* and self-written reduction. Several tests have shown that the *cub* library implementation was the fastest with *thrust* library being the slowest: it is around 6 times slower for the maximum problem size. Results are provided in Fig. 1. These results can be different for modern GPUs since relatively old GPUs are available at the author's disposal. So all tests are using *cub* library, though *thrust* is also applicable.

4   Test Problems

All problems are tested on the following hardware: Intel Xeon E5-2697V2 Ivy Bridge with up to six K40 Nvidia GPUs installed on one chassis. All computations are performed in the double-precision floating-point operations mode.

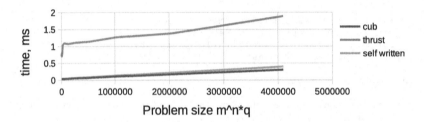

Fig. 1. Reduce by key operation absolute time of one operation for different problem sizes on a single GPU

4.1 Acceleration Tests for CPU and GPU Implementations

In this subsection, acceleration and performance tests of individual algorithms are provided. All algorithms benchmarked for consistency between CPU and GPU versions were found to be consistent. The algorithms provided above for simple operations like addition and substitution show acceleration ranging from 0.5 to 15 times vs single CPU thread, shown in Figs. 2 and 4 (left). This is expected since such operations are not computationally extensive.

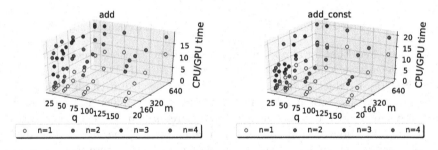

Fig. 2. CPU vs GPU time for $TM_1 - TM_2$ (left) and TM_1TM_2 (right) algorithms as function of m variables, dimension q and TM order n

A substantial acceleration was obtained for such computationally extensive operations as multiplication (in Fig. 3, right) and especially integration and convolution, Fig. 4. One can observe acceleration raging from 15 to 100–160 vs CPU single thread.

The authors are grateful to the anonymous reviewer for fruitful remarks that led to the revision and profiling of the CPU code resulting in its substantial optimization. When applied to intervals, all switching of rounding operations is minimized to prevent the processor from running idle cycles. Optimization of the algorithm and implementation of particular operation extended the usage of SSE registers in operations like convolution, integration and multiplication. This was verified during profiling with the *perf* utility.

The test of a single GPU vs CPU multithreaded implementation shows a perfectly linear scale, so that for 8 CPU threads, the best acceleration is around

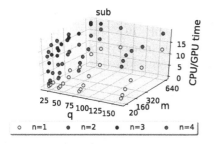

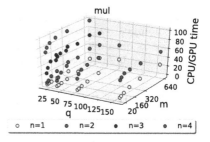

Fig. 3. CPU vs GPU time for $TM_1 - TM_2$ (left) and $TM_1 TM_2$ (right) algorithms as a function of m variables, dimension q and TM order n

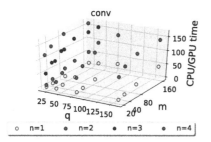

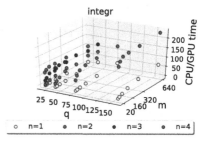

Fig. 4. CPU vs GPU time for $TM_1 * TM_2$ (left) and integration (right) algorithms as a function of m variables, dimension q and TM order nCPU vs GPU time for $TM_1 * TM_2$ (left) and integration (right) algorithms as a function of m variables, dimension q and TM order n

Table 1. Measurement of FLOPS in double precision on the convolution operation for different TM orders.

TM order	GPU, GFLOPS	CPU, GFLOPS	FLOPS ratio
1	3.347	0.769	4.348
2	30.735	0.928	33.109
3	94.551	0.917	103.036
4	140.196	0.953	147.139

3 to 20 for computationally extensive calculations and around 0.3 to 2 times for addition and subtraction.

The measurement of efficiency in terms of the floating point operations per second (FLOPS) is conduced with two utilities: *nvprof* used for GPU and *perf* used for CPU profiling and FLOPS estimation. For the *perf* utility the following event codes were used: $r530110$, $r538010$ and $r531010$. The results for the convolution operation on a TM with $q = 80$ and $m = 20$ are provided in Table 1 as a function of a TM polynomial degree. The CPU is found to reach its maximum performance relatively fast while GPU is fully loaded only when the fourth-order polynomials are used. It is observed that only 10.8% of the peak GPU performance is obtained. However, this problem is memory-bounded and algorithmically complex, and such performance can be considered to be satisfactory.

4.2   ODE System Test Problem

In the next test case the 2D Kolmogorov flow problem [25] of a viscous incompressible fluid on a 2D stretched periodic domain $\mathbb{T}^2(\alpha) := [2\pi/\alpha \times 2\pi]$ is considered, with $\alpha < 1$ being a stretching factor in the first direction. In the current case, α is selected such, that $1/\alpha$ in an integer. The problem is described with the Navier-Stokes equations. The vorticity formulation of these equations in two-dimensional spatial case looks as follows:

$$w_t + (\mathbf{u} \cdot \nabla)\, w = R^{-1}\, \triangle w + \beta \cos\left(\beta y\right) \tag{2}$$

with

$$w = \nabla \times \mathbf{u}, \; u_1 = (-\triangle^{-1}w)_x, \; u_2 = (\triangle^{-1}w)_y,$$

where $(\cdot)_j$ is a partial derivative in j coordinate, $\mathbf{u} = (u_1, u_2)^T$ and $\beta \geq 1$ being an integer. The nonlinear dynamics of this problem with different aspect ratios and values of β was considered in [26] where the limited cycles of a different period and invariant tori in the phase space are revealed using high order Galerkin spline and spectral numerical methods. Currently, the reduced system of ODEs is considered to be derived from the Galerkin approximation of (2) in the space of trigonometric polynomials:

$$(\hat{w}_{j,k})_t = -(\alpha^2 j^2 + k^2)\hat{w}_{j,k} + \beta/2\delta_\beta^{|k|} + \mathcal{B}(w), \tag{3}$$

for all $\{j, k\} \in [M/\alpha \times M]$ with M being a finite integer and

$$\mathcal{B}(w) := \sum_{l=-\frac{M/\alpha}{2}}^{\frac{M/\alpha}{2}-1} \sum_{m=-\frac{M}{2}}^{\frac{M}{2}-1} \left[\frac{\alpha m(j-l)}{\alpha^2 l^2 + m^2} \hat{w}_{l,m}\hat{w}_{(j-l),(k-m)} \right.$$
$$\left. - \frac{\alpha l(m-k)}{\alpha^2 l^2 + m^2} \hat{w}_{l,m}\hat{w}_{(j-l),(k-m)} \right],$$

is the convolution term. The main aim objective of the test case is to use the developed parallel TM methods together with the theory in [21] and to prove the existence of a periodic orbit in the finite-dimensional reduced system (3). The Picard iterations algorithm was implemented using TM operations from previous sections to get rigorous estimations for the ODE system (3). TM-extension of the right-hand side function is arithmetic operations with direct real numbers arithmetic being substituted for their TM counterparts. The sum (convolution $\mathcal{B}(w)$) part is substituted for the TM convolution rather than for direct combination of summation and multiplication algorithms only due to the performance issues.

In this paper, this problem is considered as a performance and demonstration test only.

Example of the calculation results is presented in Fig. 5.

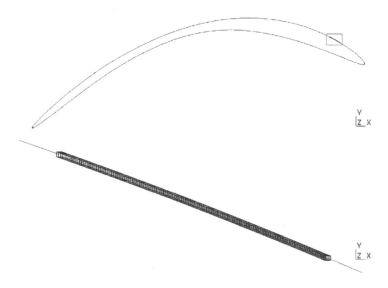

Fig. 5. Projection of a periodic cycle into 2D subspace and zoom-in of TM evolution using Picard integration

Multiple GPU implementation acceleration for the considered problem is presented in Fig. 6. This test is a combination of all computationally extensive algorithms considered above. One can see that the acceleration on 4 GPUs is close to linear, and 5 GPUs feature a rather poor acceleration (4.34 times). The reason for such behaviour is a lack of GPU loading due to the small chunk size on each GPU. For a single GPU implementation, the acceleration against a single-threaded CPU version constituted 37.5 times for the same problem size parameters as denoted in the caption show in Fig. 6.

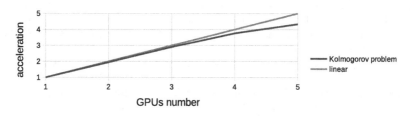

Fig. 6. MultiGPU acceleration for Kolmogorov flow problem, $m = 80$, $q = 79$, $n = 2$

5   Conclusion

The present paper provides the details of the parallel implementation of the algorithms designed in [20,21]. To our best knowledge, the provided detailed GPU implementation description is the first to be published. It is shown that the problem distribution which decreases the register pressure and application of the efficient reduction operation allows one to obtain up to 150 times acceleration in the GPU vs CPU test. Applied to the Kolmogorov flow problem in [21] the method provided an acceleration of up to 37.5 times compared to the CPU single-thread operations. The next goal is to use the provided parallel methods for the rigorous proves of periodic orbits in the ODE and PDE systems.

References

1. Makino, K.: Rigorous analysis of nonlinear motion in particle accelerators. Ph.D. thesis, MSU, vol. 3518, pp. 1–157 (1999)
2. Berz, M., Makino, K., Hoefkens, J.: Verified integration of dynamics in the solar system. Nonlinear Anal. **47**, 179–190 (2001). https://doi.org/10.1016/S0362-546X(01)00167-5
3. Burgmeier, P., Jahn, K.U., Plochov, A.G.: An interval computational method for approximating controllability sets. Computing **44**, 35–46 (1990). https://doi.org/10.1007/BF02247963
4. de Aguiar, M.S., Dimuro, G.P., da Rocha Costa, A.C.: ICTM: an interval tessellation-based model for reliable topographic segmentation. Numer. Algorithms **37**, 3–11 (2004). https://doi.org/10.1023/B:NUMA.0000049453.95969.41
5. Michel, A.N., Pai, M.A., Sun, H.F., Kulig, C.: Interval-analysis techniques in linear systems: an application to power systems. Circuits Syst. Signal Process. **12**(1), 51–60 (1993). https://doi.org/10.1007/BF01183147
6. Bogle, D., Johnson, D., Balendra, S.: Handling uncertainty in the development and design of chemical processes. In: Muhanna, R.L., Mullen, R.L. (eds.) Proceedings of the NSF workshop on reliable engineering modelling. September 15–17, 2004, Savannah, Georgia, USA, pp. 235–250. Georgia Institute of Technology (2004). https://doi.org/10.1007/s11155-006-9012-7
7. Snyder, J.M.: Interval analysis for computer graphics. Comput. Graph. **26**(2), 121–130 (1992). https://doi.org/10.1145/133994.134024
8. Ashokaraj, I., Tsourdos, A., Silson, P., White, B.A.: Sensor based robot localization and navigation: using interval analysis and uncertain Kalman filter. In: Proceedings of 2004 IEEE/RSJ International Conference on Intelligent Robots and Systems, September 28–October 2, Sendai, Japan, pp. 7–12 (2004). https://doi.org/10.1109/IROS.2004.1389887
9. Jerrell, M.: Interval arithmetic for input-output models with inexact data. Comput. Econ. **10**, 89–100 (1997). https://doi.org/10.1023/A:1017987931447
10. Appel, K., Haken, W.: Every planar map is four colorable. Bull. Am. Math. Soc. **82**(5), 711–712 (1976). https://doi.org/10.1215/ijm/1256049012
11. Babenko, K.I.: On a problem of Gauss. Dokl. Akad. Nauk SSSR **238**(5), 1021–1024 (1978)
12. Babenko, K.I., Vasiliev, M.M.: On demonstrative computations in the problem of stability of plane Poiseuille flow. Sov. Math. Dokl. **28**, 763–768 (1983)

13. Tucker, W.: The Lorenz attractor exists. Comptes Rendus de l'Académie des Sci.-Ser. I-Math. **328**(12), 1197–1202 (1999). https://doi.org/10.1016/S0764-4442(99)80439-X

14. Moore, R.E., Kearfott, R.B., Cloud, M.J.: Introduction to Interval Analysis. SIAM, Philadelphia (2009). https://doi.org/10.1137/1.978089871771

15. Sharyi, S.P.: Konechnomernyi interval'nyi analiz (Finite-Dimensional Interval Analysis). Electronic Book, Novosibirsk (2010)

16. Kearfott, R.: Interval computations: introduction, uses, and resources. Euromath Bull. **2**(1), 95–112 (1996). https://doi.org/10.1007/978-1-4613-3440-8_1

17. Berz, M., Hoffstatter, G.: Computation and application of Taylor polynomials with interval remainder bounds. Reliab. Comput. **4**, 83–97 (1998). https://doi.org/10.1023/A:1009958918582

18. Berz, M., Makino, K.: Verified integration of ODEs and flows using differential algebraic methods on high-order Taylor models. Reliab. Comput. **4**, 361–369 (1998). https://doi.org/10.1023/A:1024467732637

19. Makino, K., Berz, M.: COSY INFINITY Version 9. Nucl. Instrum. Methods Phys. Res. Sect. A: Accel. Spectrom. Detect. Assoc. Equip. **558**(1), 346–350 (2006). https://doi.org/10.1016/j.nima.2005.11.109

20. Evstigneev, N.M., Ryabkov, O.I.: Applicability of the interval Taylor model to the computational proof of existence of periodic trajectories in systems of ordinary differential equations. Differ. Equ. **54**(4), 525–538 (2018). https://doi.org/10.1134/S0012266118040092

21. Evstigneev, N.M., Ryabkov, O.I.: Algorithms for constructing isolating sets of phase flows and computer-assisted proofs with the use of interval Taylor models. Differ. Equ. **55**(9), 1198–1217 (2019). https://doi.org/10.1134/S0012266119090088

22. Flórez, J., Sbert, M., Sainz, M.A., Vehí, J.: Improving the interval ray tracing of implicit surfaces. In: Nishita, T., Peng, Q., Seidel, H.-P. (eds.) CGI 2006. LNCS, vol. 4035, pp. 655–664. Springer, Heidelberg (2006). https://doi.org/10.1007/11784203_63

23. Bagoczki, Z., Banhelyi, B.: A parallel interval arithmetic-based reliable computing method on a GPU. Acta Cybern. **23**, 491–501 (2017). https://doi.org/10.14232/actacyb.23.2.2017.4

24. Lessard, J.P.: Computing discrete convolutions with verified accuracy via banach algebras and the FFT. Appl. Math. **63**(3), 219–235 (2018). https://doi.org/10.21136/AM.2018.0082-18

25. Arnold, V.I., Meshalkin, L.D.: Mathematical life in USSR: A. N. Kolmogorov sem. on selected problems of analysis (1958–1959). Uspekhi Math. Nauk **15**(1), 247–250 (1960)

26. Evstigneev, N.M., Magnitskii, N.A., Silaev, D.A.: Qualitative analysis of dynamics in Kolmogorov's problem on a flow of a viscous incompressible fluid. Differ. Equ. **51**(10), 1292–1305 (2015). https://doi.org/10.1134/S0012266115100055

Combining Local and Global Search in a Parallel Nested Optimization Scheme

Konstantin Barkalov$^{(\boxtimes)}$ ⓘ, Ilya Lebedev ⓘ, Maria Kocheganova ⓘ,
and Victor Gergel ⓘ

Lobachevsky State University of Nizhni Novgorod, Nizhni Novgorod, Russia
{konstantin.barkalov,ilya.lebedev,maria.rachinskaya}@itmm.unn.ru,
gergel@unn.ru

Abstract. This paper considers global optimization problems and numerical methods to solve them. The following assumption is made regarding the dependence of the objective function on its parameters: it is multiextremal concerning selected variables only, and the dependence on the other variables has a local character. Such problems may arise when identifying unknown parameters of the mathematical models based on experimental results. A parallel computational scheme, which accounts for this feature, is being proposed. This novel scheme is based on the idea of nested (recursive) optimization, which suggests optimizing the significant variables at the upper recursion level (using the global optimization method) and solving the local optimization problems at the lower level (using the local method). The generated local subproblems will not affect each other and can be solved in parallel. The efficiency of the proposed parallel computation scheme has been confirmed by the numerical experiments conducted on several hundreds of test problems.

Keywords: Global optimization · Local optimization · Recursive optimization scheme · Parallel algorithms

1 Introduction

Global optimization problems arise frequently in various fields of modern science and technology. For example, the problems of finding the configurations of various chemical compounds corresponding to the minimum interaction energy are of current interest [1]. In particular, such problems arise in the development of new medical drugs [12]. Another example is the problem of finding optimal control parameters for complex processes [11,13]. Many other applications also generate global optimization problems (see the review [18]).

The application of global optimization methods for identifying the parameters of the mathematical models based on the data of experiments has become a conventional practice. These problems require such a search for values of the

This study was supported by the Russian Science Foundation, project No. 16-11-10150.

L. Sokolinsky and M. Zymbler (Eds.): PCT 2020, CCIS 1263, pp. 100–112, 2020.
https://doi.org/10.1007/978-3-030-55326-5_8

unknown model parameters, which makes the computational results based on the model close to the ones obtained experimentally.

The number of parameters which are required to be identified in such a way may reach tens and hundreds in complex mathematical models [2,15]. In the problems of such dimensionality, regular search methods for a global solution cannot be applied because of the extremely large computational costs for covering the search domain by the trial points. This remains true even with the use of efficient algorithms (for example, [5,10,17]) constructing essentially nonuniform coverage.

However, in the problems of model identification, it is usually possible to select the parameters, which will affect the result most significantly. As a rule, the number of such parameters is small. The rest of the parameters either cause an insufficient effect (in this case, multiextremality can be neglected) or dependence on them is local.

Currently, the development of parallel algorithms which would minimize the essentially multidimensional functions (hundreds of variables) and account for the dependence of the objective function on different groups of parameters is relevant. Specifically, the objective function is assumed to be multiextremal for some of the variables. The rest of the variables have a local effect, i.e., the objective function is unimodal for them. In this case, the problem-solving can be organized according to the nested (recursive) optimization scheme. The multiextremal subproblem, which requires the use of complex global search algorithms, will be solved at the upper level of the recursion. The unimodal subproblems (each corresponding to a fixed set of values of the first part of the parameters) will be solved at the lower level. To solve these subproblems, efficient methods of local optimization, linear programming or linear algebra can be applied (subject to the character of the local subproblems).

Below you will find a description of this paper structure. Section 2 describes the nested optimization scheme taking into account different properties of the objective function at different nesting levels. A particular example of such a problem is presented, along with a general scheme of the parallel computation organization. Section 3 is devoted to the description of the parallel global search algorithm using space-filling curves. The computational scheme of the parallel algorithm is presented, and its main properties are described. Section 4 discusses the methods to solve the local subproblems within the framework of the nested optimization scheme. In addition, this section briefly describes the properties pertaining to the local extremum search problems and outlines the known methods for this type of problems. Section 5 contains the results of a numerical experiment. The scalability of the proposed parallel computation scheme for solving a series of test problems was evaluated. Section 6 concludes the paper.

2   Nested Optimization Scheme

Let us consider an optimization problem of the type

$$f(x^*, y^*) = \min \{f(x, y) : x \in S, y \in D\},$$
$$S = \{x \in R^M : a_i \leq x_i \leq b_i, 1 \leq i \leq M\},$$
$$D = \{y \in R^N : c_i \leq y_i \leq d_i, 1 \leq i \leq N\}. \tag{1}$$

Let us assume that at any fixed set of values $\overline{x}$ the function $f(\overline{x}, y)$ is a multiextremal one and satisfies the Lipschitz condition

$$|f(\overline{x}, y') - f(\overline{x}, y'')| \leq L\|y' - y''\|, \ y', y'' \in D, \ 0 < L < \infty,$$

with the constant L unknown a priori. At the same time, at any fixed set of values $\overline{y}$, the function $f(x, \overline{y})$ is a unimodal one, i.e. the function $f(x, \overline{y})$ has a single minimum point x^* (generally depending on $\overline{y}$).

Accounting for such a peculiarity of the considered problem class can essentially reduce the computational costs of the optimum search. Indeed, according to the known nested optimization scheme [4], the initial problem can be reduced to the problem of finding the global minimum of the function $\varphi(y)$

$$\varphi(y^*) = \min_{y \in D} \varphi(y), \tag{2}$$

where

$$\varphi(y) = \min_{x \in S} f(x, y). \tag{3}$$

According to (2), computing a single value of the function $\varphi(y)$ (let us call this process a search trial) implies unimodal problem solution (3), which can be performed with one of the local search methods. The efficiency and sophistication of the local optimization methods allow solving subproblems (3) with the precision essentially exceeding that of the global optimization methods. Correspondingly, the problem (2) can be considered a global optimization problem, where the objective function can be computed with high precision but this operation is time-consuming.

The following approximation problem can be considered a particular example of the problem, where a different character of dependence on different groups of parameters takes place. Let us assume m values $u_j = u(t_j), 1 \leq j \leq m$, of the function $u(t)$ have been obtained in an experiment and the analytical expression for $u(t)$ is known to take the form

$$u(t) = u(t, \omega, c) = \sum_{i=0}^{n} [c_{2i+1} \sin(\omega_i t) + c_{2i+2} \cos(\omega_i t)]. \tag{4}$$

In this expression, $c = (c_1, \ldots, c_{2n+2})$, $\omega = (\omega_0, \omega_1, \ldots, \omega_n)$ are the unknown parameters defining a particular function $u(t)$.

Let us introduce a measure of function $u(t, w, c)$ deviation from the experiential data as a sum of squares

$$\Delta(w, c) = \sum_{j=1}^{m} [u_j - u(t_j, w, c)]^2 .$$

Now, following the idea of least squares fitting, it is possible to present the approximation problem as the problem of minimizing the objective function

$$\Delta(w^*, c^*) = \min_{w,c} \Delta(w, c). \tag{5}$$

The solution (w^*, c^*) of this problem will define the best approximation.

Problem (5) can be written in a recursive form

$$\varphi(w^*) = \min_{w} \varphi(w), \tag{6}$$

$$\varphi(w) = \min_{c} \Delta(w, c). \tag{7}$$

Nested subproblem (7) is a classical linear least squares problem, and its solution can be obtained by solving a system of linear algebraic equations regarding the unknown c, which can be done, for example, by Gaussian elimination. At the same time, external problem (6) will be, in general, a multiextremal one, and its solving will require considerable computational resources.

So far, to solve problem (1) in nested scheme (2), (3) one can apply a parallel global search algorithm for solving external subproblems (2). At that, a set of independent subproblems (3) will be generated at every iteration of the algorithm, the solving of which can be performed in parallel using the local methods. A general scheme of the computation organization using several processors is presented in Fig. 1.

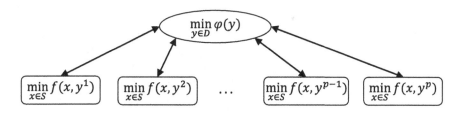

Fig. 1. A tree of subproblems in the parallel nested optimization scheme

The parallel program processes will form a tree. The root process will solve problem (2) and distribute subproblems (3) among the child processes. Each subproblem is solved by a separate process; the data exchange between the processes can be organized using MPI. The data transmission between the processes will be minimal: it is required to send the coordinates of points $y^1, ..., y^p$ from the root to the child processes and to receive the found minimal values back

$$\varphi(y^1) = \min_x f(x, y^1), \quad ..., \quad \varphi(y^p) = \min_x f(x, y^p).$$

A detailed discussion of the algorithms used at different levels of the process tree is given in the next sections.

3   Parallel Global Search Algorithm

An efficient parallel global optimization algorithm intended for solving the problems of kind (2) has been developed in Lobachevsky University of Nizhni Novgorod (UNN) [3,8,23]. The main idea of the parallelization consists in such an organization of computations during which several trials are performed in a parallel way. This approach is featured by a high efficiency (the part of the computational process with the major amount of computations is parallelized) and generality (it is applicable for a wide class of algorithms).

In the developed global search algorithm, a novel method to reduce the dimensionality of the problems to be solved is applied. The reduction of the dimensionality is based on the fundamental fact, according to which an N-dimensional hyperparallelepiped D and the interval $[0, 1]$ of the real axis are equal-power sets, and the interval $[0, 1]$ can be mapped continuously into D using the Peano curve $y(x)$, i.e. $D = \{y(x) : 0 \le x \le 1\}$.

Utilizing this fact, one can reduce the minimization problem of a multidimensional function $\varphi(y)$ to the minimization of a one-dimensional function $\varphi(y(x))$

$$\varphi(y(x^*)) = \min \{\varphi(y(x)) : x \in [0, 1]\}, \tag{8}$$

where the function $\varphi(y(x))$ will satisfy a uniform Hölder condition

$$|\varphi(y(x')) - \varphi(y(x''))| \le H |x' - x''|^{1/N}$$

with the Hölder constant H linked to the Lipschitz constant L by the relation $H = 2L\sqrt{N+3}$ and $y(x)$ being the Peano curve from $[0, 1]$ onto D.

The algorithms for the numerical construction of the Peano curve approximations (called *evolvents*) are given in [21,24]. As an illustration, two evolvents are shown in Fig. 2. The figure demonstrates that the precision of the evolvent depends on the density level m used in the construction.

It is worth noting that many well-known global optimization algorithms are based implicitly on the idea of the dimensionality reduction and adaptation of one-dimensional algorithms for solving multidimensional problems [6,20,26].

According to [23,24], the rules of the global search algorithm, where p trials are performed in parallel at every iteration, are as follows. At the first iteration of the method, the trials are performed in parallel in two boundary points $x^1 = 0$, $x^2 = 1$ as well as in $(p-2)$ arbitrary internal points of the interval $[0, 1]$, i.e. $x^i \in (0, 1), 3 \le i \le p$.

Supposing $n \geq 1$ iterations of the method have been carried out, where the trials were performed at $k = k(n)$ points x^i, $1 \leq i \leq k$, then the points $x^{k+1}, \ldots, x^{k+p}$ of the search trials at the next iteration are determined according to the following rules.

Rule 1. Renumber the points $x^1, \ldots, x^k$ of the preceding trials in lowercase in ascending order of the coordinate values, i.e.

$$0 = x_1 < x_2 < \cdots < x_{k-1} < x_k = 1,$$

and juxtapose them with the values $z_i = \varphi(y(x_i))$, $1 \leq i \leq k$, computed at these points.

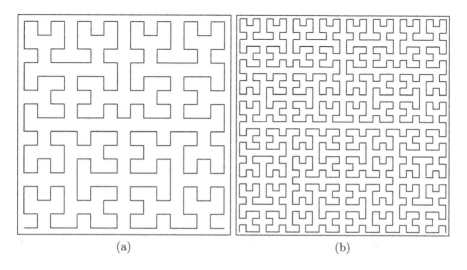

(a)   (b)

Fig. 2. The evolvents in two dimensions with (a) $m = 4$ and (b) $m = 5$

Rule 2. Compute the current lower estimates

$$\mu = \max \left\{ \frac{|z_i - z_{i-1}|}{\Delta_i}, \ 2 \leq i \leq k \right\}, \tag{9}$$

where $\Delta_i = (x_i - x_{i-1})^{1/N}$. If μ is equal to zero, then assume $\mu = 1$.

Rule 3. For each interval (x_{i-1}, x_i), $2 \leq i \leq k$, compute the *characteristics* $R(i)$:

$$R(i) = \Delta_i + \frac{(z_i - z_{i-1})^2}{r^2 \mu^2 \Delta_i} - 2\frac{z_i + z_{i-1}}{r\mu}, \tag{10}$$

where $\Delta_i = (x_i - x_{i-1})^{1/N}$. The value $r > 1$ is an algorithm parameter. An appropriate selection of r allows considering the product $r\mu$ as an estimate of the Lipschitz constant L of the objective function.

Rule 4. Arrange characteristics $R(i)$, $2 \leq i \leq k$, in decreasing order

$$R(t_1) \geq R(t_2) \geq \cdots \geq R(t_k) \geq R(t_k) \tag{11}$$

and select p maximum characteristics with interval numbers $t_j, 1 \leq j \leq p$.

Rule 5. Carry out new trials at points $x^{k+j} \in (x_{t_j-1}, x_{t_j})$, $1 \leq j \leq p$, calculated using the formula

$$x^{k+j} = \frac{x_{t_j} + x_{t_j-1}}{2} - \frac{\text{sign}(z_{t_j} - z_{t_j-1})}{2r} \left[\frac{|z_{t_j} - z_{t_j-1}|}{\mu} \right]^N. \tag{12}$$

The algorithm terminates if the condition $\Delta_{t_j} < \epsilon$ is performed at least for one number t_j, $1 \leq j \leq p$; $\epsilon > 0$ is the predefined accuracy.

This method of parallelization has the following justification. The characteristics of intervals (10) used in the method can be considered as probability measures of the global minimum point location in these intervals. Inequalities (11) arrange intervals according to their characteristics, and the trials are carried out in parallel in the first p intervals with the largest probabilities.

Various modifications of this algorithm and the corresponding theory of convergence are given in [3, 21, 23, 24]. The results of the experimental comparison (see, for example, [22]) show that the global search algorithm is superior to many known methods of similar purpose, including both deterministic and heuristic algorithms.

4   Local Search Algorithms

In this section, let us consider the issues related to solving subproblem (3), which (in the notation $f(x) = f(x, \overline{y})$ for the fixed value $\overline{y} \in D$) corresponds to the problem of finding a local extremum

$$f(x) \to \min, x \in S. \tag{13}$$

To date, a huge number of various local optimization methods for the problems of kind (13) have been developed. They include such methods as the gradient ones, Newton, and quasi-Newton methods, the methods of conjugate directions, etc. The majority of them utilizes the principle of local descent when the method progressively passes the points with smaller values of the objective function at every step. Almost all these methods can be represented in a form of the iterative relation

$$x^{k+1} = x^k + s^k d^k,$$

where x^k are the points of main trials consisting in computing a set of some local characteristics of the objective function $I^k = I(x^k)$ at the point x^k, where d^k are the shift directions from the points x^k computed on the base of the main trial results, and s^k are the coefficients defining the magnitudes of the shifts along the directions chosen.

To define the shift magnitudes s^k along the directions d^k, the methods can include some auxiliary (working) steps. This results in additional measuring of the objective function local characteristics along the direction d^k. The transitions from the points x^k to the points x^{k+1} is performed in such a way that the function $f^k = f(x^k)$ value essentially decreases as a result of this step.

The set of the computed local characteristics $I^k = I(x^k)$ may include: the function values $f^k = f(x^k)$, the gradient $\nabla f^k = \nabla f(x^k)$, and the matrix of the second derivatives (Hessian) $\Gamma_k = \Gamma^f(x^k)$. A particular set of the characteristics to be measured depends on the properties of the problem being solved as well as on the optimization method chosen.

In the local optimization problems arising in the applications, the a priori information on the objective functions is usually quite limited. For example, a local character of the function dependence on the parameters may only be suggested where the gradient values (and Hessian) are unknown. In this case, the applied methods are the direct search ones, which do not utilize any suggestions on the objective function smoothness. When searching for the minimum, these methods measure the function values only. The rules of the placement of the iteration points in these ones are based on some heuristic logical schemes.

Hooke–Jeeves method [9] and Nelder–Mead method [14] belong to the popular methods of the direct search. Despite their apparent simplicity and theoretical non-substantiation of the direct search methods, these two methods are well established in the practical computations. This can be explained as follows. Actually, many smooth optimization methods are very sensitive to computational errors in the function values transforming the theoretically smooth function into the non-smooth one. Because of this, in the practical computations, these methods lose positive properties, which are often promised by theory. In these conditions, direct search methods allow achieving better results.

5 Numerical Experiments

The scheme of parallel computations described in previous sections was implemented in the Globalizer software system developed in UNN [8,25]. The numerical experiments were carried out using the Lomonosov supercomputer (Lomonosov Moscow State University). Each supercomputer node included two quad-core processors Intel Xeon X5570 2.93 GHz and 12 Gb RAM. To build the Globalizer system for running on the Lomonosov supercomputer, we used the GCC 5.5.0 compiler and Intel MPI 2017.

To simulate the applied problems, where the locally-affecting parameters can be selected, a test function of the kind

$$f(x, y) = G(y) + R(x) \tag{14}$$

was used when conducting the experiments. Here, $G(y)$ is a multiextremal function of the dimensionality $N = 2$ and $R(x)$ is a unimodal function of the dimensionality $N \gg 2$. As the multiextremal part of the problem, the functions

$$G(y) = -\left\{ \left(\sum_{i=1}^{7} \sum_{j=1}^{7} A_{ij}g_{ij}(y) + B_{ij}h_{ij}(y) \right)^2 \right.$$
$$\left. + \left(\sum_{i=1}^{7} \sum_{j=1}^{7} C_{ij}g_{ij}(y) + D_{ij}h_{ij}(y) \right)^2 \right\}^{1/2}, \qquad (15)$$

were considered, where

$$y = (y_1, y_2) \in R^2, 0 \le y_1, y_2 \le 1,$$
$$g_{ij}(y) = \sin(i\pi y_1)\sin(j\pi y_2),$$
$$h_{ij}(y) = \cos(i\pi y_1)\cos(j\pi y_2),$$

and the coefficients $A_{ij}, B_{ij}, C_{ij}, D_{ij}$ are the random numbers uniformly distributed in the interval $[-1, 1]$. This class of functions is often used for testing global optimization algorithms. As the local part of the problem, the modified Rosenbrock function

$$R(x) = \sum_{i=1}^{N-1} \left[(1 - x_i)^2 + 100(x_{i+1} - x_i^2)^2 \right], \quad -2 \le x_i \le 2, \, 1 \le i \le N,$$

was used, where the extremum point was randomly shifted in the search domain using the linear coordinate transform.

As an example, contour plots of a pair of functions $G(y)$ and $R(x)$ are presented in Fig. 3. These functions are the complex ones for the corresponding global and local optimization methods since the function $G(y)$ is essentially multiextremal, and the function $R(x)$ has a ravine clearly manifested structure.

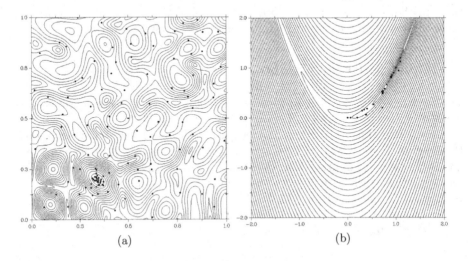

(a)   (b)

Fig. 3. Contour plot of the functions (a) $G(y)$ and (b) $R(x)$

In Fig. 3, the trial points performed by the global search algorithm and Hooke–Jeeves method when solving the corresponding problems are also marked.

The trial point positions clearly demonstrate the function $G(y)$ multiextremality (the trial points form a nonuniform coverage of the search domain) as well as the ravine character of the function $R(x)$ (the trial points form an elongated group along the ravine bottom).

The problem was considered to be solved if in the course of minimizing the function $G(y)$, the global search algorithm generates the next trial point y^k in the ϵ-neighbourhood of the global minimum y^*, i.e. $\|y^* - y^k\| \leq \epsilon$, where $\epsilon = 10^{-2}$. At that, the precision of finding the minimum when minimizing the unimodal function $R(x)$ was selected 1000 times lower, i.e. $\delta = 10^{-5}$. In the global search algorithm, the parameter $r = 4.5$ from (10) and the evolvent building parameter $m = 10$ were used; the employed local Hooke–Jeeves method had no additional parameters.

In the first series of experiments, 100 problems of kind (14) were solved. Their local subproblems dimensionality was $N = 50$. The problems were solved using 1, 4, and 8 cluster nodes, which employed from 2 to 32 MPI-processes. According to the parallelization scheme presented in Fig. 1, at the upper nesting level the global optimization problem was solved in a single (master) process, which initiated parallel solving of the lower-level local problems (from 1 up to 31 problems, respectively). So far, the minimum possible number of the employed processes equalled 2, one per each nesting level.

Table 1 presents the average number of iterations K_{av} performed by the global search algorithm when minimizing the function $G(y)$, the average time of solving the whole problem T_{av}, and the speed-up S subject to the number of employed MPI processes p.

Table 1. Results of solving a series of problems of dimensionality $N = 52$

p	K_{av}	T_{av}	S
2	349.4	9.14	—
16	25.6	0.70	13.2
32	12.9	0.37	24.6

In the second series of experiments, 10 problems of kind (14) were solved, in which the dimensionality of the local subproblems was equal to $N = 100$. The running parameters were the same as in the previous series of experiments. The results of solving are presented in Table 2.

Table 2. The results of solving a problems series with dimensionality $N = 102$

p	K_{av}	T_{av}	S
2	299.1	41.19	—
16	22.0	2.92	14.1
32	11.7	1.50	27.4

In the last series of experiments, 10 problems were solved, in which three-dimensional functions from GKLS class [7] were used as a multiextremal part. This class of functions is often used for investigating global optimization algorithms [3,16,19]. The dimensionality of the local subproblems was equal to $N = 100$. The problems were solved at 1, 8, 16, and 32 cluster nodes, on which from 4 to 128 MPI processes were used. The running parameters were the same as in the previous series of experiments. The results of solving are presented in Table 3, where speed-up is calculated for running on a single node using 4 MPI processes.

Table 3. The results of solving a problems series with dimensionality $N = 103$

p	K_{av}	T_{av}	S
4	730.8	63.4	—
32	87.4	7.37	8.6
64	37.0	3.77	16.8
128	16.5	1.78	35.7

6   Conclusion

The results of the experiments performed on a series of optimization problems demonstrate that the parallel computation scheme proposed in this work has a good potential for the parallelization. Specifically, when solving a series of 100 problems with a total number of variables equal to 52 (from which the first two variables caused a global effect on the objective function and the other fifty affected the one locally), the obtained speed-up was 24.6 when using 32 processes. When solving problems with more complex local subproblems (with the local variables number $N = 100$), the speed-up increased up to 35 using 128 processes. In the last case, the speed-up was calculated subject to using 4 processes on one node. The developed parallel computation scheme can be applied in identifying complex mathematical models, which feature a large number (tens and hundreds) of unknown parameters.

References

1. Abgaryan, K.K., Posypkin, M.A.: Optimization methods as applied to parametric identification of interatomic potentials. Comput. Math. Math. Phys. **54**(12), 1929–1935 (2014). https://doi.org/10.1134/S0965542514120021
2. Akhmadullina, L., Enikeeva, L., Gubaydullin, I.: Numerical methods for reaction kinetics parameters: identification of low-temperature propane conversion in the presence of methane. Procedia Eng. **201**, 612–616 (2017)
3. Barkalov, K., Strongin, R.: Solving a set of global optimization problems by the parallel technique with uniform convergence. J. Glob. Optim. **71**(1), 21–36 (2017). https://doi.org/10.1007/s10898-017-0555-4

4. Carr, C., Howe, C.: Quantitative Decision Procedures in Management and Economic: Deterministic Theory and Applications. McGraw-Hill, New York (1964)
5. Evtushenko, Y., Malkova, V., Stanevichyus, A.A.: Parallel global optimization of functions of several variables. Comput. Math. Math. Phys. **49**(2), 246–260 (2009)
6. Evtushenko, Y., Posypkin, M.: A deterministic approach to global box-constrained optimization. Optim. Lett. **7**, 819–829 (2013)
7. Gaviano, M., Kvasov, D., Lera, D., Sergeyev, Y.: Software for generation of classes of test functions with known local and global minima for global optimization. ACM Trans. Math. Softw. **29**(4), 469–480 (2003)
8. Gergel, V., Barkalov, K., Sysoyev, A.: A novel supercomputer software system for solving time-consuming global optimization problems. Numer. Algebra Control Optim. **8**(1), 47–62 (2018)
9. Hooke, R., Jeeves, T.A.: "Direct search" solution of numerical and statistical problems. J. ACM **8**(2), 212–229 (1961)
10. Jones, D.R.: The DIRECT global optimization algorithm. In: The Encyclopedia of Optimization, pp. 725–735. Springer, Heidelberg (2009). https://doi.org/10.1007/978-0-387-74759-0
11. Kalyulin, S., Shavrina, E., Modorskii, V., Barkalov, K., Gergel, V.: Optimization of drop characteristics in a carrier cooled gas stream using ANSYS and Globalizer software systems on the PNRPU high-performance cluster. Commun. Comput. Inf. Sci. **753**, 331–345 (2017)
12. Kutov, D., Sulimov, A., Sulimov, V.: Supercomputer docking: investigation of low energy minima of protein-ligand complexes. Supercomput. Front. Innov. **5**(3), 134–137 (2018)
13. Modorskii, V., Gaynutdinova, D., Gergel, V., Barkalov, K.: Optimization in design of scientific products for purposes of cavitation problems. In: AIP Conference Proceedings, vol. 1738 (2016)
14. Nelder, J., Mead, R.: A simplex method for function minimization. Comput. J. **7**(4), 308–313 (1965)
15. Nurislamova, L.F., Gubaydullin, I.M., Koledina, K.F., Safin, R.R.: Kinetic model of the catalytic hydroalumination of olefins with organoaluminum compounds. React. Kinet. Mech. Catal. **117**(1), 1–14 (2015). https://doi.org/10.1007/s11144-015-0927-z
16. Paulavičius, R., Sergeyev, Y., Kvasov, D., Žilinskas, J.: Globally-biased DISIMPL algorithm for expensive global optimization. J. Glob. Optim. **59**(2–3), 545–567 (2014)
17. Paulavičius, R., Žilinskas, J., Grothey, A.: Parallel branch and bound for global optimization with combination of Lipschitz bounds. Optim. Method Softw. **26**(3), 487–498 (2011)
18. Pinter, J.: Global Optimization: Scientific and Engineering Case Studies. Springer, New York (2006). https://doi.org/10.1007/0-387-30927-6
19. Sergeyev, Y., Kvasov, D.: A deterministic global optimization using smooth diagonal auxiliary functions. Commun. Nonlinear Sci. Numer. Simul. **21**(1–3), 99–111 (2015)
20. Sergeyev, Y.D., Kvasov, D.E.: Global search based on efficient diagonal partitions and a set of Lipschitz constants. SIAM J. Optim. **16**(3), 910–937 (2006)
21. Sergeyev, Y.D., Strongin, R.G., Lera, D.: Introduction to Global Optimization Exploiting Space-Filling Curves. Springer Briefs in Optimization. Springer, New York (2013). https://doi.org/10.1007/978-1-4614-8042-6

22. Sovrasov, V.: Comparison of several stochastic and deterministic derivative-free global optimization algorithms. In: Khachay, M., Kochetov, Y., Pardalos, P. (eds.) MOTOR 2019. LNCS, vol. 11548, pp. 70–81. Springer, Cham (2019). https://doi.org/10.1007/978-3-030-22629-9_6

23. Strongin, R., Gergel, V., Barkalov, K., Sysoyev, A.: Generalized parallel computational schemes for time-consuming global optimization. Lobachevskii J. Math. **39**(4), 576–586 (2018)

24. Strongin, R.G., Sergeyev, Y.D.: Global Optimization with Non-Convex Constraints. Sequential and Parallel Algorithms. Kluwer Academic Publishers, Dordrecht (2000)

25. Sysoyev, A., Barkalov, K., Sovrasov, V., Lebedev, I., Gergel, V.: Globalizer – a parallel software system for solving global optimization problems. In: Malyshkin, V. (ed.) PaCT 2017. LNCS, vol. 10421, pp. 492–499. Springer, Cham (2017). https://doi.org/10.1007/978-3-319-62932-2_47

26. Žilinskas, J.: Branch and bound with simplicial partitions for global optimization. Math. Model. Anal. **13**(1), 145–159 (2008)

SART Algorithm Parallel Implementation with Siddon and MBR Methods for Digital Linear Tomosynthesis Using OpenCL

N. S. Petrov[1]([✉]), V. S. Linkov[1], and R. N. Ponomarenko[2]

[1] SFedU Scientific and Technical Center "Technocenter", Taganrog, Russia
nspetrov@sfedu.ru, linkov.vs@fltrrf.ru
[2] LLC Sevkavrentgen-D, Maysky, Russia
r_ponomarenko@skrz.ru

Abstract. Most practical tomographic reconstruction methods in a conical beam are based on the filtered backprojection (FBP) method, even though such iterative methods as SART can reconstruct the volume with high accuracy for sufficiently large conical angles due to the high computational complexity of iterative methods. However, the rapid development of the graphic processors that can be also used for non-graphic computing as hardware accelerators, compensate this disadvantage. The paper describes a parallel implementation of the SART algorithm for the task of digital linear tomosynthesis. In the context of the problem, the algorithm consists of two consecutive parts: forward and backprojection. For forward projection, the Siddon method for the tomosynthesis unit linear geometry was modified, while for backprojection, the minimum bounding rectangle (MBR) method was used backproject. The parallel algorithm was implemented on a heterogeneous CPU-GPU system using the OpenCL platform and allowed a significant increase in the reconstruction speed compared to single and multi-threaded implementation on the CPU.

Keywords: Tomosynthesis · SART · Heterogeneous system · OpenCL

1 Introduction

Digital tomosynthesis (DTS) is a three-dimensional x-ray method for acquiring layered images of an object from a set of two-dimensional projections. Tomosynthesis is based on a multi-angle radiography scheme, which provides for the acquisition of projection images of the object examined at limited irradiation angles and the subsequent restoration of its internal structure by solving a system of integral projection equations. This technique relates to reconstructive visualization methods [1].

The radiation exposure of digital tomosynthesis does not exceed that of computerized tomography, since reconstruction requires fewer projections.

© Springer Nature Switzerland AG 2020
L. Sokolinsky and M. Zymbler (Eds.): PCT 2020, CCIS 1263, pp. 113–130, 2020.
https://doi.org/10.1007/978-3-030-55326-5_9

The advantage of the method is that by minimizing projection overlays, it allows visualizing structures inaccessible with classical fluoroscopy, for example, 1–2 cervical vertebrae, skull bones, etc. Digital tomosynthesis is used primarily for chest examination [2].

During tomosynthesis, each image (projection) carries information not only about the intensity of a particular x-ray beam passing through the object, but also about its trajectory. The entire volume of the object under examination is divided into volume elements referred to as voxels. For each projection, a system matrix is calculated to present information about the passage of a particular ray (its trajectory) through a specific voxel. The result of a ray passing (intensity) through a set of voxels on its trajectory is contained in a detector pixel, which is illuminated by a particular ray, a set of pixels for each projection forms a matrix of the ray sums [3].

Thus, given the information as of the X-ray beams trajectories and as of the intensity of the beams passing each of the trajectories of each of the projections, a three-dimensional view of the object can be reconstructed (with a certain discreteness resulting from geometrical dimensions of the pixel detector and the voxel height). The image reconstruction accuracy is the highest in the area where more rays pass through the object [4].

The purpose of the work is to modify and implement the methods for reconstruction of the object in an acceptable time for linear digital tomosynthesis installation. The second part describes the selected method and sets the reconstruction task, the third part gives a mathematical description of the methods modified for parallel calculations, the fourth part demonstrates the computational experiment results using the developed methods and algorithms.

2   Simultaneous Algebraic Reconstruction Technique (SART)

The filtered backprojections (FBP) method is the most popular for the reconstruction of three-dimensional objects from x-ray projections [5]. However, for the of linear tomosynthesis installation with limited radiation angles, algebraic iterative methods are more preferable [6,7]. Conversely, the difficulty of algebraic methods implementation lies in the large amount of computer calculations associated with operations on the system matrix [8].

Tomographic reconstruction using algebraic methods reduces to solving the following system of algebraic equations [9]:

$$p_i = \sum_{j=1}^{N} w_{ij} v_j \tag{1}$$

The v_j value must be reconstructed in $N = n^3$ voxels belonging to the reconstruction area, where the index j takes values from 1 to N, using p_i values in pixels with the current index i, belonging to the projection images P_α, where α is the angle that characterizes the geometric orientation of the source-detector pair. In Eq. (1), w_{ij} is the weight coefficient with which the j-indexed voxel contributes to the brightness value of the i-indexed pixel. SART uses an iterative way to solve this system of equations, with the value correction in the jth voxel for all voxels belonging to the reconstruction area being performed simultaneously at the current k-numbered iteration and λ relaxation coefficient, as indicated below in formula (2) [10]:

$$v_j^{(k+1)} = v_j^{(k)} + \lambda \frac{\sum_{i=1}^{M} \left(\frac{p_i - \sum_{n=1}^{N} w_{in} v_j^{(k)}}{\sum_{j=1}^{N} w_{in}} \right) w_{ij}}{\sum_{i=1}^{M} w_{ij}}. \tag{2}$$

Major problems has been encountered while implementing this algorithm include calculation, storage and reading of weight coefficients w_{ij} from the memory [11,12]. A direct approach to the implementation of this algorithm assumes using a CPU and one of the simplest weight coefficient w_{ij} calculation methods based on rasterization, such as Bresenham or Xiaolin Wu. This approach involves a significant amount of additional memory allocated for storing intermediate values of the numerator and denominator of formula (2), it does not provide acceptable reconstruction accuracy, or fully parallelize calculations, not even on a multi-core CPU, which, in total, results in a long reconstruction time.

In this regard, reconstruction time reduction and accuracy improvement for the linear tomosynthesis installation requires the following:

1) separate the calculations by formula (2) into forward and backprojection;
2) do independent calculations for each pixel of the projection and voxel of the reconstructed volume to achieve maximum parallelism;
3) perform calculations for the linear installation geometry and the conical beam;
4) exclude competitive access of the memory threads for writing;
5) minimize exchange operations through PCIe.

When separated into forward and backprojections, formula (2) is divided into 2 parts:

$$c_i = \frac{p_i - \sum_{n=1}^{N} w_{in} v_j^{(k)}}{\sum_{j=1}^{N} w_{in}} \tag{3}$$

$$v_j^{(k+1)} = v_j^{(k)} + \lambda \frac{\sum\limits_{i=1}^{M} c_i w_{ij}}{\sum\limits_{i=1}^{M} w_{ij}}. \tag{4}$$

First, an array of correction values is calculated by (3). Each c_i value for the current projection P_α can be calculated in parallel and independently of each other. Then, according to (4), the values of voxels v_j of the reconstructed volume are updated concurrently and independently from each other. Thus, for each projection P_α from the entire P projection set, calculations are performed in 2 stages, each one providing parallel processing by pixels and voxels, respectively.

A common problem with this approach is choosing the calculation methods for w_{ij} weighting coefficients. Since storing a pre-calculated full system matrix will result in excessive memory use and a significant increase in access time to matrix elements, which is unacceptable for quick execution of the algorithm, it is necessary to choose methods for calculating and temporary storage of the coefficients immediately while running the SART algorithm. Besides, it should be taken into consideration that this approach implies deleting the weighting coefficients from the local RAM areas, which means that they have to be recalculated during the backprojection. So the w_{ij} coefficients values must be the same for (3) and (4). As indicated in [13] identical w_{ij} are obtained when using Siddon's methods (ray driven) [14] or ray casting by Basu - De Man [15–17] for forward projection and Minimum Bounding Rectangle (MBR) method for backprojection. The Siddon method was selected to be the first version of the implementation as it was simpler for calculations.

Thus, the subjects of the research and development are the following:

1) adaptation of the Siddon and MBR methods for the linear geometry of the tomosynthesis unit for calculating the weight coefficients of the SART algorithm;
2) parallel implementation of the SART algorithm on a heterogeneous software and hardware platform to speed up the reconstruction result.

3   Methods for Obtaining Weighting Coefficients

3.1   Installation Geometry Description

Figure 1 provides a graphic representation of the installation geometry of a digital linear tomosynthesis (with the example of "Cosmos-D" manufactured by Sevkavrentgen-D LLC), where:

src is the point source of x-ray radiation (in the calculations, the radiation is considered to propagate from the point source located at the focal spot point *src* and cover the entire surface of the detector in all positions during scanning) or the focal spot of the source;

COR is the rotation center or cutting plane height;
COD is the center of a detector;
L is the distance from the detector surface to the src point or focal length.
H is the distance from the detector surface to the table surface.
G is the distance from the table surface to COR.

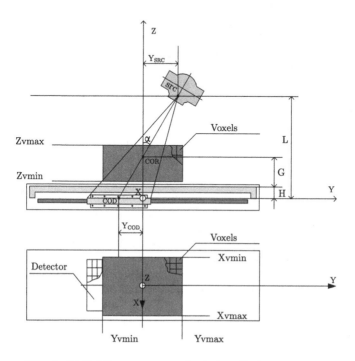

Fig. 1. Digital linear tomosynthesis installation geometry

All calculations are carried out in a single Cartesian coordinate system. In this case, the center of the coordinate system O is located in the plane of motion of the detector surface at the point COD at the initial position of the complex.

Scanning with Digital Linear Tomosynthesis. During scanning, the detector and emitter move on parallel planes in the opposite directions, so that the line connecting src and COD always intersects the OZ axis at one point which is the COR point. The angle α between the OZ axis and the line connecting src and COD is called the projection angle. The angle α will be considered positive when the source moves in the positive direction of the OY axis. P is the total number of projections.

The angle α between the OZ axis and the line connecting src and COD is called the projection angle. The angle α will be considered positive when the source moves in the positive direction of the OY axis. P is the total number of projections.

Uniform scanning (usually during modeling) means scanning with a uniform step symmetrical to the OZ axis. For uniform scanning, the step of changing α between the adjacent projections is $\Delta\alpha$ and ranges from $-\alpha_b$ to α_b with a step $\Delta\alpha$, where α_b is the boundary value of the projection angle (5, 6).

In real complexes, random scanning is usually performed to obtain projections for an arbitrary set of angles. An array of scan angles $\alpha[P]$ is acquired.

$$\alpha_p = -\alpha_b + \Delta\alpha p, \ p \in [0; P-1], \tag{5}$$

$$\Delta\alpha = \frac{2\alpha_b}{P-1}. \tag{6}$$

The radiation source and the detector center coordinates (7) are described respectively as follows:

$$\begin{aligned} x_{src} = 0, \ y_{src} = (L - (G-H))\tan\alpha, \ z_{src} = L \\ x_{COD} = 0, \ y_{COD} = -(G+H))\tan\alpha, \ z_{COD} = 0. \end{aligned} \tag{7}$$

Detector Parameters. The following are the basic parameters of the detector.

Dx is the physical size of the detector along the OX axis;
Dy is the physical size of the detector along the OY axis;
Mx is resolution (number of detector elements) along the OX axis;
My is resolution (number of detector elements) along the OY axis;
$M = Mx \cdot My$ is the total number of the detector elements (pixels).

Figure 2 shows the points on the detector element used in various algorithms for obtaining a partial system matrix.

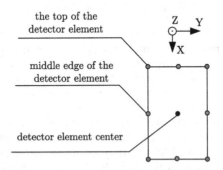

Fig. 2. Detector element

The detector element dimensions are determined:

$$Bx = Dx/Mx, \ By = Dy/My. \tag{8}$$

The detector extreme points coordinates are determined:

$$X_{Dmin} = X_{COD} - Dx/2, \ X_{Dmax} = X_{COD} + Dx/2$$
$$Y_{Dmin} = Y_{COD} - Dy/2, \ Y_{Dmax} = Y_{COD} - Dy/2. \tag{9}$$

Then, using (8) and (9), we acquire the detector element center coordinates:

$$Xbin_{ix} = X_{Dmin} + (ix + 0.5)Bx, \ ix \in [0, M_x - 1]$$
$$Ybin_{iy} = Y_{Dmin} + (iy + 0.5)By, \ iy \in [0, M_y - 1]$$
$$Zbin = 0.$$

Parameters of the Reconstructed Volume. Below are the main parameters of the reconstructed volume and voxels:

X_{Vmin} is the left border of the reconstructed volume along the OX axis;
X_{Vmax} is the right border of the reconstructed volume along the OX axis;
Y_{Vmin} is the left border of the reconstructed volume along the OY axis;
Y_{Vmax} is the right border of the reconstructed volume along the OY axis;
Z_{Vmin} is the left border of the reconstructed volume along the OZ axis;
Z_{Vmax} is the right border of the reconstructed volume along the OZ axis;
Nx, Ny, Nz is resolution along the OX, OY, OZ axes respectively;
$N = Nx \cdot Ny \cdot Nz$ is the total number of voxels.

Figure 3 shows the points on the voxel used in various algorithms for obtaining a partial system matrix.

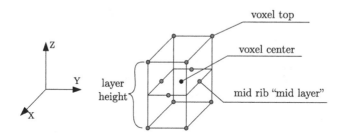

Fig. 3. Points on the voxel

The dimensions of the detector element are determined:

$$Vx = (X_{Vmax} - X_{Vmin})/N_x$$
$$Vy = (Y_{Vmax} - Y_{Vmin})/N_y$$
$$Vz = (Z_{Vmax} - Z_{Vmin})/N_z.$$

Voxel number along the OX axis: $jx \in [0, N_x - 1]$.
Voxel number along the OY axis: $jy \in [0, N_y - 1]$.
Voxel number along the OZ axis: $jz \in [0, N_z - 1]$.

The boundary values of the coordinates of the volume occupied by the voxel can be calculated by the formulas:

$$Xvoxmin_{jx} = X_{Vmin} + jx \cdot Vx, \ Xvoxmax_{jx} = X_{Vmin} + (jx + 1) \cdot Vx$$
$$Yvoxmin_{jy} = Y_{Vmin} + jy \cdot Vy, \ Yvoxmax_{jy} = Y_{Vmin} + (jy + 1) \cdot Vy$$
$$Zvoxmin_{jz} = Z_{Vmin} + jz \cdot Vz, \ Zvoxmax_{jz} = Z_{Vmin} + (jz + 1) \cdot Vz.$$

The coordinates of the voxel vertices can be calculated using the formulas:

$$
\begin{aligned}
&Xvox_{jx}^0 = X_{Vmin} + jx \cdot Vx \\
&Yvox_{jy}^0 = X_{Vmin} + jy \cdot Vy \\
&Zvox_{jz}^0 = X_{Vmin} + jz \cdot Vz \\
&Xvox_{jx}^1 = X_{Vmin} + (jx + 1) \cdot Vx, \ Yvox_{jy}^1 = Yvox_{jy}^0, \ Zvox_{jz}^1 = Zvox_{jz}^0 \\
&Xvox_{jx}^2 = Xvox_{jx}^1, \ Yvox_{jy}^2 = Y_{VOLmin} + (jy + 1) \cdot Vy, \ Zvox_{jz}^2 = Zvox_{jz}^0 \\
&Xvox_{jx}^3 = Xvox_{jx}^0, \ Yvox_{jy}^3 = Yvox_{jy}^2, \ Zvox_{jz}^3 = Zvox_{jz}^0 \\
&Xvox_{jx}^4 = Xvox_{jx}^0, \ Yvox_{jy}^4 = Yvox_{jy}^0, \ Zvox_{jz}^4 = Z_{Vmin} + (jz + 1) \cdot Vz \\
&Xvox_{jx}^5 = Xvox_{jx}^1, \ Yvox_{jy}^5 = Yvox_{jy}^1, \ Zvox_{jz}^5 = Zvox_{jz}^4 \\
&Xvox_{jx}^6 = Xvox_{jx}^2, \ Yvox_{jy}^6 = Yvox_{jy}^2, \ Zvox_{jz}^6 = Zvox_{jz}^4 \\
&Xvox_{jx}^7 = Xvox_{jx}^3, \ Yvox_{jy}^7 = Yvox_{jy}^3, \ Zvox_{jz}^7 = Zvox_{jz}^4.
\end{aligned}
\tag{10}
$$

3.2 Siddon's Ray Casting Forward Projection

We adapt the original Siddon algorithm [14, 18] into the developed complex. The algorithm begins with the formation of the parametric equation of the segment that intersects the volume with voxels. In our case, the equation is:

$$
\begin{cases}
X(t) = X_{src} + (Xbin_{ix} - X_{src})t \\
Y(t) = Y_{src} + (Ybin_{iy} - Y_{src})t \\
Z(t) = Z_{src} + (Zbin_{iz} - Z_{src})t.
\end{cases}
$$

We formulate equations for three orthogonal sets of equally spaced parallel planes, with the number of planes on each axis being one more than the number of voxels – $(N_x + 1)(N_y + 1)(N_z + 1)$, and the planes numbered by indices lx, ly, lz:

$$
\begin{cases}
Xpl_{lx} = X_{Vmin} + lx \cdot Vx, \ lx \in [0, 1, \dots, N_x] \\
Ypl_{ly} = Y_{Vmin} + ly \cdot Vy, \ ly \in [0, 1, \dots, N_y] \\
Zpl_{lz} = Z_{Vmin} + lz \cdot Vz, \ lz \in [0, 1, \dots, N_z].
\end{cases}
$$

We shall calculate the parameter t value for the intersection points of the segment with the extreme planes of the orthogonal sets of planes (left along the corresponding axis L, right R):

$$
\begin{cases}
tx_L = \dfrac{X_{Vmin} - X_{src}}{Xbin_{ix} - X_{src}} \\
tx_R = \dfrac{X_{Vmax} - X_{src}}{Xbin_{ix} - X_{src}}, \ if \ (Xbin_{ix} - X_{src}) \neq 0.
\end{cases}
$$

The values ty_L and ty_R, tz_L and tz_R are calculated in the same way as tx_L and tx_R.

Next, the minimum and maximum value of the parameter t for each set of planes are to be found (it is necessary to consider the equivocating of the values tx_L and tx_R when the coordinates $Xbin_{ix}$ and X_{src} coincide, the same way for other sets):

$$tx_{min} = \begin{cases} -10^{38}, \text{ if } (Xbin_{ix} - X_{src}) = 0 \\ tx_L, \text{ if } txL < txR \\ tx_R, \text{ other} \end{cases}$$

$$tx_{max} = \begin{cases} 10^{38}, \text{ if } (Xbin_{ix} - X_{src}) = 0 \\ tx_L, \text{ if } txL > txR \\ tx_R, \text{ other.} \end{cases}$$

The values ty_{min} and ty_{max}, tz_{min} and tz_{max} are calculated in the same way as tx_{min} and tx_{max}.

We obtain the value for the entry and exit points from the reconstructed volume according to the formulas (the values 0 and 1 are excluded from the formulas, since in transmission tomography the source and receiver cannot be included into the reconstructed volume):

$$t_{min} = max\{tx_{min}, ty_{min}, tz_{min}\}$$
$$t_{max} = min\{tx_{max}, ty_{max}, tz_{max}\}.$$

If $t_{max} \le t_{min}$, the segment does not intersect the reconstructed volume, we get the result of $RayCnt = 0$ and the algorithm execution ends. For all the points inside the reconstructed volume, the value of t lies in the interval $[t_{min}, t_{max}]$. The segment intersection with planes is determined independently by three sets of planes. We shall calculate the starting and ending indices of the planes by the sets intersected by the segment (operation $\lfloor val \rfloor$ - the floor function)

$$\begin{cases} lx_{min} = \left\lfloor N_x - \dfrac{X_{Vmax} - (Xbin_{ix} - X_{src})t_{min} - X_{src}}{Vx} \right\rfloor \\ lx_{max} = \left\lfloor \dfrac{X_{src} + (Xbin_{ix} - X_{src})t_{max} - X_{Vmin}}{Vx} \right\rfloor \end{cases}, \text{ if } (Xbin_{ix} - X_{src}) > 0$$

$$\begin{cases} lx_{min} = \left\lfloor N_x - \dfrac{X_{Vmax} - (Xbin_{ix} - X_{src})t_{max} - X_{src}}{Vx} \right\rfloor \\ lx_{max} = \left\lfloor \dfrac{X_{src} + (Xbin_{ix} - X_{src})t_{min} - X_{Vmin}}{Vx} \right\rfloor \end{cases}, \text{ if } (Xbin_{ix} - X_{src}) < 0.$$

The values ly_{min} and ly_{max}, lz_{min} and lz_{max} are calculated in the same way as lx_{min} and lx_{max}.

For each set of planes, we shall calculate an array of t values:

$$tx_{lx} = \frac{X_{Vmin} + lx \cdot Vx - X_{src}}{Xbin_{ix} - X_{src}},$$

$$\begin{cases} lx = \{lx_{min}, \ldots, lx_{max}\}, & \text{if } (Xbin_{ix} - X_{src}) > 0 \\ lx = \{lx_{max}, \ldots, lx_{min}\}, & \text{if } (Xbin_{ix} - X_{src}) < 0. \end{cases}$$

The values ty_{ly} and tz_{lz} are calculated in the same way as tx_{lx}.

Next, the arrays tx, ty, tz need to be combined into one array t so that the values in the array t are arranged in the increasing order. When combining, the matching values, if any, shall be discarded. If, after combining, the t_{min} and t_{max} values (or one of them) are absent in the array, the array needs to be replenished with these values. The number of elements in the $RayCntDot$ array is determined during the merge process.

We calculate the number of crossed voxels: $RayCnt = RayCntDot - 1$.

We calculate the array of lengths of the segment intersection and the voxels:

$$D = \sqrt{(Xbin_{ix} - X_{src})^2 + (Ybin_{iy} - Y_{src})^2 + (Zbin_{iz} - Z_{src})^2},$$

$$RayW_{cnt} = D(t_{cnt+1} - t_{cnt}), \quad cnt = \{0, \ldots, RayCnt - 1\}$$

We calculate the voxel numbers along the axes:

$$\begin{cases} RayVoxelNums_{cnt,0} = \left\lfloor \frac{X_{src} + (t_{cnt+1} + t_{cnt})(Xbin_{ix} - X_{src})/2 - X_{Vmin}}{Vx} \right\rfloor \\ RayVoxelNums_{cnt,1} = \left\lceil \frac{Y_{src} + (t_{cnt+1} + t_{cnt})(Ybin_{iy} - Y_{src})/2 - Y_{Vmin}}{Vy} \right\rceil \\ RayVoxelNums_{cnt,2} = \left\lceil \frac{Z_{src} + (t_{cnt+1} + t_{cnt})(Zbin_{iz} - Z_{src})/2 - Z_{Vmin}}{Vz} \right\rceil. \end{cases}$$

Next, using the obtained indices of crossed voxels and the length of the intersections, we calculate the value of c_i with formula (3).

3.3   Minimum Bounding Rectangle Backprojection (MBR Backprojection) Partially Using Siddon's Method

When backprojecting under the SART algorithm, each voxel of the reconstructed volume (working voxel) should be adjusted to the weighted average sum over all detector elements. Moreover, for most detector elements, the value of the coefficients of the partial system matrix is zero, and, therefore, the weighted average sum must be calculated with non-zero elements.

To "reject" most non-zero elements, we use the minimum bounding rectangle method (MBR, Fig. 4). To do this, we need to:

- draw rays from *scr* through eight voxel vertices;
- find the coordinates of the points of rays' intersection with the detector plane;
- on the axes *OX* and *OY* determine the minimum and maximum value from the coordinates of the intersection points;
- determine the boundaries of the detector element numbers found in the minimum bounding rectangle area.

After this, the partial system matrix coefficient is determined and the numerator and denominator of formula (4) are calculated in the cycle for all the detector elements located in the MBR.

Let us consider the method described above in more detail.

The coordinates of the voxel vertices are calculated using formulas (10).

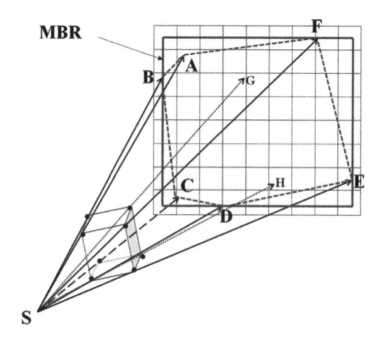

Fig. 4. Minimum bounding rectangle backprojection

To find the intersection points with the detector plane for each of the eight rays, we make the parametric equation of the line:

$$\begin{cases} X(t)^v = X_{src} + (Xvox_{jx}^v - X_{src})t \\ Y(t)^v = Y_{src} + (Yvox_{jy}^v - Y_{src})t, \ v \in [1,\dots,8] \\ Z(t)^v = Z_{src} + (Zvox_{jz}^v - Z_{src})t. \end{cases}$$

In our case, the detector plane coincides with the *XOY* plane, respectively, for all intersection points $Zinters^v = 0$ and at the intersection point

$$t_{inters}^v = -\frac{Z_{src}}{Zvox_{jz}^v - Z_{src}},$$

we obtain the coordinates of the intersection of the rays with the detector plane by substituting in the equations for the coordinates X and

$$
\begin{cases}
X^v_{inters} = X_{src} - \dfrac{Z_{src}(Xvox^v_{jx} - X_{src})}{Zvox^v_{jz} - Z_{src}} \\[2mm]
Y^v_{inters} = Y_{src} - \dfrac{Z_{src}(Yvox^v_{jy} - Y_{src})}{Zvox^v_{jz} - Z_{src}}, \quad v = [1,\dots,8] \\[2mm]
Z^v_{inters} = 0.
\end{cases}
$$

We shall determine the minimum and maximum values of the coordinate

$$
\begin{aligned}
X_{intersmin} &= min_{v\in[1\dots8]}\{X^v_{inters}\} \\
X_{intersmax} &= max_{v\in[1\dots8]}\{X^v_{inters}\} \\
Y_{intersmin} &= min_{v\in[1\dots8]}\{X^v_{inters}\} \\
Y_{intersmax} &= max_{v\in[1\dots8]}\{X^v_{inters}\}.
\end{aligned}
$$

Determine MBR:

$$
\begin{aligned}
ix_{min} &= \left\lfloor \frac{X_{intersmin} - X_{Dmin}}{Bx} \right\rfloor \\
ix_{max} &= \left\lfloor \frac{X_{intersmax} - X_{Dmin}}{Bx} + 1 \right\rfloor \\
iy_{min} &= \left\lfloor \frac{Y_{intersmin} - Y_{Dmin}}{By} \right\rfloor \\
iy_{max} &= \left\lfloor \frac{Y_{intersmax} - Y_{Dmin}}{By} + 1 \right\rfloor.
\end{aligned}
$$

To calculate the coefficients of a particular system matrix for each element of the detector located in the MBR ($ix = [ix_{min},\dots,ix_{max}]$, $iy = [iy_{min},\dots,iy_{max}]$), we carry out the calculations similarly to the initial part of the Siddon's algorithm (until t_{min} and t_{max} are obtained). Next, we calculate the partial system matrix coefficient for each detector element in the MBR:

$$
w_{ix,iy} = (tmax - tmin)D,
$$

where

$$
D = \sqrt{(Xbin_{ix} - X_{src})^2 + (Ybin_{iy} - Y_{src})^2 + (Zbin_{iz} - Z_{src})^2}.
$$

Using $w_{ix,iy}$ we calculate the voxel value by the formula (4).

4   Parallel Implementation of the SART Algorithm on a Heterogeneous CPU-GPU System Using OpenCL

4.1   Algorithm Description

As the hardware platform for implementing the reconstruction algorithm, a heterogeneous system consisting of a central processor and a graphic processor was selected as it has the optimal performance, price, power consumption and compactness ratio. So, the personal computer of the tomosynthesis unit operator, can both serve as a calculator and carry out reconstruction within an acceptable time.

The most popular heterogeneous computing software platforms are CUDA and OpenCL. The advantage of CUDA is the widespread support provided by the scientific community, as well as slightly higher optimization and performance on NVIDIA GPUs compared to OpenCL. OpenCL has the advantage of being a cross-platform, able to perform calculations not only on GPUs of NVIDIA and AMD, but also on multi-core processors and even FPGA. For this reason, the OpenCL framework was chosen for implementation.

An OpenCL application is a collection of software codes executed on a host and OpenCL devices. A host is usually understood as the Central processor (CPU) of a computing device (computer, tablet, phone, etc.), and OpenCL devices are a set of computing units corresponding to a graphics processor (GPU), multicore CPU, or other parallel architecture processors available from a host program. Execution of an OpenCL application thus involves execution of a host program (on the host) and execution of a special code on one or more OpenCL devices running the host program.

The general scheme of the application performing the reconstruction consists of three parts:

1) a top-level function that performs data preparation and runs a function an OpenCL host;
2) the OpenCL host function executed on the CPU controls the data exchange between the host and the device, runs kernel functions for execution on the device; this function can be performed in a separate dynamic library and called from a top-level function written in any programming language;
3) a kernel function that runs in multiple instances on the device and each instance runs in parallel in a separate thread. The forward projection kernel is based on the calculations in Sect. 3.2 and the backprojection kernel in Sect. 3.3.

Below is a generalized algorithm for implementing SART using the ray-driven (ray casting) method, adapted for parallel computing on multi-core CPUs and GPUs.

Features of the algorithm:

– does not involve significant allocation of additional memory (except for the memory consumed by ray casting and MBR backprojection algorithms), all coefficients are calculated "on the fly";
– in forward projection, the coefficients are calculated by any of the ray casting methods (as in the original algorithm); in the current implementation, the Siddon's method is used;
– for backprojection, the minimum bounding rectangle approach is used; in the current version, the approach is implemented partially using the Siddon's method;

– this algorithm is also suitable for implementation the Basu - De Man method (distance-driven) [15] when replacing the Ray Casting function (Sect. 3.2) and MBR backporojection with those corresponding to the Basu - De Man method.

Generalized algorithm (host function).

1. Preparatory operation.
   (a) Read from file into string variables with the text of the kernels of the forward and backward projection.
   (b) Determine the number of platform and device on which the calculations will be performed (clGetPlatformIDs and clGetDeviceIDs).
   (c) Create context for the selected device (clCreateContext).
   (d) Compile kernel programs for the selected device (clBuildProgramWith-Source, clBuildProgram and clCreateKernel).
   (e) Create a command queue for the selected device (clCreateCommandQueue).
   (f) Create memory objects for data exchange between host and device (clCreateBuffer).
   (g) Set the correspondence between kernel function arguments and memory objects and host variables (clSetKernelArg).
   (h) Set the global number of threads for forward and backprojection and the size of local groups used on the device (globalItemSize and localItemSize).
   (i) Rearrange the projections.
   (j) Write the initial arrays and initial geometry parameters to the created memory objects on the device (clEnqueueWriteBuffer).
2. SART algorithm.
   (a) At each step of the cycle iterations of the algorithm START to execute:
       i. Set the current relaxation coefficient as an argument of the backprojection kernel function (clSetKernelArg).
       ii. At each step of the cycle on projections $P_\alpha \in P$ to execute:
           A. Calculate current installation geometry parameters.
           B. Select the current projection P_α from the projection array and write it to the corresponding device memory object.
           C. Run the forward projection kernels using the Siddon method (clEnqueueNDRangeKernel). The result is an array of correction coefficients c_i stored in the device memory to be used when starting the backprojection kernels.
           D. Run the backprojection kernels using the MBR method (clEnqueueNDRangeKernel).
   (b) Read array of voxels from device memory to host memory (clEnqueueReadBuffer).
   (c) Clear all created memory objects (clReleaseMemObject).

4.2   Computational Experiment

As a hardware platform, a personal computer was used in the following configuration:

- CPU: Intel Core i7-8700K, 6 cores, 12 threads.
- GPU No. 1: AMD Radeon RX Vega 64, 4096 universal processors, 8 GB of HBM memory with a bandwidth of 484 GB/s, maximum FP32 performance is 12.7 TFLOPS.
- GPU No. 2: NVIDIA GTX 1060 6 GB, 1280 universal processors, 6 GB of GDDR5 memory with a bandwidth of 192 GB/s, maximum FP32 performance is 3.9 TFLOPS.
- RAM: DDR4 32 GB.

The calculations were based on the detector resolution $M = 1440^2$ and the physical size of the reconstructed volume of 350^3 mm, the range of variation of the scan angles was adopted from $-30°$ to $30°$ in increments of $1°$, which amounted to 61 projections. As the initial data the projections obtained by simulating the transmission of the Shepp–Logan mathematical phantom with a volume of 896^3 voxels by the Siddon method were used.

Table 1 below shows the time of one iteration of reconstruction of volume 128^3 for various devices using OpenCL.

Table 1. Reconstruction time for various devices

Device	i7-8700K	GTX 1060 6 GB	RX Vega 64
Time, s	28	7,4	2,9

Figure 5 shows the dependence of the time of one iteration of reconstruction on the dimension of the reconstructed volume for the GPU Vega 64. Figure 6

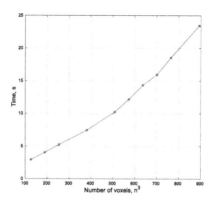

Fig. 5. Reconstruction time on RX Vega 64

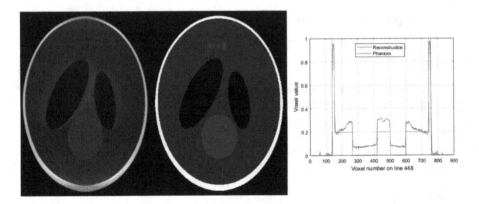

Fig. 6. Result of reconstruction

shows the result of the reconstruction of the 896^3 volume and a comparison with the Shepp–Logan phantom.

Thus, we got a relatively short reconstruction time with a satisfactory quality of reconstruction using affordable devices. The acceleration of calculations on the GPU compared to the CPU was ~10 times at $N = 128^3$, and with an increase in the size of the reconstructed volume, this gap increases significantly.

5   Conclusion

The result of the work is accelerated calculation of volume reconstruction for the problem of linear tomosynthesis by the SART method. To do this, the Siddon method for forward projection was modified and the MBR method for backprojection was used. At this stage of the work, the reconstruction was obtained only for projections based on the Shepp–Logan phantom. In the next stages, reconstruction results based on real images shall be obtained, the accuracy shall be improved, but at the same time more computationally complex algorithms such as those based on distance-driven [16] and trapezoid footprint methods [19, 20] shall be implemented.

Acknowledgement. The study was carried out with the financial support of the Ministry of Education and Science of the Russian Federation in the framework of the decree of the Government of the Russian Federation No. 218 in cooperation with LLC Sevkavrentgen–D.

References

1. Dobbins, J.T., Godfrey, D.J.: Digital x-ray tomosynthesis: current state of the art and clinical potential. Phys. Med. Biol. **48**(19), R65–106 (2003)
2. Dobbins, J.T., et al.: Digital tomosynthesis of the chest for lung nodule detection: interim sensitivity results from an ongoing NIH-sponsored trial. Med. Phys. **35**(6), 2554–2557 (2008)

3. Hu, Y.-H., Zhao, W., Mertelmeier, T., Ludwig, J.: Image artifact in digital breast tomosynthesis and its dependence on system and reconstruction parameters. In: Krupinski, E.A. (ed.) IWDM 2008. LNCS, vol. 5116, pp. 628–634. Springer, Heidelberg (2008). https://doi.org/10.1007/978-3-540-70538-3_87

4. Jacobs, F., Sundermann, E., De Sutter, D., Christiaens, M., Lemahieu, I.: A fast algorithm to calculate the exact radiological path through a pixel or voxel space. J. Comput. Inf. Technol. **6**, 89–94 (1998)

5. Claus, B.E.H., Eberhard, J.W., Schmitz, A., Carson, P., Goodsitt, M., Chan, H.-P.: Generalized filtered back-projection reconstruction in breast tomosynthesis. In: Astley, S.M., Brady, M., Rose, C., Zwiggelaar, R. (eds.) IWDM 2006. LNCS, vol. 4046, pp. 167–174. Springer, Heidelberg (2006). https://doi.org/10.1007/11783237_24

6. Zhang, Y., Chan, H., Sahiner, B., et al.: A comparative study of limited-angle cone-beam reconstruction methods for breast tomosynthesis. Med. Phys. **33**(10), 3781–3795 (2006)

7. Mueller, K.: Fast and accurate three-dimensional reconstruction from cone-beam projection data using algebraic methods. Ph.D. thesis, Graduate School of the Ohio State University (1998)

8. Andersen, A.H.: Algebraic reconstruction in CT from limited views. IEEE Trans. Med. Imaging **8**(1), 50–55 (1989)

9. Reiser, I., Bian, J., Nishikawa, R.M., Sidky, E.Y., Pan, X.: Comparison of reconstruction algorithms for digital breast tomosynthesis. In Proceedings of the 9th International Meeting on Fully Three Dimensional Image Reconstruction in Radiology and Nuclear Medicine, pp. 155–158 (2007)

10. Andersen, A.H., Kak, A.C.: Simultaneous algebraic reconstruction technique (SART): a superior implementation of the ART algorithm. Ultrason. Imaging **6**(1), 81–94 (1984)

11. Zhao, H., Reader, F.J.: Fast ray-tracing technique to calculate line integral paths in voxel arrays. In: IEEE Nuclear Science Symposium Conference Record, vol. 4, pp. 2808–2812 (2003)

12. Du, Y., Yu, G., Xiang, X., Wang, X.: GPU accelerated voxel-driven forward projection for iterative reconstruction of cone-beam CT. BioMed. Eng. OnLine (2017). https://doi.org/10.1186/s12938-016-0293-8

13. Pang, W., Qin, J., Lu, Y., et al.: Accelerating simultaneous algebraic reconstruction technique with motion compensation using CUDA-enabled GPU. Int. J. CARS **6**, 187–199 (2011)

14. Siddon, R.L.: Fast calculation of the exact radiological path for a 3-dimensional CT array. Med. Phys. **12**(2), 252–255 (1985)

15. De Man, B., Basu, S.: Distance-driven projection and backprojection in three dimensions. Phys. Med. Biol. **49**(11), 2463–2475 (2004)

16. Levakhina, Y.M., Duschka, R.L., Barkhausen, J., Buzug, T.M.: Digital tomosynthesis of hands using simultaneous algebraic reconstruction technique with distance-driven projector. In 11th REFERENCES 181 International Meeting on Fully Three-Dimensional Image Reconstruction in Radiology and Nuclear Medicine, pp. 167–170 (2011)

17. Choi, S., Lee, D., et al.: Development of a chest digital tomosynthesis R/F system and implementation of low-dose GPU-accelerated compressed sensing (CS) image reconstruction. Med. Phys. **45**, 1871–1888 (2018)

18. Gao, H.: Fast parallel algorithms for the X-ray transform and its adjoint. Med. Phys. **39**(11), 7110–7120 (2012)

19. Wu, M., Fessler, J.A.: GPU acceleration of 3D forward and backward projection using separable footprints for X-ray CT image reconstruction. In: 3rd Workshop on High Performance Image Reconstruction, pp. 53–56 (2011)
20. Long, Y., Fessler, J.F., Balter, J.M.: 3D forward and back-projection for X-ray CT using separable footprints. In: 11th International Meeting on Fully Three-Dimensional Image Reconstruction in Radiology and Nuclear Medicine, pp. 146–149 (2011)

Developing an Efficient Vector-Friendly Implementation of the Breadth-First Search Algorithm for NEC SX-Aurora TSUBASA

Ilya V. Afanasyev[1]([envelope]) [ORCID], Vladimir V. Voevodin[1] [ORCID], Kazuhiko Komatsu[2] [ORCID], and Hiroaki Kobayashi[2] [ORCID]

[1] Research Computing Center of Moscow State University, Moscow 119234, Russia
afanasiev_ilya@icloud.com, voevodin@parallel.ru
[2] Tohoku University, Sendai, Miyagi 980-8579, Japan
{komatsu,koba}@tohoku.ac.jp

Abstract. Breadth-First Search (BFS) is an important computational kernel used as a building-block for many other graph algorithms. Different algorithms and implementation approaches aimed to solve the BFS problem have been proposed so far for various computational platforms, with the direction-optimizing algorithm being the fastest and the most computationally efficient for many real-world graph types. However, straightforward implementation of direction-optimizing BFS for vector computers can be extremely challenging and inefficient due to the high irregularity of graph data structure and the algorithm itself. This paper describes the world's first attempt aimed to create an efficient vector-friendly BFS implementation of the direction-optimizing algorithm for NEC SX-Aurora TSUBASA architecture. SX-Aurora TSUBASA vector processors provide high-performance computational power together with a world-highest bandwidth memory, making it a very interesting platform for solving various graph-processing problems. The implementation proposed in this paper significantly outperforms the existing state-of-the-art implementations both for modern CPUs (Intel Skylake) and NVIDIA V100 GPUs. In addition, the proposed implementation achieves significantly higher energy efficiency compared to other platforms and implementations both in terms of average power consumption and achieved performance per watt.

Keywords: Breadth-First Search · Graph algorithms · NEC SX-Aurora TSUBASA · Vectorisation

1 Introduction

Developing efficient implementations of various graph algorithms is an extremely important problem of modern computer science, since graphs are frequently used to model many real-world objects in different fields of application. These

© Springer Nature Switzerland AG 2020
L. Sokolinsky and M. Zymbler (Eds.): PCT 2020, CCIS 1263, pp. 131–145, 2020.
https://doi.org/10.1007/978-3-030-55326-5_10

fields include social network and web-graphs analysis, solving infrastructure and socio-economic problems, and many others. In all mentioned fields of application, all graphs may have a very large size, which requires using supercomputers to accelerate computations. Traditionally, two types of supercomputer systems are used for graph processing: distributed-memory and shared memory systems. Distributed-Memory systems are typically used for extremely large-scale graph processing (having about 2^{40} edges), while shared-memory architectures are typically used for processing much smaller graphs (about 2^{32}–2^{34} edges). However, shared-memory architectures are generally significantly more efficient on a per core, per dollar, and per joule basis than distributed memory systems [17], while shared-memory algorithms tend to be simpler compared to their distributed counterparts. In addition, shared memory systems are frequently designed around using GPUs and other co-processors, which provide higher performance and better energy efficiency.

So far, many graph algorithm implementations have been proposed for different classes of modern shared-memory systems. In particular, several high-performance graph processing libraries and frameworks have been developed for multicore ×86 CPUs and NVIDIA GPUs. However, possibility of implementing graph algorithms for vector computers belonging to another important family of shared-memory systems are studied to a much lesser extent. The newest generation of NEC vector processors with a codename SX-Aurora TSUBASA has a great potential for efficient graph-processing, since it is equipped with world's highest bandwidth memory (1.2 TB/s), while many graph algorithms are memory-bandwidth bound. At the time of this writing, there are no known attempts of implementing the BFS algorithm for SX-Aurora system.

2   NEC SX-Aurora TSUBASA Architecture

NEC SX-Aurora TSUBASA is the latest SX vector supercomputer with dedicated vector processors [13,19]. SX-Aurora TSUBASA inherits the design concepts of the vector supercomputer and enhances its advantages to achieve higher sustained performance and higher usability. Different from its predecessors in the SX supercomputer series [8,12], the system architecture of SX-Aurora TSUBASA mainly consists of vector engines (VEs) equipped with a vector processor and a vector host (VH) of an x86 node. The VE is used as a primary processor for executing applications while the VH is used as a secondary processor for executing basic operating system (OS) functions offloaded from the VE. The VE has eight powerful vector cores. As each core provides 537.6 GFlop/s of single-precision performance with 1.40 GHz frequency, the peak performance VE reaches 4.3 TFlop/s.

Each SX-Aurora vector core consists of three components: scalar processing unit (SPU), vector processing unit (VPU), and memory subsystem. Most computations are performed by VPUs, while SPUs provide typical CPU functionality. Since SX-Aurora is not just a typical accelerator, but rather a self-sufficient processor, SPUs are designed to provide relatively high performance on scalar computations. VPU of each vector core has its own relatively simple instruction pipeline aimed at decoding and reordering incoming vector instructions from SPU. Decoded instructions are executed on vector-parallel pipelines (VPP). To store the results of intermediate calculations, each vector core is equipped with 64 vector registers with a total register capacity of 128 KB. Each register is designed to store a vector of 256 double precision elements (DP). On the memory subsystem side, six HBM modules in the vector processor can deliver the world's highest memory bandwidth of 1.22 TB/s [8]. This high memory bandwidth contributes to higher sustained performance, especially in memory-bound applications.

3   BFS Problem Description

Let us consider an undirected non-weighted graph $G(V, E)$ where V is the set of vertices and E is the set of edges. Hereinafter, the following notation will be used $|V|$ - total number of vertices, $|E|$ - total number of edges in graph G. Edge $(u, v) \in E$ connects vertices u and v. Consider a given source vertex $s \in V$ is selected in the graph. A path from vertex s to arbitrary vertex $v \in V$ in graph G is a finite sequence of edges and vertices $S(s, v) = (a_0, E_0, a_1, E_1, ...a_{n-1}, E_{n-1}, a_n)$, so that each two adjacent edges E_i and E_{i-1} have a common vertex a_I, and $s = a_0, v = a_n$. Length of the path $|S(s, v)|$ is defined as a number of edges included into this path. An output of the breadth-first search algorithm assigns a number to each vertex v of the graph $G(V, E)$, equal to the minimum number of edges in path $S(s, v)$, where all possible paths between the vertices s and v are considered.

The BFS algorithm has a linear computational complexity equal to $O(|V| + |E|)$. However, this computational complexity is optimal for sequential implementations only. Since such sequential implementations as Dijkstra's algorithm [7] include various data dependencies, different variations of the BFS algorithms have been proposed to provide more convenient parallelization. If implemented efficiently, the performance of many BFS algorithms is limited by memory bandwidth of target architecture. Therefore, different optimizations aimed at improving the memory hierarchy use have to be implemented in order to achieve high performance.

4   BFS Algorithms and Existing Implementations

At the moment of this writing, the highest performance is achieved by hybrid BFS algorithms, which combine properties of many other BFS algorithms. Two most basic algorithms, top-down and bottom-up, will be described in this section together with the main ideas of the hybrid algorithm, which forms a basis for the proposed implementation for the NEC SX-Aurora TSUBASA architecture. This section discusses only the general structure and basic mathematical ideas of BFS algorithms, while all implementation and optimization details (including data structures, parallelization strategies, memory access patterns) are discussed further. At the end, this section provides a review of the existing best-known BFS implementations demonstrating the highest performance on various modern architectures, including multicore CPUs and NVIDIA GPUs.

4.1   Conventional Top-Down BFS Algorithm

The top-down BFS algorithm (Listing 1.1) has the following structure. At the beginning of each iteration, a frontier of vertices is constructed from the vertices visited in the previous iteration (for the first iteration, the frontier consists of the source vertex). In each following iteration, the frontier is expanded by visiting all vertices reachable from the current frontier: each vertex investigates all of its neighbors and adds to the next frontier vertices which have not been visited yet. The presence of each vertex inside the frontier can be represented either via a bit mask or a queue. Usually, vertices visited in each iteration are stored via a bit mask converted into a queue an the end of the iteration.

Listing 1.1. Single iteration of the top-down BFS algorithm

```
top_down_step(vertices, frontier, next, parents)
for v ∈ frontier do
    for n ∈ neighbors[v] do
        if parents[n] = -1 then
            parents[n] ⟵ v
            next ⟵   next ∪ {n}
        end if
    end for
end for
```

4.2   Bottom-Up BFS Algorithm

When the frontier is large, it is better to use another algorithm called bottom-up BFS (Listing 1.2). Instead of traversing all adjacent edges of each frontier vertex, this algorithm investigates unvisited vertices only. For each unvisited vertex, it is necessary to find at least one already visited neighbor and set it as a parent of the vertex. As soon as a visited neighbor is found for a particular vertex, there is no need to further traverse other neighbors, which drastically reduces the number of checked edges, and, consequently, the amount of indirect memory accesses.

Listing 1.2. Single iteration of the bottom-up BFS algorithm

```
bottom−up−step(vertices, frontier, next, parents)
for v ∈ vertices do
    if parents[v] = -1 then
        for n ∈ neighbors[v] do
            if n ∈ frontier then
                parents[v] ⟵n
```

```
                    next←─next ∪ {v}
                    break
              end if
          end for
      end if
end for
```

The bottom-up BFS algorithm does not necessarily reduce the amount of the edges traversed for any given iteration. For example, in the first iteration when only the source vertex is visited, the bottom-up algorithm will examine $V - 1$ unvisited graph vertices. If the source vertex does not have many outgoing connections, almost all the neighbors of each graph vertex will be investigated, resulting into $O(|E|)$ vertices traversed. In the meantime, the top-down algorithm will only traverse vertices adjacent to the source vertex. Thus, it is necessary to develop a hybrid algorithm, combining the properties of both top-down and bottom-up algorithms.

4.3 Hybrid (Direction-Optimizing) BFS

The idea of hybrid (direction-optimizing) BFS was proposed in [3]. Advantageous for low-diameter graphs, this algorithm combines the conventional top-down algorithm with the bottom-up algorithm, resulting into the reduced overall number of edges examined, which in turn accelerates the search in a whole. The proposed algorithm makes it possible to decide, whichever top-down or the bottom-up algorithm should be used in the given iteration based on the number of vertices and edges processed at two adjacent iterations of the algorithm, as well as two coefficients, adjustable to a specific of input graph type. This algorithm with some adjustment is used for our SX-Aurora TSUBASA implementation, as well as for many third-party implementations for various platforms.

4.4 BFS Implementations

At the time of writing, no known attempts of developing vectorized BFS implementations for NEC SX-Aurora TSUBASA architecture have been made. Research [1] describes implementations of several graph algorithms for previous generation (SX-ACE) of the NEC vector architecture. However, only comparable per-socket and slightly better per-core performance has been achieved compared to Intel Skylake processors.

Several studies propose attempts of vectorizing the BFS algorithm using the AVX-512 vector extension for Intel multi-core CPUs. Paper [5] describes a vectorized "SlimSel" graph storage format to efficiently vectorize the BFS algorithm based on the sparse matrix-vector multiplication. However, this approach has a significantly larger computational complexity compared to the direction-optimized algorithm for many important types of graphs, and therefore cannot be applied in many application fields.

Multiple non-vectorized parallel BFS implementations have been proposed for multi-core processors (mainly the ×86 architecture), including implementations based on the direction-optimizing algorithm [20,21]. In addition, several graph-processing frameworks and libraries, developed for multicore architectures, include BFS implementations: Ligra [17], the GAP Benchmark Suite [4], and Galois [16].

Various high-performance BFS implementations have been designed for NVIDIA GPUs [10,15] demonstrate the possibility of achieving significant acceleration and greater energy efficiency compared to CPU implementations. Many graph-processing libraries and frameworks have been developed for the NVIDIA GPU architecture, including Gunrock [18], Enterprise [14], MapGraph [9] and CuSHA [11]. Among existing GPU implementations, Gunrock is reported to have the highest performance.

5   Hardware Characteristics

All the experiments have been conducted using a single SX-Aurora TSUBASA Vector Engine equipped with a 48 GB RAM, which achieves the peak performance of 2.45 TFLOP/s. To generate vectorized code, the NCC compiler of version 2.4.1 has been used.

6   Input Graphs

In this paper we use synthetic RMAT [6] and random uniform graphs to investigate the developed implementation performance. Synthetic graphs can be freely scaled by changing their size, average degree and edge distribution properties. Thus, with the synthetic graphs it is possible to easily model different situations: for example, various graph data can be placed in the certain cache level of target platform's memory hierarchy. All graphs used in this paper have an average connection count equal to 32, and the number of vertices in the range $[2^{18}, 2^{25}]$. The largest graph has 33 million of vertices and 1 billion edges.

7   Implementation Details

The proposed BFS implementation of the direction-optimizing algorithm for SX-Aurora TSUBASA architecture consists of the following blocks:

- BFS iterations using the bottom-up algorithm
- BFS iterations using the top-down algorithm
- changing the state between the top-down and the bottom-up algorithm
- converting bitmask to queue and vice versa.

Each block has significant algorithmic and implementation differences, compared to the existing implementations of the direction-optimizing algorithm for multicore systems, since it is extremely important to efficiently vectorize each algorithm part.

7.1   Graph Storage Format

The storage format of the input graph defines the memory access pattern, together with the organization of parallel threads and vector instruction workload. Memory access patterns for NEC SX-Aurora TSUBASA should generally consist of different vector cores accessing independent non-overlapping memory regions, while memory access from a single vector core should have a sequential (or localized) pattern within the same vector instruction. In addition, SX-Aurora TSUBASA provides relatively efficient implementations of gather and scatter instructions, through which indirect memory accesses are implemented.

The implementation of the direction-optimizing BFS algorithm requires storing information about neighbors of each graph vertex, since in each iteration only a part of adjacent edges of each frontier vertex is traversed. Thus, the proposed implementation uses traditional Compressed Sparse Rows (CSR) format with several adjustments. For each graph vertex, there is a stored list of vertices adjacent to it; all adjacent lists are stored in a single array of size $|E|$, which is supplemented by an array of offsets of size —V + 1—. Since BFS does not require information about weights of edges, each edge is represented only by a destination ID of its neighbor vertex.

To increase the graph processing efficiency, several modifications to the traditional CSR format have been proposed. First, all input graph vertices are sorted based on the number of connections of each vertex. This sorting allows storing the most frequently accessed vertices separately from the vertices with a medium and low connectivity degree that are rarely accessed during graph traversals. To assure that the resulting graph is equivalent to the input one, vertices and their neighbors are renumbered using new indices acquired after the sorting. Second, all neighbors of each vertex are sorted in the ascending order for each vertex to store information about the connection with more "significant" vertices in the beginning of its adjacency list.

To allow efficient vector-friendly graph traversals during the bottom-up stage, a certain group of graph edges is stored separately inside the vector extension. For each vertex, only the first $max(4, 0.2 * |E|/|V|)$ adjacent edges are stored inside the vector extension. The vector extension is designed to efficiently provide information about adjacent edges, when each group of 256 graph vertices is processed using the single vector instruction. The modified CSR format and its vector extension are illustrated in Fig. 1.

7.2   Top-Down Iteration

At the beginning of each top-down iteration, it is necessary to extract IDs of the vertices visited during the previous iteration. A vectorized algorithm converting an array of levels into a list of vertex IDs belonging to a specified level is described below. This algorithm generates a list of vertex IDs, sorted in the descending order of the connections count. After such list is generated, it is divided into 3 independent subsets.

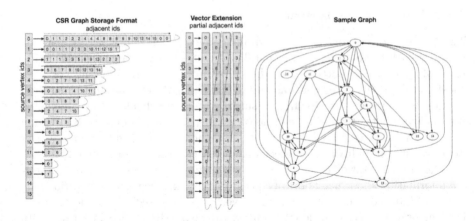

Fig. 1. Preprocessed CSR graph storage format (left) and its vector extension (middle) for a sample graph, with loops and multiple arc removed for simplicity (right)

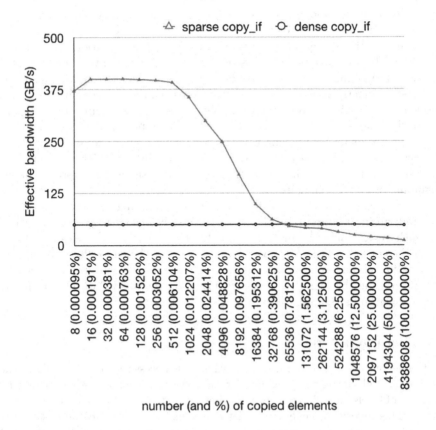

Fig. 2. Achieved bandwidth (in GB/s) for sparse and dense implementations of copy_if operation

- The first subset contains vertices with a large number of connections (over $8 * 256$, where 8 is the number of vector cores, 256 is the maximum vector length). Each vertex from this subset is processed collectively by all vector cores, while each vector core operates with batches of 256 adjacent edges using a single vector instruction of the maximum length.
- The second subset contains vertices, with a moderate number of connections (under $8 * 256$, over 256). Each vertex from this subset is processed by a separate vector core, while all adjacent edges of the vertex are processed in portions of 256 elements.
- The third subset contains vertices with several connections (less than $8 * 256$). These vertices are processed collectively by all vector cores, when each vector core processes portions of 256 vertices with a single vector instruction, one adjacent edge for each vertex at a time.

All three lists can be later processed separately using a standard top-down algorithm, which can be efficiently parallelized and vectorized for each type of vertices.

7.3 Bottom-Up Iteration

In order to improve memory access pattern during the bottom-up iterations, two types of the bottom-up algorithm have been implemented. At the beginning of each bottom-up iteration, the number of unvisited vertices is calculated to allow choosing a more efficient algorithm. If the percentage of the unvisited vertices exceeds a predetermined threshold (10% of the total number of vertices), then the beginning of bottom-up steps is performed using the previously constructed vector extension of the graph. In this case, only the edges included into the vector extension are examined first: all graph vertices are processed collectively by all vector cores, while each vector core processes portions of 256 consequent vertices. Since the same number of adjacent edges are stored for each vertex inside the vector extension, this algorithm part has an extremely efficient parallel workload balancing and a good memory access pattern.

Since traversing a vector extension alone is not enough to complete a bottom-up BFS iteration, a list of the remaining unvisited vertices is created. Vertices included in this list are divided into 3 separate groups based on the count of their connections as implemented in the top-down algorithm. Thereafter, for the remaining vertices all adjacent edges are examined using the whole CSR representation of the input graph.

If the number of unvisited vertices does not exceed the specified threshold value, the vector extension is not used. In this case, the algorithm immediately generates a list of unvisited vertices and processes these vertices using the CSR representation.

7.4 Converting Bitmap to Queue

Both top-down and bottom-up algorithms heavily rely on the procedure which generates lists of currently active vertices based on the current level of each vertex. This operation can be viewed as a copy_if operation designed to copy indices of elements with a given value (activity flag) from a given input array of size $|V|$. Implementing copy_if operation for SX-Aurora TSUBASA vector processors can be difficult, since main computational core of this operation (the removal of the specified elements from an array of fixed length equal to 256) demonstrates low performance. Thus, a straightforward implementation of copy_if operation may significantly bottleneck the performance of the whole BFS algorithm.

In the current work, we developed two vectorized parallel copy if implementations for SX-Aurora architecture. The first implementation operates with an input array of an arbitrary structure, from which a large number of elements have to be copied - named dense copy_if. The second implementation operates with vectors, from which only a very small number of elements have to be copied - named sparse copy_if. For these implementations, the kernel, removing elements from a fixed-size array, is invoked in different conditions. For the dense copy_if implementation, the removal of elements is performed for each 256 consequent elements of the input array. In the case of sparse copy_if implementation, all necessary elements are pre-copied into an intermediate buffer via scatter operation, while the removal of a small number of non-required elements is performed later. The performance difference between two proposed implementations is demonstrated in Fig. 2. The required algorithm version (sparse or dense) can be selected via preliminary scanning of the array containing information about BFS levels and estimating the number of elements to be copied. This operation is based on a parallel vectorized reduction, which can be implemented very efficiently on the SX-Aurora TSUBASA [2].

7.5   Changing Traversal Type

For the SX-Aurora architecture, the algorithm switching the traversal type between top-down and bottom-up, can be slightly altered. For power-law graphs, an early switch to the bottom-up algorithm is performed as soon as at least several vertices with a high connectivity degree are marked as visited. Particularly, for RMAT graphs, switch to the bottom-up algorithm occurs immediately after the first iteration. Since the proposed graph representation stores vertices with high connectivity degree consequently in the memory, this condition can be efficiently evaluated.

7.6   Optimizing Indirect Memory Accesses

Computational time of many graph algorithms is mostly spent for processing indirect memory accesses. Implementing indirect memory accesses on the SX-Aurora TSUBASA architecture has several fundamental differences compared to other architectures, such as Intel CPUs or NVIDIA GPUs. First, indirect accesses to 4-byte data achieve maximum throughput of 450–500 GB/s only in the case when the requested data can be fully placed inside the Last Level Cache (LLC) of 16 MB size [2]. The BFS algorithm stores level values for each graph vertex, which are indirectly accessed during graph traversals. When the graph scale increases, the size of array levels quickly exceeds the LLC size, resulting in a significantly lower performance for large-scale graphs. However, in the case of power-law graphs, most of the indirect accesses are performed to the vertices with the highest connectivity degree adjacent to a large number of graph edges. Since the proposed implementation stores information about such vertices consequently in memory, only the first part of levels array can be prefetched into the cache, which can be done on the NEC SX-Aurora TSUBASA architecture using a special prefetch directive from the NCC compiler.

However, a naive implementation of this approach demonstrates extremely low performance: when vector cores frequently request data from the same (or adjacent) memory regions, many cache conflicts occur, resulting in a rapid performance degradation. To avoid this problem, the following optimization is proposed: at the beginning of each BFS iteration, the first 3500 values from the shared array of levels are copied into private independent arrays of each parallel thread. Thus, each copied array is cached independently in LLC. Values in these private arrays are placed with an interval of 7 elements in order to avoid conflicts at the single cache line level. This optimization allows a throughput of 400–450 GB/s when traversing power-law graphs.

8   Performance Evaluation

Figures 3 and 4 present the performance comparison of the developed implementation for SX-Aurora TSUBASA with the existing third-party high-performance graph processing frameworks and libraries involving BFS implementations. Performance data of all third-party implementations has been collected on multicore Intel(R) Xeon(R) Gold 6126 processors and NVIDIA V100 GPUs. All the performance values are measured in TEPS (Traversed Edges Per Second), which can be defined as the number of edges in a graph divided by the execution time of the implementation. Despite the fact that the direction-optimizing algorithm does not necessarily traverse all graph edges, TEPS value is calculated based on the total number of graph edges (unlike TEPS value reported by, for example, Gunrock). The presented comparison demonstrates that the developed BFS implementation for the SX-Aurora TSUBASA architecture has significantly higher performance on both RMAT and uniform-random graphs compared to multicore CPU and NVIDIA GPU implementations. This significant acceleration is achieved due to the optimal computational complexity of the implemented direction-optimizing algorithm, appropriate vector-friendly graph representation and the efficient usage of SX-Aurora computational resources, including memory bandwidth and vectorization potential.

8.1   Energy Efficiency Evaluation

In addition to the performance comparison, this paper estimates energy efficiency of the developed implementation, and compares it with energy efficiency of other reviewed implementations for different platforms (Intel CPUs and NVIDIA

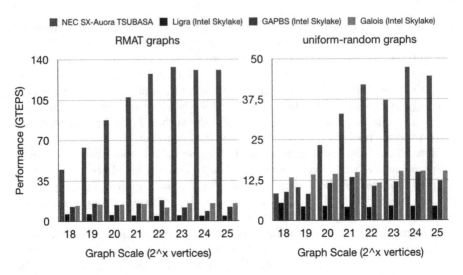

Fig. 3. Performance comparison (in GTEPS) between proposed SX-Aurora TSUBASA implementation and CPU frameworks

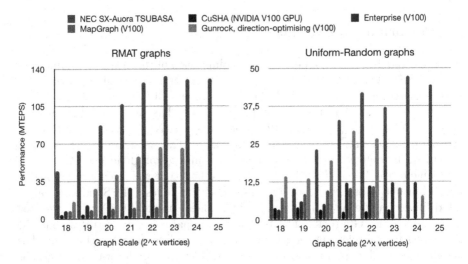

Fig. 4. Performance comparison (in GTEPS) between proposed SX-Aurora TSUBASA implementation and GPU frameworks

GPUs). In this section, two metrics are used: average power of the system during computations and performance per watt (in GTEPs/w) of various implementations. The platform used for energy efficiency evaluation consisted of Intel Xeon Vector Host, Vector Engine and NVIDIA V100 GPU, installed on a single server. The power was measured with an IPMI (Intelligent Platform Management Interface) tool and proprietary tools of SX-Aurora TSUBASA and Tesla V100. For the power of the whole system that includes Xeon, VE, and V100, the IPMI tools are used. For the powers of VE and V100, SX-Aurora TSUBASA Monitoring and Maintenance Manager (MMM) and NVIDIA System Management Interface are used, respectively, as shown in Fig. 5 (left).

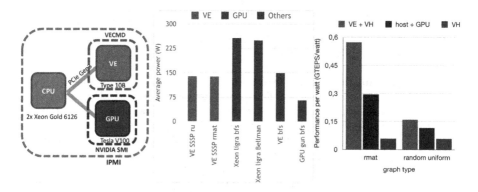

Fig. 5. Power measurement tools (left), average power comparison (middle), performance per watt comparison (right)

Figure 5 (middle) examines the average power consumption of different system components when executing three most high-performance implementations for each of the reviewed platforms: Gunrock BFS for NVIDIA V100 GPU, Ligra for Intel Xeon Vector Host, and the proposed BFS implementation for vector engine. The power consumption of different system components has been measured separately for Vector Engine, GPU, and others that include Xeon, SSD, Fan, etc. Since vector engines and GPUs cannot operate without a host, the power consumption of all system components was summed up. Figure 5 demonstrates that the power consumed by the entire system is comparable for all three implementations, while power consumption of the components involved in actual BFS computation on GPU and VE, is significantly lower. When doing a BFS, VE consumes 83 W (with idle power 41 W), GPU - 65 W (with idle power 39 W), while Intel Xeon Vector Host consumes 255 W. Figure 5 (right) compares performance per watt for the same three implementations, demonstrating significantly better energy efficiency of the developed implementation for VE.

9   Conclusions

This paper proposes an efficient vector-friendly implementation of the direction-optimizing BFS algorithm for modern vector NEC SX-Aurora TSUBASA architecture. Several microarchitecture optimization techniques have been discussed, including a specialized vector-friendly graph format, optimized cache-aware indirect memory accesses, and improved memory access pattern during adjacency lists' traversals. The proposed implementation executed on a single vector engine significantly outperforms (5–10 times) other state-of-the art CPU BFS implementations (Ligra, Galois, GAPBS), executed on 12-core Intel(R) Xeon(R) Gold 6126 processor, and demonstrates 2–10 times better performance than state-of-the-art GPU implementations (Gunrock, Enterprise, CuSHA, MapGraph), executed on NVIDIA GPU V100. In addition, in terms of performance per watt the proposed SX-Aurora TSUBASA implementation demonstrates slightly lower average power consumption and significantly better energy-efficiency than both Intel Xeon CPU and NVIDIA V100 GPUs.

Acknowledgement. This project was partially supported by JSPS Bilateral Joint Research Projects program, entitled "Theory and Practice of Vector Data Processing at Extreme Scale: Back to the Future" and by MEXT Next Generation High-Performance Computing Infrastructures and Applications R&D Program, entitled "R&D of A Quantum-Annealing-Assisted Next Generation HPC Infrastructure and its Applications". The reported study was funded by RFBR, project number 19-31-27001. The reported study was supported by the Russian Foundation for Basic Research, project No. 18-29-03230. The research is carried out using the equipment of the shared research facilities of HPC computing resources at Lomonosov Moscow State University.

References

1. Afanasyev, I.V., et al.: Developing efficient implementations of bellman-ford and forward-backward graph algorithms for nec SX-ACE. Supercomput. Front. Innov. **5**(3), 65–69 (2018)
2. Afanasyev, I.V., Voevodin, V.V., Voevodin, V.V., Komatsu, K., Kobayashi, H.: Analysis of relationship between SIMD-processing features used in NVIDIA GPUs and NEC SX-Aurora TSUBASA vector processors. In: Malyshkin, V. (ed.) PaCT 2019. LNCS, vol. 11657, pp. 125–139. Springer, Cham (2019). https://doi.org/10.1007/978-3-030-25636-4_10
3. Beamer, S., Asanović, K., Patterson, D.: Direction-optimizing breadth-first search. Sci. Program. **21**(3–4), 137–148 (2013)
4. Beamer, S., Asanović, K., Patterson, D.: The gap benchmark suite. arXiv preprint arXiv:1508.03619 (2015)
5. Besta, M., Marending, F., Solomonik, E., Hoefler, T.: Slimsell: a vectorizable graph representation for breadth-first search. In: 2017 IEEE International Parallel and Distributed Processing Symposium (IPDPS), pp. 32–41. IEEE (2017)
6. Chakrabarti, D., Zhan, Y., Faloutsos, C.: R-mat: a recursive model for graph mining. In: Proceedings of the 2004 SIAM International Conference on Data Mining, pp. 442–446. SIAM (2004)
7. Dijkstra, E.W.: A note on two problems in connexion with graphs. Numerische Mathematik **1**(1), 269–271 (1959)
8. Egawa, R., et al.: Potential of a modern vector supercomputer for practical applications: performance evaluation of SX-ACE. J. Supercomput. **73**(9), 3948–3976 (2017)
9. Fu, Z., Personick, M., Thompson, B.: MapGraph: a high level API for fast development of high performance graph analytics on GPUs. In: Proceedings of Workshop on GRAph Data management Experiences and Systems, pp. 1–6. ACM (2014)

10. Hiragushi, T., Takahashi, D.: Efficient hybrid breadth-first search on GPUs. In: Aversa, R., Kołodziej, J., Zhang, J., Amato, F., Fortino, G. (eds.) ICA3PP 2013. LNCS, vol. 8286, pp. 40–50. Springer, Cham (2013). https://doi.org/10.1007/978-3-319-03889-6_5

11. Khorasani, F., Vora, K., Gupta, R., Bhuyan, L.N.: CuSHA: vertex-centric graph processing on GPUs. In: Proceedings of the 23rd International Symposium on High-Performance Parallel and Distributed Computing, pp. 239–252. ACM (2014)

12. Komatsu, K., Egawa, R., Isobe, Y., Ogata, R., Takizawa, H., Kobayashi, H.: An approach to the highest efficiency of the HPCG benchmark on the SX-ACE supercomputer. In: Proceedings of the Conference on High Performance Computing Networking, Storage and Analysis (SC15), Poster, pp. 1–2 (2015)

13. Komatsu, K., et al.: Performance evaluation of a vector supercomputer SX-Aurora TSUBASA. In: Proceedings of the International Conference for High Performance Computing, Networking, Storage, and Analysis, SC 2018, pp. 54:1–54:12. IEEE Press, Piscataway (2018)

14. Liu, H., Huang, H.H.: Enterprise: breadth-first graph traversal on GPUs. In: Proceedings of the International Conference for High Performance Computing, Networking, Storage and Analysis, SC 2015, pp. 1–12. IEEE (2015)

15. Merrill, D., Garland, M., Grimshaw, A.: Scalable GPU graph traversal. In: ACM Sigplan Notices, vol. 47, pp. 117–128. ACM (2012)

16. Nguyen, D., Lenharth, A., Pingali, K.: A lightweight infrastructure for graph analytics. In: Proceedings of the Twenty-Fourth ACM Symposium on Operating Systems Principles, pp. 456–471. ACM (2013)

17. Shun, J., Blelloch, G.E.: Ligra: a lightweight graph processing framework for shared memory. In: ACM Sigplan Notices, vol. 48, pp. 135–146. ACM (2013)

18. Wang, Y., Davidson, A., Pan, Y., Wu, Y., Riffel, A., Owens, J.D.: Gunrock: a high-performance graph processing library on the GPU. In: ACM SIGPLAN Notices, vol. 51, p. 11. ACM (2016)

19. Yamada, Y., Momose, S.: Vector engine processor of NECs brand-new supercomputer SX-Aurora TSUBASA. In: International symposium on High Performance Chips (Hot Chips 2018) (2018)

20. Yasui, Y., Fujisawa, K., Sato, Y.: Fast and energy-efficient breadth-first search on a single NUMA system. In: Kunkel, J.M., Ludwig, T., Meuer, H.W. (eds.) ISC 2014. LNCS, vol. 8488, pp. 365–381. Springer, Cham (2014). https://doi.org/10.1007/978-3-319-07518-1_23

21. Zhang, Y., Hansen, E.A.: Parallel breadth-first heuristic search on a shared-memory architecture. In: AAAI-06 Workshop on Heuristic Search, Memory-Based Heuristics and Their Applications (2006)

Synchronous and Asynchronous Parallelism in the LRnLA Algorithms

Anastasia Perepelkina$^{(\boxtimes)}$ⓘ and Vadim Levchenkoⓘ

Keldysh Institute of Applied Mathematics, Moscow, Russia
mogmi@narod.ru, lev@keldysh.ru

Abstract. In modern computers, parallelism is essentially multilevelled. This is considered in the theory of Locally Recursive non-Locally Asynchronous (LRnLA) algorithms. LRnLA algorithms are constructed by analyzing the causal structure of the dependency graph to find areas of dependence, influence, and asynchronous domains. The base set for subdivision of the task into subtasks is constructed by intersecting these 3D1T space-time conoids, and the asynchronous subtasks may be distributed to cluster nodes. The subdivision continues to engage all available parallelism in a recursive manner. Thus, the problem of load imbalance is solved by decreasing the number of synchronization events, and, at the same time, increasing the number of asynchronous subtasks. On the other hand, in data-level parallelism, vectorization requires synchronized and aligned data access. LRnLA method provides a method to construct data structure and parallel access for non-local vectorization. Thus, LRnLA theory accounts for all the levels of parallelism.

Keywords: AVX512 · LRnLA algorithms · Temporal blocking

1 Introduction

Presently, the supercomputer race drives researchers towards new insights and achievements [3]. The breakthrough to exascale computing is not just a quantitative, but a qualitative goal, as it involves a new approach to scientific computing in general. With more computing power, computer modeling becomes more detailed, abandons legacy approximations, provides more reliable results. The latter is an important factor in the range of the applicability of computer models. Another qualitative change lies in the computing software itself. The development and implementation of new algorithms is essential in the HPC technology transition. Let us focus on several known challenges for the numerical modeling codes in the context of the supercomputer race [7].

The Parallel Scalability problem is the main topic of the present paper. The exponential growth of the degree of parallelism of supercomputers has been going on for over than 20 years. At this moment, there is a qualitative shift, when parallel processing becomes the main source of performance power. In the desired figure of 10^{18} Flops, at least 10^9 comes from hierarchical parallelism, and

© Springer Nature Switzerland AG 2020
L. Sokolinsky and M. Zymbler (Eds.): PCT 2020, CCIS 1263, pp. 146–161, 2020.
https://doi.org/10.1007/978-3-030-55326-5_11

Table 1. Parallel degree of supercomputers [1]

Computer	Peak perf., Flops	Total parallel degree	Nodes	Chips per node	Cores per chip	FPUs per core	Flop per FPU * latency
Summit	$0.2 \cdot 10^{18}$	$0.6 \cdot 10^9$	4608	6GPU	80	32	2 * 4
				2CPU	22	4	2 * 5
Frontera	$38 \cdot 10^{15}$	$86 \cdot 10^6$	8008	2CPU	28	2 * 8	2 * 6
Christofari	$8.8 \cdot 10^{15}$	$25 \cdot 10^6$	75	16GPU	80	32	2 * 4
				2CPU	24	2 * 8	2 * 6
Historical							
Jaguar 2008	$1.4 \cdot 10^{15}$	$2.4 \cdot 10^6$	18769	2CPU	4	2 * 2	1 * 4
Lomonosov 2009	$0.4 \cdot 10^{15}$	$0.56 \cdot 10^6$	4420	2CPU	4	2 * 2	1 * 4
ASCI Red 1997	$1.8 \cdot 10^{12}$	$37 \cdot 10^3$	4576	2CPU	1	1	1 * 4

the rest is one chip clock rate. Table 1 [1] shows the parallel components of the best supercomputers from the Top500 list.

During the transition from TeraFlops to PetaFlops, the leading supercomputers won by increasing the node count.

At the time, the TeraFlops figure had been surpassed due to the drastic increase in the node count of clusters (Cluster Level Parallelism - CLP). For the first computer of such type, ASCI Red (Table 1), apart from the pipe-lining of the computing devices, the parallel degree had exclusively come from this level. For such a homogenous parallel system, for more than 20 years, various algorithms and methods of their efficient implementation for mathematical modeling have been intensively developing, being applied in many purposes. Among the relevant issues in this area, there are load balancing and fault tolerance problems.

CLP may continue to rise (as in, e.g. Jaguar); however the increase in parallelism has been shifted inside of the cluster node, and it gave rise to new levels of parallelism (PLP, TLP, DLP, ILP) and the development of parallelism hierarchy:

PLP — Processor Level Parallelism, that corresponds to the NUMA systems and systems with coprocessors with shared host memory and separate device memory;

TLP — Thread Level Parallelism, that corresponds to the multi- and many-core processors and may overlap with the NUMA level;

DLP — Data Level Parallelism, or SIMD, that corresponds to the hardware vectorization (altivec, AVX, SVE, and other)

ILP — Instruction Level Parallelism, which comes from pipelining speedups. In Table 1, last column, it is estimated as a number of pipeline stages, which can be approximated by latency, if one stage is assumed to take one cycle.

For GPU devices the term SIMT, which corresponds to a combination of ILP and DLP, is used as well.

The CLP remains an important part of the acceleration, but at the moment less than half of the parallelism orders of magnitude come from it.

Table 1 illustrates that none of these levels dominate in the generation. Moreover, none of these reaches the parallelism degree of the CLP.

However, if any level is ignored in the implementation, the computing efficiency is limited. Additionally, thoughtful use of all levels of parallelism may lead to overcoming the Memory Wall challenge.

The essence of the *Memory Wall* problem is that the growth of the memory bandwidth falls behind the growth of the processing power. The most successful illustration for this is the Roofline model [21]. The prominent feature of the stencil computing omnipresent in the simulation codes is low operational intensity. Therefore, the performance scales with memory bandwidth, and the continuation of the Moore law seems less important from this point of view.

There is a solution to the problem in the increasing memory hierarchy complexity. The efficient use of memory caches is an algorithmic issue, which can be solved with cache-aware [4,22] and cache-oblivious [5] algorithms.

Heterogeneity of the modern computers is the most prominent issue since the introduction of GPUs; however, simultaneous use of processors from different generations is also quite common.

Code complexity becomes an issue when, taking into account the hierarchy of cache and the parallelism, the computer itself is complex. One solution is to entrust the work to automatic optimization of the compiler and automatic parallelization frameworks. However, in the context of increasing computer complexity, the task of code optimizing is too difficult under the current circumstances. On the other hand, development of the abstraction level that describes algorithms in a uniform way both for data localization in levels of cache and computation distribution among all levels of parallelism could also create a cleaner code.

To summarize, the challenge is the development of a framework for the unified description of algorithms, that would provide flexibility to adjust the algorithm to the specifics of every level of the parallel hierarchy while solving the memory wall problem at the same time. As the cache subsystem is also hierarchical, the algorithm construction should also have an underlying hierarchy. To achieve a cleaner code, the description must be uniform on all levels.

For example, there is a variety of temporal blocking algorithms [24], the most classical of which being wavefront [23], also known as time skewing in simulation codes. Another option is Diamond Blocking [14], and overlapped tiling. These algorithms help covering the memory exchange operations at the prominent bottleneck sites [19].

Locally Recursive non-Locally Asynchronous (LRnLA) algorithms [9,13] provide a theory of algorithm construction, giving a bird's-eye view of all the hardware hierarchies, and provides both cache localization strategies and distribution of tasks on all levels of parallelism. During the evolution of the LRnLA algorithm methods of construction, a range of codes in applied computing has been developed [8,26,27].

The use of data access locality to overcome the bottlenecks in the memory bandwidth was demonstrated in previous publications [10,11,17]. An efficient parallel scaling is inherent in the algorithm construction and has also been demonstrated [27]. In this work, we describe the theoretical part of the LRnLA algorithm application to different levels of parallelism (Sects. 2.3, 3). We propose a new algorithm for unified CLP and DLP, which connects the circle of large-scale and small-scale space and time domain decomposition (Sect. 3.4). The new algorithm has been implemented for a Computational Fluid Dynamics (CFD) code base on the Lattice Boltzmann Method (LBM), and the efficiency of AVX vectorization is shown in Sect. 4.

2   Algorithms

The LRnLA algorithms [13] exist to make more efficient calculation of memory-bound big data problems. The method is not limited to the explicit numerical schemes on cartesian grids, but these are a common illustration of the method capabilities. In the current work, the scope of possible applications is limited to the cube stencil numerical schemes with fixed time step on regular grids.

2.1   Sample Problem

LBM [20] is a numerical scheme for CFD, which is robust and highly paralleliz-able. The cell update is

$$f_i(\boldsymbol{x} + \boldsymbol{c}_i, t + 1) = f_i^*(\boldsymbol{x}, t); \quad i = 1, ..Q; \tag{1}$$

$$f_i^*(\boldsymbol{x}, t) = \Omega(f_1(\boldsymbol{x}, t), f_2(\boldsymbol{x}, t), ...f_Q(\boldsymbol{x}, t)). \tag{2}$$

Here c_i are discrete values of velocity. For the purpose of the performance benchmark, the most common 3D variation of LBM, with $Q = 19$ velocities (commonly denoted as D3Q19 in LBM texts), collision term (2) in BGK form, and an equilibrium function as a polynomial of order 2 are taken. Thus, $c_i = (0, 0, 0), (\pm 1, 0, 0), (0, \pm 1, 0), (0, 0, \pm 1), (\pm 1, \pm 1, 0), (\pm 1, 0, \pm 1), (0, \pm 1, \pm 1)$. All stencil dependencies are within a $3 \times 3 \times 3$ cube. On the other hand, it is a stencil scheme, which makes the locality wall problem quite prominent.

The operational intensity for stepwise update in case of ideal caching is $1/4$ lattice update per one memory exchange of the whole cell data. All problems of stencil code parallelization are present as well.

- The data amount per cell is at least Q values, which creates a challenge for data localization on smaller cache levels.
- The stencil is multidimensional, which necessarily leads to misalignment in data access that impedes automatic vectorization.
- Data race condition may arise, since the update in one cell depends on the pre-update data of its neighbors.

Since the stencil shape is a cube, the better fitting LRnLA algorithms subdivide space into cubes [13]. The description of the LRnLA algorithms with a cube base may be split in coordinates, so the illustrations in this paper are of lower dimensionality.

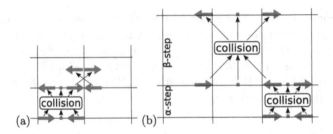

Fig. 1. (a) In the swap pattern, the DFs are collided and swapped at the first stage, and swapped between neighboring cells at the second stage. (b) In the AA pattern, the DFs are collided in the even steps, and pulled in, collided, and pushed out at the odd stage

The LBM scheme has two stages in the time update. Since the stencil half-width for one update is 1, the stencil half-width for each stage may be assumed to be equal to 1/2. When two stages are merged for one cell update, the data race condition occurs for parallel processing. The solution for this issue is using two array copies, or one of the advanced propagation patterns, such as AA [2], EsoTwist [6]. These allow to save memory space by reading and writing data from the same location in each elementary update. The swap algorithm described in [17] has this property at both stages of the time step update. In this paper, the AA and swap patterns are relevant (Fig. 1).

2.2   Notation

The LRnLA algorithms are introduced in the space of the dependency graph of the problem. The nodes are operations of cell data update in a numerical scheme. They are allocated to coordinates in time and space. When d-dimensional space and time is considered, it is denoted as dD1T.

The algorithm is a shape that covers some nodes in the dD1T space, and a rule of its subdivision. The nodes inside the shape are executed in the algorithm. The subdivision rule shows the division of this task into sub-tasks. This definition is recursive.

The subdivision plane requires all dependencies between the nodes across the plane to be unilateral. These hyperplanes in dD1T may be flat in time so that they define a synchronization time instant. If all dependency graph nodes below the horizontal plane are executed, the whole domain is updated to a fixed time iteration. This means that the solution of the problem may be saved to a disk, visualized, or analyzed with various tools. Another type of subdivision plane corresponds with the faces of the dependency or influence conoids in dD1T.

Thus, after several levels of recursive subdivision, there exists a set of shapes with defined dependencies between them. The set can be analyzed to find asynchronous tasks among them, and distribute them between parallel processors.

In the following, the simulation domain is regarded as a cube with a linear size 2^R, where R is an integer number. The task is to progress the simulation by

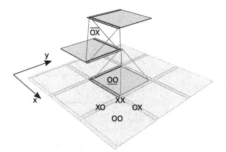

Fig. 2. Subdivision of space at $t = 0$ and $t = 2^R$, and a dual subdivision in between. One $\underline{00}$ is built as an intersection of the domain of influence of OO at $t = 0$ and the domain of influence of XX at $t = 2^{R-1}$. One $\overline{0X}$ is built as an intersection of the domain of dependence of OX at $t = 2^R$ and the domain of influence of XO at $t = 2^{R-1}$

2^R time steps, so the whole task is a cube of $2^{R(d+1)}$ dependency graph nodes. For a non-cubic domain, the task may be constructed from such cubes.

2.3   ConeTur and ConeFold

For cube stencils, LRnLA algorithms ConeTur and ConeFold are used.

ConeTur [9,13] is built by defining a dD space subdivision on the 0-th and 2^R-th time steps, and a dual subdivision at the 2^{R-1}-th time step.

The elements of the basic 1D subdivision are a line segment and a point. The elements of the basic 3D subdivision are a cube, three types of faces, three types of edges, and a vertex. Let O denote a line and X denote a point. OOO can denote a cube, while XOO, OXO, OOX denote three types of faces and XXO, XOX, OXX denote three types of edges, and XXX denote a vertex. The first, the second and the third symbol are the extent of the shape in the x, y or z-axis correspondingly. The linear size of the shapes is 2^R if O, and 0 if X.

In dual subdivision, in the point where the OOO cube center was located, there is a XXX vertex. All other pairs also have dual notations, such as XOO is dual to OXX.

Between $t = 0$ and $t = 2^{R-1}\Delta t$ we construct dependence regions of shapes at $t = 2^{R-1}$ and influence regions of shapes at $t = 0$. The intersections of the influence and dependence regions of the dual shapes correspond to the first half of the ConeTur algorithms. These ConeTurs may be coded by the symbols of their lower base with an underline, such as $\underline{0XX}$.

The second half are the intersections of the dependence regions of shapes at $t = 2^R$ and influence regions of shapes on $t = 2^{R-1}\Delta t$. These ConeTurs are coded by the symbols of their lower base, such as $\overline{X00}$. In 2D1T, $\underline{00}$ is a pyramid, $\underline{X0}$ and $\underline{0X}$ are two tetrahedra. $\underline{XX}$ may be merged with $\overline{XX}$ to form an octahedron (Fig. 2). In general dimensionality, the notation is as in $\underline{0_iX_j}$, where index shows the number of repeated symbols.

Thus, a set of ConeTur shapes at rank R is defined.

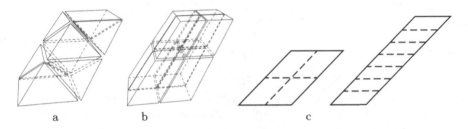

Fig. 3. ConeFold in 2D1T as a merge of ConeTurs (a) and its subdivision (b). ConeFold vs ConeTorre subdivision in 1D1T (c)

For a subdivision of these shapes, the edges of dD shapes at all time instants are subdivided in halves. In time steps $t = 0$, $t = 2^{R-1}\Delta t$, $t = 2^{R}\Delta t$ the linear size of cubes becomes 2^{R-1}, and, after this, the base and the dual subdivisions match. In time steps $t = 2^{R-2}$, $t = 3 \cdot 2^{R-2}$, the dual subdivisions are introduced. The ConeTurs ranked $R - 1$ are built between these time instants analogous to the previous step. The ConeTurs ranked r are subdivided into a set of ConeTurs ranked $r - 1$ which are inside it after this subdivision. Thus, the shape and its subdivision are defined.

ConeFold [13,17,18,25] shape ranked r may be constructed as a merge of the ConeTurs ranked r, one ConeTur of each type, so that they form a prism (Fig. 3). The ConeFold ranked r is subdivided into 2^{d+1} ConeFolds ranked $r - 1$ by subdividing each of its edges in halves.

There is another way to construct a ConeFold. On $t = 0$, take a cube with linear size 2^R. On $t = 2^R$, take a cube with same size, shifted by 2^R to the right on all coordinate axes. ConeFold is an intersection of the region of influence of the first cube and the region of dependence of the second one.

The ConeTorre algorithm is constructed when ConeFolds are stacked onto each other and merged into one prism (Fig. 3,(c)). Alternatively, the same shape may be acquired if a cube is taken at a later time step, N_T, and shifted by N_T cells from the base cube. ConeTorre is subdivided with planes at several time instants. Since the ConeTorre construction does not rely on the binary subdivision, its parameters may be chosen differently. The pieces formed by this subdivision are referred to as floors.

The floor is subdivided into flat layers of one time step each.

Stepwise is a type of subdivision of any shape into flat layers with the height of one full time step only. If the whole problem domain from $t = 0$ to $t = N_T$ undergoes stepwise subdivision, the resulting algorithm has no temporal blocking. Any LRnLA algorithm shape with height more than 1 may be subdivided in a stepwise way. The floor of a ConeTorre is subdivided in a stepwise fashion.

LRnLA cell is the closure of the recursive subdivision of ConeFold. The example for the LBM method is shown in Fig. 1(a). It may also appear as a subdivision of a flat layer, formed by stepwise subdivision, into elementary updates.

3   Parallelism

3.1   Synchnorous and Asynchronous Parallelism

LRnLA subdivision recursively decomposes a problem into sub-tasks, and on each level of subdivision the subdivision type may vary. With this, the subdivision rules may be adjusted so that for every level of the parallel hierarchy of a computer there is a level of algorithm subdivision fitted to it. At this level of abstraction, they can be processed with any type of parallelism.

On the other hand, all levels of the parallel hierarchy have distinct features, so there is a need for some guidelines. For this purpose, we introduce a distinction between synchronous and asynchronous types of parallelism. The metric for the distinction is the number of synchronizations present in the algorithm and its implementation.

An example of *synchronous* parallelism is vectorized processing. In CPU, it is the AVX2/AVX512 instructions; in GPU, it is the execution of one warp. For the construction of the algorithm, we may assume that operations in one hardware vector are executed synchronously. Thus, the characteristics of synchronous parallelism are low synchronization cost (synchronization performed by vector shuffle operations) and small amount of data. The data for synchronous processing is often localized in the register file or on higher levels of cache.

Asynchornous parallelism is represented by PLP. Synchronization is performed by message passing, or by writing and reading to/from the slower levels of memory. This way, the synchronization may be pairwise, even one-side in the extreme case of MPI, or occur in small portions of processors only. For this type of parallelism, large asynchronous computation blocks are detected, and the synchronization of some portions may be asynchronous.

For simplicity, one may say that the synchronous parallelism is performed with barrier synchronization, and the asynchronous parallelism is performed with pairwise synchronization. However, this distinction is vague since at the DLP, the threads may be synchronized either way with little change in synchronization cost. Actual implementations may be placed on a spectrum of synchronous to asynchronous parallelism.

3.2   Parallelism in LRnLA Algorithms

The parallelism hierarchy generally suggests gradual transition from asynchronous to synchronous when going from larger to smaller scales. This hierarchy can be found in the LRnLA subdivision of the dependency graph. Table 2 is presented to show how the switch from asynchronous to synchronous parallelism works in the LRnLA method.

Let us assume that starting at rank R, the subdivision has been performed $R - r$ times, $0 < r < R$. As a result, there are ConeFolds or ConeTurs with linear size of the base equal to 2^r. Among them, one can find asynchronous algorithms. Let us assume that asynchronous parallelism is applied between such shapes, and synchronous parallelism is applied to the shapes inside it, that arise

Table 2. Parallel algorithm properties. The values that correspond to the desired properties are highlighted in blue. The values that lead to complex parallelization patterns and issues with load balancing are highlighted in red. Green indicates excessively high values. In the last column red, blue and green show low, high and excessively high operational intensity respectively.

Algorithm	Asynchrony		Synchrony		Locality
	Degree	LUp's before sync	Degree	Local data	$I[\frac{o}{s}]$
DD SW	$2^{d(R-r)}$	2^{dr}		2^{dr}	$1/4$
ConeTur	$(1...d)\cdot 2^{d(R-r)}$	$2^{dr}\frac{2^r}{d+1}$	$2^l\binom{d}{l}$	2^{dr}	$\frac{2^r-1}{d+1}$
ConeFold	$1...\frac{2^{d(R-r)}}{d+1}$	$2^{(d+1)r}$	$\binom{d+1}{l}$	$(d+1)2^{dr}$	$\frac{2^r}{d+1}$
TorreFold	$1...2^{(d-1)(R-r)}$		$(\frac{2^R-n_t}{dn_t}+1)^d$	$2^{dr}(1+\frac{n_t}{2^r}d)$	
ReFold	$2^{d(R-r-1)}$	$2^{(d+1)r}$	$(\frac{2^{r-r'}-1}{d}+1)^d$	$2^{dr'}(d+1)$	$2^{r'}/d$
			2^d	2^d	

from its subdivisions. In the applied codes, we start with a cube-shaped domain. However, for the purpose of the current discussion let us assume that at rank R there exist a full ConeFold, or a full ConeTur of the $\underline{0}_d$ type.

Both traditional and LRnLA algorithms decompose problems into sub-tasks for parallel processing. In *asynchronous* parallelism, the general guidelines are: each task should process its own data; the synchronization time should be covered by the time of computation in one sub-task; there should be more sub-tasks than there are parallel processors to ensure load-balancing.

Thus, we aim for higher degree of parallelism, and for large number of lattice updates (LUp's) in each sub-task. This, in turn, shows that asynchronous parallelism is targeted towards weak parallel scaling.

In *synchronous* parallelism, data is localized in higher levels of cache. This is the requirement both for the low synchronization cost and for the alleviating the memory bandwidth bottleneck in the sequential processing. Thus, it is better to store less data for each core, so that it fits higher levels of its cache. The thread in TLP work with shared data; the same portion of data is updated by one thread, and then by the next thread and so on. This way, synchronous parallelism targets at strong parallel scaling.

When several processors per node are used, parallelism may be synchronous or asynchronous. It is fitted with consideration of the specific NUMA parameters and the algorithm parameters. LRnLA subdivision allows enough flexibility to fit the hardware parameters. Weak scaling inevitably saturates for parallelism inside one node, since the amount of available memory and cache is limited. With the increase in the node number, weak scaling efficiency may stay close to 100% almost indefinitely [9].

The *domain decomposition* stepwise (DD SW) algorithm illustrates the common approach to parallel execution without LRnLA dD1T subdivision. The domain is decomposed into blocks with 2^{dr} cells. All of them are updated once, and then the processors exchange data. Communication between the nodes is pairwise: one node is waiting for messages from its neighbors to continue computation.

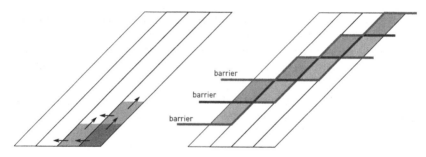

Fig. 4. Asynchronous TorreFold for 4 parallel threads with semaphore synchronization (left) and synchronous TorreFold with barrier synchronization (right)

Thus, node parallelism is asynchronous, with degree of parallelism equal to $2^{d(R-r)}$. In one domain, cells may be processed in parallel, allowing for maximal degree of freedom equal to 2^{dr}. Both vectorization and barrier thread synchronization act as synchronous parallelism for this purpose. On the other hand, thread synchronization may also be implemented with mutexes, thus some of the asynchronous parallelism may occur in one node as well.

ConeTurs ranked r form a set of different types of shapes. The dependency graph of ConeTurs contains several tiers. For example, in 2D1T, the first tier is $2^{d(R-r)}$ <u>00</u> shapes. The next tier is twice bigger, since it contains $2^{d(R-r)}$ <u>0X</u> shapes and $2^{d(R-r)}$ of <u>X0</u> shapes. The number of shapes in a tier show the maximal possible parallelism degree.

One ConeTur is subdivided into $d + 1$ tiers, their degree of parallelism on l-th tier is $2^{l}\binom{d}{l}$, where $\binom{d}{l}$ is a binomial coefficient. Even when this value is at maximum, it is hard to take advantage of, since the shape of the slice is different at each time instant, and the code becomes too complex.

With the ConeFold subdivision, the first tier of the dependency graph contains one ConeFold, the next has d ConeFolds. The number increases up to $2^{d(R-r)}/(d + 1)$. At the starting tiers and at the tiers close to the end, the parallelism degree is small, so the issue of load balancing may limit parallel efficiency.

TorreFold algorithm (Fig. 4) is a special way of processing ConeTorres. The ConeFold shape ranked R is decomposed into ConeTorres with base size 2^{dr}. One ConeTorre is assigned to a processing core. In the asynchronous interpretation, the cores are synchronized with one-sided semaphores. After each core computes one floor (stepwise execution of n_t time steps in the ConeTorre), it allows the d neighboring ConeTorres to progress one floor up. In the synchronous interpretation, the dependency graph tier with a maximal degree of parallelism is detected. It is possible that several such tiers exist. These tiers are processed one by one with barrier synchronization. For parallelism inside one node, synchronous processing is better. The barrier synchronization does not allow the first ConeTorre to progress much further than the others. This way, the projection of parallel ConeFolds to dD is smaller, and the shared data for the synchronization is less. The maximal degree of parallelism and the data for synchronous processing are presented for synchronous TorreFold in Table 2. On the other hand, some of the ConeTorres in case $d > 1$ are completely asynchronous, so that $2^{dr} \cdot 2^R$ operations in them may be processed without data exchange. After the first ConeTorre, the next d ConeTorres are asynchronous, up to $2^{(d-1)(R-r)}$ at the diagonal of the starting ConeFold.

Since all LRnLA algorithms have the 'LR' side for localization of data in higher level of cache which helps overcome the Roofline of the main data storage site [11,13,17], the operational intensity is also shown in Table 2 for comparison.

3.3 Synchronous Processing of Non-local Data

Localization of the processed data with its dependencies on the smaller levels of hierarchy is an important feature of the LRnLA subdivision. Smaller algorithms, ranked 0 and 1 are naturally localized in the register file, therefore synchronous parallelism such as vectorization may be applied. This approach leads to local vectorization.

The communication between locally situated cells occurs more often in time. Even with synchronous parallelism, communication cost remains the bottleneck for parallel scalability, as shuffle operations may produce a significant overhead. Thus, non-local vectorization may be more efficient.

In LRnLA codes, non-local vectorization may be applied independently from the LRnLA subdivision. For this purpose, the $(d-1)$D1T subdivision of the dD1T task is performed. In this case, the LRnLA cell contains the update of all cells along one of the coordinate axes. It is implemented with a vectorized loop in the code. It was done previously both for AVX vectorization [18] and for CUDA-threads [26,27]. The disadvantage of this approach is the loss of locality.

Non-local vectorization may also be implanted into the LRnLA subdivision with more locality. For this, the dependency graph of the problem is transformed before the LRnLA subdivision. For a vector length of N_v, N_v pieces of the dependency graph are folded onto each other. To preserve the locality at the boundaries, the overlaying parts are mirrored. After the subdivision of the dependency graph, that is folded onto itself, the LRnLA cell contains the update for N_v cells that are situated non-locally in space. At the boundaries of the folded domain, the LRnLA cell contains the communication of the mirrored data.

In [15] the domain folding was used to apply wavefront blocking simultaneously with periodic boundary. Indeed, this technique allows implementation of periodic boundary with the ConeFold algorithm, which is impossible otherwise. In [17] this method was used for the first time for efficient vectorization of operations.

3.4 Refolding for Asynchronous Parallelism

Let us fold the dependency graph for the $2 \cdot 2^{(d+1)r}$ cube in two in each of the coordinate axes, and perform ConeFold subdivision at rank r. After this, let us unfold the dependency graph. The subdivision that is seen on the unfolded

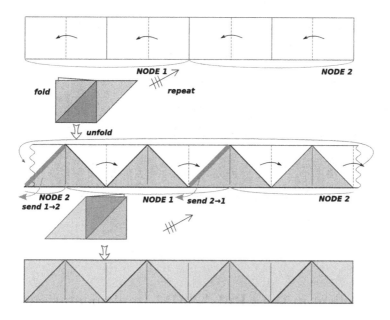

Fig. 5. The ReFold algorithm in 1D1T

domain is similar to the lower half of the ConeTur subdivision of rank $r + 1$ on the periodic domain [16].

Now, let us use this idea to update the domain with the size $2 \cdot B \cdot 2^r$ (Fig. 5). First, fold pairs of cubes with numbers $\{0, 1\}$, $\{2, 3\}$,... $\{2B - 2, 2B - 1\}$. The execution of rank R ConeFolds in these folded domains is asynchronous. Next, unfold the pairs to the original position, and form the pairs $\{2B - 1, 0\}$, $\{1, 2\}$,... $\{2B - 3, 2B - 2\}$. The unprocessed dependency graph nodes in these folded cubes are inside the left boundary ConeFolds ranked R. These are asynchronous as well. The execution of these ConeFolds, at the same time, ensures the communication between cube-shaped domains.

Thus, it is possible to distribute $2B/N_p$ blocks to each node. For communication, the nodes need to send only a portion of the boundary block to the neighboring node.

To compare the parallelism of the resulting algorithm (denoted ReFold) to the parallelism of the preceding algorithms (Table 2), let us start with the domain with size 2^R, decompose it into cubes with size 2^r. There are 2^{R-r} such cubes along each axis, and thus 2^{R-r-1} pairs. The ConeTurs constructed by executing ConeFolds on each pair correspond to the asynchronous part of the algorithm. They may be distributed between the nodes. Each ConeFold ranked r may be decomposed with the TorreFold rule into ConeTorres ranked r' to be processed by several cores, and vectorization works on 2^d folds of the algorithm. Thus, two rows are shown for the algorithm in Table 2, assuming $n_t = 2^{r'}$.

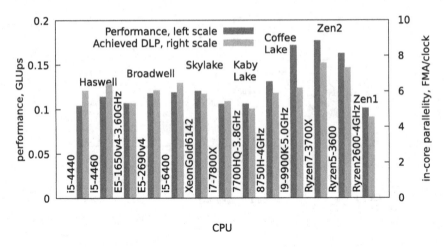

Fig. 6. One-thread performance of the vectorized code on different CPU

4   Parallelism in the LRnLA LBM Implementations

The GPU code [11,12] has the following 3D1T LRnLA subdivision. The domain is subdivided into ConeTorres between $t = 0$ and $t = N_T$. The ConeTorre is subdivided into floors with height `CFsteps`. The floor is processed stepwise with the AA pattern (Fig. 1(b)). The ConeTorres are distributed between SMs, the communication is implemented with TorreFold. The 6^3 cells in the ConeTorre base are distributed between CUDA threads with barrier synchronization.

The CPU code [12,17] has the following 3D1T LRnLA subdivision. The domain is a cube with linear size $2 \cdot 2^R$. It is folded once in each axis for mirrored non-local vectorization, resulting in a cube of vectorized data with linear size 2^R. It is subdivided into ConeTorres with linear size of the base equal to 2^{nLArank}. The ConeTorre is subdivided into ConeFolds ranked `nLArank`. The ConeFold subdivision progresses down to the LRnLA cell, which is illustrated in Fig. 1(a). The ConeTorres are distributed between CPU cores, the communication is implemented with TorreFold.

Additionally we have implemented the ReFold algorithm in this code. The one-thread performance (Fig. 6) illustrates the efficiency of vectorization on various processors. The domain shape is a cube, the problem size is equal to 2GB. The mean performance in FMA per second divided by the clock frequency is plotted on the second scale. The ideal goal would be 16 FMA per clock (except Zen1), since there are 2 FPUs which can perform 8 FMA in a vectorized operation. It is not achieved due to the memory bandwidth limits (see Roofline graph in [17]). Nevertheless, at least 40% of the vectorization efficiency has been reached.

5   Conclusion

In this paper, we have shown how the LRnLA method of algorithm construction provides a solution to implementing efficient parallel stencil computation in the environment of complex hierarchical parallelism.

The advantage of temporal blocking for intra-node communications is a well-established fact. Since data transfer events happen less often, there are no issues with parallel scalability in LRnLA, which has been demonstrated in various codes over the years. This is the 'nLA' side of the method for asynchronous parallelism.

Now, we have shown that synchronous parallelism may be also described in the same terms. The locally recursive subdivision and non-locally asynchronous property of the resulting shapes are the two sides of the LRnLA method. These two sides do not contradict each other, but appear to be highly intertwined both with parallelism ability and information propagation in real physics. With non-local mirrored vectorization, the synchronous parallelization falls into the nLA domain. With the new ReFold algorithm, the small-scale and large-scale asynchrony connect in a circle, and the algorithm description is uniform throughout the entire parallel hierarchy.

Moreover, the proposed vectorization is easy to implement. For symmetrical stencils, computation in the mirrored domain differ only in the sign of some values. This is done by introducing vector constants like $\{1, -1, ...\}$ into the formula. The vector reshuffle operations are only applied in the boundaries of the folded domain; thus, the complexity of the implementation is not higher than applying any boundary condition. The approach is implemented for the LBM numerical scheme both on CPU and GPU.

Acknowledgements. The work is supported by the Russian Science Foundation, grant #18-71-10004.

References

1. Top500 Supercomputer Sites. https://www.top500.org/. Accessed 12 Feb 2019
2. Bailey, P., Myre, J., Walsh, S.D., Lilja, D.J., Saar, M.O.: Accelerating lattice Boltzmann fluid flow simulations using graphics processors. In: International Conference on Parallel Processing, 2009, ICPP 2009, pp. 550–557. IEEE (2009). https://doi.org/10.1109/ICPP.2009.38
3. Bergman, K., et al.: Future high performance computing capabilities: summary report of the advanced scientific computing advisory committee (ASCAC) sub-committee (2019). https://doi.org/10.2172/1570693
4. Datta, K., et al.: Stencil computation optimization and auto-tuning on state-of-the-art multicore architectures. In: Proceedings of the 2008 ACM/IEEE Conference on Supercomputing, Austin, Texas–November 15–21, 2008, p. 4. IEEE Press (2008). https://doi.org/10.1109/SC.2008.5222004
5. Frigo, M., Strumpen, V.: Cache oblivious stencil computations. In: Proceedings of the 19th Annual International Conference on Supercomputing, ICS 2005, pp. 361–366. ACM, New York (2005). https://doi.org/10.1145/1088149.1088197

6. Geier, M., Schönherr, M.: Esoteric twist: an efficient in-place streaming algorithms for the lattice Boltzmann method on massively parallel hardware. Computation **5**(2), 19 (2017). https://doi.org/10.3390/computation5020019
7. Geist, A., Lucas, R.: Major computer science challenges at exascale. Int. J. High Perform. Comput. Appl. **23**(4), 427–436 (2009). https://doi.org/10.1177/1094342009347445
8. Korneev, B., Levchenko, V.: Detailed numerical simulation of shock-body interaction in 3D multicomponent flow using the RKDG numerical method and "DiamondTorre" GPU algorithm of implementation. J. Phys.: Conf. Ser. **681**, 012046 (2016). https://doi.org/10.1088/1742-6596/681/1/012046
9. Levchenko, V.: Asynchronous parallel algorithms as a way to archive effectiveness of computations. J. Inf. Tech. Comp. Syst. **1**, 68–87 (2005). (in Russian)
10. Levchenko, V., Perepelkina, A., Zakirov, A.: DiamondTorre algorithm for high-performance wave modeling. Computation **4**(3), 29 (2016). https://doi.org/10.3390/computation4030029
11. Levchenko, V., Zakirov, A., Perepelkina, A.: GPU implementation of ConeTorre algorithm for fluid dynamics simulation. In: Malyshkin, V. (ed.) PaCT 2019. LNCS, vol. 11657, pp. 199–213. Springer, Cham (2019). https://doi.org/10.1007/978-3-030-25636-4_16
12. Levchenko, V., Zakirov, A., Perepelkina, A.: LRnLA lattice Boltzmann method: a performance comparison of implementations on GPU and CPU. In: Sokolinsky, L., Zymbler, M. (eds.) PCT 2019. CCIS, vol. 1063, pp. 139–151. Springer, Cham (2019). https://doi.org/10.1007/978-3-030-28163-2_10
13. Levchenko, V.D., Perepelkina, A.Y.: Locally recursive non-locally asynchronous algorithms for stencil computation. Lobachevskii J. Math. **39**(4), 552–561 (2018). https://doi.org/10.1134/S1995080218040108
14. Malas, T., Hager, G., Ltaief, H., Stengel, H., Wellein, G., Keyes, D.: Multicore-optimized wavefront diamond blocking for optimizing stencil updates. SIAM J. Sci. Comput. **37**(4), C439–C464 (2015). https://doi.org/10.1137/140991133
15. Osheim, N., Strout, M.M., Rostron, D., Rajopadhye, S.: Smashing: folding space to tile through time. In: Amaral, J.N. (ed.) LCPC 2008. LNCS, vol. 5335, pp. 80–93. Springer, Heidelberg (2008). https://doi.org/10.1007/978-3-540-89740-8_6
16. Perepelkina, A., Levchenko, V.: Enhanced asynchrony in the vectorized ConeFold algorithm for fluid dynamics modelling. Math. Model. **3**(2), 52–54 (2019)
17. Perepelkina, A., Levchenko, V.: LRnLA algorithm ConeFold with non-local vectorization for LBM implementation. In: Voevodin, V., Sobolev, S. (eds.) RuSCDays 2018. CCIS, vol. 965, pp. 101–113. Springer, Cham (2019). https://doi.org/10.1007/978-3-030-05807-4_9
18. Perepelkina, A.Y., Levchenko, V.D., Goryachev, I.A.: Implementation of the kinetic plasma code with locally recursive non-locally asynchronous algorithms. J. Phys.: Conf. Ser. **510**, 012042 (2014). https://doi.org/10.1088/1742-6596/510/1/012042
19. Shimokawabe, T., Endo, T., Onodera, N., Aoki, T.: A stencil framework to realize large-scale computations beyond device memory capacity on GPU supercomputers. In: Cluster Computing (CLUSTER), pp. 525–529. IEEE (2017). https://doi.org/10.1109/CLUSTER.2017.97
20. Succi, S.: The Lattice Boltzmann Equation: For Fluid Dynamics and Beyond. Oxford University Press, Oxford (2001)
21. Williams, S., Waterman, A., Patterson, D.: Roofline: an insightful visual performance model for multicore architectures. Commun. ACM **52**(4), 65–76 (2009). https://doi.org/10.1145/1498765.1498785

22. Wolf, M.E., Lam, M.S.: A data locality optimizing algorithm. In: ACM Sigplan Notices, vol. 26, pp. 30–44. ACM (1991). https://doi.org/10.1145/113446.113449
23. Wolfe, M.: Loops skewing: the wavefront method revisited. Int. J. Parallel Program. **15**(4), 279–293 (1986). https://doi.org/10.1007/BF01407876
24. Wonnacott, D.G., Strout, M.M.: On the scalability of loop tiling techniques. IMPACT **2013**, 3 (2013)
25. Zakirov, A.: Application of the locally recursive non-locally asynchronous algorithms in the full wave modeling. Ph.D. thesis, MIPT, Moscow (2012). (in Russian)
26. Zakirov, A., Levchenko, V., Ivanov, A., Perepelkina, A., Levchenko, T., Rok, V.: High-performance 3D modeling of a full-wave seismic field for seismic survey tasks. Geoinformatika **3**, 34–45 (2017)
27. Zakirov, A., Levchenko, V., Perepelkina, A., Zempo, Y.: High performance FDTD algorithm for GPGPU supercomputers. J. Phys.: Conf. Ser. **759**, 012100 (2016). https://doi.org/10.1088/1742-6596/759/1/012100

Parallel Combined Chebyshev and Least Squares Iterations in the Krylov Subspaces

Yana Gurieva[1] and Valery Il'in[1,2(✉)]

[1] Institute of Computational Mathematics and Mathematical Geophysics,
SB RAS, Novosibirsk, Russia
`yana@lapasrv.sscc.ru`, `ilin@sscc.ru`
[2] Novosibirsk State University, Novosibirsk, Russia

Abstract. The combined Chebyshev−Least Squares iterative processes in the Krylov subspaces to solve symmetric and non-symmetric systems of linear algebraic equations (SLAEs) are proposed. This approach is a generalization of the Anderson acceleration of the Jacobi iterative method as an efficient alternative to the Krylov methods. The algorithms proposed are based on constructing some basis of the Krylov subspaces and a minimization of the residual vector norm by means of the least squares procedure. The general process includes periodical restarts and can be considered to be an implicit implementation of the Krylov procedure which can be efficiently parallelized. A comparative analysis of the methods proposed and the classic Krylov approaches is presented. A parallel implementation of the iterative methods on multi-processor computer systems is discussed. The efficiency of the algorithms is demonstrated via the results of numerical experiments on a set of model SLAEs.

Keywords: Chebyshev iterative algorithms · Least squares · Krylov subspaces · Anderson acceleration · Numerical stability · Convergence of iterations · Numerical experiments

1 Introduction

Let us consider a SLAE

$$Au = \sum_{l' \in \omega_\ell} a_{l,l'} u_{l'} = f, \quad A = \{a_{l,l'}\} \in \mathcal{R}^{N,N},$$
$$u = \{u_l\}, \ f = \{f_l\} \in \mathcal{R}^N, \tag{1}$$

with a large real sparse matrix resulting from some approximations of multi-dimensional boundary value problems by the finite element or finite volume, or some other methods, on a non-structured grid in a computational domain with

V. Il'in—The reported study was funded by RFBR according to the research project No. 18-01-00295 A.

L. Sokolinsky and M. Zymbler (Eds.): PCT 2020, CCIS 1263, pp. 162–177, 2020.
https://doi.org/10.1007/978-3-030-55326-5_12

contrast material properties in its subdomains. We suppose that, in general, this matrix is a non-symmetric and ill-conditioned. In (1), ω_ℓ denotes a set of indices of the non-zero entries in the ℓ-th row of matrix A, and their number N_ℓ is assumed to be much smaller than N, the order of the system. Let us note that the algorithms considered below can easily be extended to a complex case.

A numerical solution of such systems is the bottle-neck of mathematical modelling in real extreme interdisciplinary problems, as it requires the computational resources which non-linearly grow if the dimension of a system increases (for example, 10^{10} d.o.f. and higher). In this case, a road map to provide high performance calculations consists in the parallel implementation of modern multi-preconditioned iterative processes in the Krylov subspaces based on the domain decomposition methods (DDM, see [1,2] and the references therein). The main achievements in this direction are based on a combination of the efficient mathematical approaches and scalable parallel technologies on multi-processor systems with distributed and hierarchical shared memory.

In [3], the authors have offered special procedures to accelerate the convergence of the Jacobi method as an "efficient alternative" to the classic Krylov methods. In order to solve a linear system, they use the Anderson acceleration, which was originally proposed in [4] to solve systems of non-linear algebraic equations. A comparative experimental analysis presented in [3], has demonstrated a considerable advantage of the original alternating Anderson-Jacobi method over the popular generalized minimal residual method (GMRES) with respect to the solution time. The idea of the alternating Anderson-Jacobi method consists in a periodical (i.e. after a prescribed number of stationary iterations) usage of an acceleration method based on a direct solution to an auxiliary least squares problem, instead of a successive orthogonalization of the direction vectors, which is typical of the Krylov variational type methods.

The objective of the present paper is to generalize and experimentally study the similar approaches. We apply the non-stationary Chebyshev iterative algorithms [5] with variable iterative parameters as a preliminary tool to construct some basis vectors in the Krylov subspaces with a subsequent minimization of the residual vector norm by means of the least squares method (LSM, [6]).

In this context, the parallel implementation of the approaches proposed is considered. We contemplate the advantage of the Chebyshev acceleration and LSM realization by means of the hybrid programming tools at the heterogeneous clusters with a distributed and hierarchical shared memory. However, the main goal of our research is to numerically analyze the convergence rate and stability of our iterative methods. Thus, we estimate the speed-up of the parallelization only qualitatively. Let us remark that an approach similar to the one mentioned above was applied by P.L. Montgomery [7] to solve special systems of linear algebraic equations over a finite field and was referred to as the block Lanczos method.

This paper is organized as follows. In Sect. 2, certain adaptive versions of the Chebyshev iterations to solve symmetric and non-symmetric SLAEs are described. In Sect. 3, we consider the two-level accelerations of the Chebyshev procedures in the Krylov subspaces. First, we present the idea of implicit, or block, least squares method with periodical restarts in the Krylov subspaces, which uses a preliminary constructed by means of the Chebyshev iterative algorithm basis vectors. We call this approach as CHEB-LSM(m), where m is the period of restarts. A special subsection is concerned with an analysis of the efficiency of the iterative algorithms parallel versions considered in comparison with a classic variational method in the Krylov subspaces. Also, we have described the upper level of the hybrid Chebyshev–Krylov iterations based on the conjugate residual (CR) or semi-conjugate residual (SCR) approaches with the polynomial preconditioning. Section 4 contains the results of numerical experiments for the algorithms offered, on a series of the test SLAEs resulting from the grid approximation of the three-dimensional boundary value problems for the convection-diffusion equation. In Conclusion, we consider the performance and efficiency of the algorithms presented and discuss some plans for the future studies.

2   The Adaptive Chebyshev Iterations

In this section, we consider three versions of the Chebyshev iterative acceleration and the least squares method. Hereafter, for simplicity we assume that the matrix A is a non-singular one and has the real positive eigenvalues $\lambda_k \in [\lambda_1, \lambda_N]$. The cases of a complex matrix spectrum will be also discussed.

First, let us consider an iterative two-step process

$$u^{n+1} = u^n + \tau_n r^n, \quad r^n = f - Au^n, \quad n = 0, 1, \ldots, \tag{2}$$

where u^0 is an arbitrary initial guess, and the iterative parameters τ_n are the roots of the Chebyshev polynomials $T_m(\lambda)$ of the first kind [5]:

$$\tau_n = 2 / \left[\lambda_N + \lambda_1 - (\lambda_N - \lambda_1) \cos \frac{(2n-1)\pi}{2m} \right], \quad n = 1, \ldots, m. \tag{3}$$

If we set $m = 1$ in (3), then process (2) will present a Jacobi, or a Richardson algorithm with the optimal constant iterative parameter $\tau_n = \tau = 1/(\lambda_N + \lambda_1)$. Here we suppose that the matrix A is diagonalizable. As by assumption it has the real positive eigenvalues, these values τ_n from (3) provide the optimal iterative process (2) in the sense of a maximum convergence rate:

$$||r^m||/||r^0|| \le \varepsilon_m = \min_{\tau_1, \ldots, \tau_m} \left\{ \max_{\lambda \in [\lambda_1, \lambda_N]} \{|P_m(\lambda)|\} \right\},$$
$$P_m(\lambda) = \prod_{k=1}^m (\lambda - \tau_k) \Big/ (-1)^m \prod_{k=1}^m \tau_k = T_m \left(\tfrac{\lambda_1 + \lambda_N - 2\lambda}{\lambda_N - \lambda_1} \right) \Big/ T_m \left(\tfrac{\lambda_N + \lambda_1}{\lambda_N - \lambda_1} \right). \tag{4}$$

Here $P_m(\lambda)$ is the polynomial of order m with a minimum of the maximal value of its modulus at $[\lambda_1, \lambda_N]$ and satisfies the conditions $P_m(0) = 1$, $|P_m(\lambda)| \le 1$ for $\lambda \in [\lambda_1, \lambda_n]$. For the given ε_m, the number of iterations which provides the fulfilment of condition (4) is estimated by the inequality

$$n(\varepsilon_m) \le 0,5|ln(\varepsilon_m/2)|\sqrt{C} + 1, \quad C = \lambda_N/\lambda_1. \tag{5}$$

After the $m - th$ iteration, the parameter values in (2) are repeated: $\tau_{m+k} = \tau_k$, $k = 1, 2, \ldots$, etc. The disadvantage of this algorithm consists in the optimality at each $l \cdot m$-th iteration only, $l = 1, 2, \ldots$. Ideally, for the given accuracy $\varepsilon \ll 1$ we should choose m from the equality $\varepsilon_m = \varepsilon$ and use formula (3) to calculate the polynomial roots, however, this is not easy to do.

Let us also note that the implementation of algorithm (2)–(3) requires a special stable ordering of the iterative parameters τ_n [5]. If we denote a set of the integer values $\theta_m = \{2n - 1 : n = 1, \ldots, m\}$, which are used in (3), then, for example, for $m = 8$ a "stable set" is $\theta_8 = \{1, 15, 7, 9, 3, 13, 5, 11\}$.

Another way to construct the stable Chebyshev iterative method is based on the three-term formulas:

$$
\begin{aligned}
&u^1 = u^0 + \tau r^0, \quad \tau = 2/(\lambda_1 + \lambda_N), \quad n = 1, 2, \ldots, \\
&r^n = f - Au^n, \quad u^{n+1} = u^n + \gamma_n \tau r^n + (\gamma_n - 1)(u^n - u^{n-1}), \\
&\gamma_n = 4/(4 - \gamma_{n-1}\gamma^2), \quad \gamma_0 = 2, \quad \gamma = (C - 1)/(1 + C).
\end{aligned}
\tag{6}
$$

These formulas satisfy the optimality condition (4) at each iteration. This follows from the recurrent relations for the polynomial $P_n(\lambda)$:

$$
P_{n+1}(\lambda) = \gamma_n(1 - \tau\lambda)P_n(\lambda) + (1 - \gamma_n)P_{n-1}(\lambda), \quad P_0(\lambda) = 1, \quad P_1(\lambda) = 1 - \tau\lambda.
\tag{7}
$$

A form of the Chebyshev acceleration which is equivalent to (6) can be presented as a set of the following coupled two-term relations:

$$
\begin{aligned}
&p^0 = r^0 = f - Au^0, \quad n = 1, 2, \ldots : \\
&u^n = u^{n-1} + \alpha_{n-1}p^{n-1}, \quad r^n = r^{n-1} - \alpha_{n-1}Ap^{n-1}, \quad p^n = r^n + \beta_n p^{n-1}, \\
&\alpha_0 = \tau, \quad \alpha_n = \gamma_n \tau, \quad \beta_n = (\gamma_n - 1)\alpha_{n-1}/\alpha_n.
\end{aligned}
\tag{8}
$$

Hereafter, we call different versions of the Chebyshev acceleration process as Cheb - 2, Cheb - 3, and Cheb - 2c keeping in mind formulas (2), (6), and (8), respectively.

Let us note that formulas (8) formally coincide, with those for the conjugate gradient (CG) or the conjugate residual (CR) methods [2], but with other values of α_n and β_n. Strictly speaking, we should also mention that the optimality condition (4), and the optimal number of iterations (5) are valid for the symmetric positive definite (s.p.d) or symmetric matrices A. But if $A \neq A^T$ has a complex spectrum with known boundaries, the considered Chebyshev acceleration can be generalized, see [1, 8, 9] and the literature therein.

If the boundaries of the matrix spectrum are not known, we should use a certain adaptive procedure to construct the Chebyshev iterative process. We offer an heuristic approach first for the real positive spectrum of the matrix A. In formulas (3), we use the upper bound as an approximate value of the maximum eigenvalue λ_N which can be calculated via the uniform matrix norm:

$$
\lambda_N^{(0)} \approx \|A\|_\infty = \max_l \left\{ \sum_{l'=1}^{N} |a_{l,l'}| \right\} \geq \lambda_N,
$$

and can be easily estimated in practice. For the minimal eigenvalues we use a rough estimation $\lambda_1^{(0)} > \lambda_1 \cong O(N^{-2})\lambda_N^{(0)}$, which can be done by means of different approaches, see [8–11], for example.

Choosing such approximations of the spectrum bounds, we perform m_0 iterations by formulas (2), or (6), or (8) (in practice it turns out that $m_0 \approx 15 \div 30$ is sufficient) and calculate the convergence ratio $\rho_{m_0} = ||r^{m_0}||/||r^0||$, as well as some values of polynomials $P_{m_0}(\nu_k)$ for $\nu_k \in [0, \lambda_1^{(0)}]$, $k = 1, 2, \ldots$, by means of (7). Because the inequality $P_{m_0}(\lambda_1) \geq \max_{\lambda \in [\lambda_1^{(0)}, \lambda_N^{(0)}]} \{|P_{m_0}(\lambda)|\}$, is fulfiled, then it is possible to take $\lambda_1^{(1)} \approx \lambda_1$ from the asymptotic equations $P_{m_0}(\lambda_1) \approx \rho_{m_0}$, using an interpolation via $P_n(\nu_k)$. In the next iterations, we can implement any kind of the Chebyshev iterations with the values $\lambda_1^{(1)}$ and $\lambda_N^{(0)}$.

3   The Least Squares Minimization in the Krylov Subspaces

In this section, we consider the level of a variational improvement of the spectral Chebyshev iterations. Also, we present the comparison of the performance of the proposed approaches and that of the classic processes in the Krylov subspaces.

3.1   The Implicit Chebyshev – LSM Approach

After $m < N$ iterations of the Chebyshev acceleration (for any its version from the previous section) we can formally write down

$$u^m = u^0 + \gamma_0 p^0 + \ldots + \gamma_{m-1} p^{m-1},$$
$$r^m = r^0 - \gamma_0 A p^0 - \ldots - \gamma_{m-1} A p^{m-1}.$$

Now we can recalculate the vectors u^m, r^m if we save the vectors $p^0, \ldots, p^{m-1}$ and $w_k = A p^k$, $k = 0, \ldots, m - 1$. The refined vectors $\hat{u}^m$, $\hat{r}^m$ are defined from a minimization of the norm $||\hat{r}^m|| = (\hat{r}^m, \hat{r}^m)^{1/2}$:

$$\hat{u}^m = u^m + c_0 p^0 + \ldots + c_{m-1} p^{m-1} = u^m + V c^m,$$
$$\hat{r}^m = r^m - c_0 A p^0 - \ldots - c_{m-1} A p^{m-1} = r^m - A V c^m, \qquad (9)$$
$$V = \{p^k\} \in \mathcal{R}^{N,m}, \quad c^m = (c_0, \ldots, c_{m-1})^T \in \mathcal{R}^m.$$

In this case, we find the new coefficients $c_0, \ldots, c_{m-1}$ instead of $\gamma_0, \ldots, \gamma_{m-1}$ from the approximate equality $r^m \approx 0$, which corresponds to the following SLAE:

$$W c^m = r^m, \quad W = AV = \{w_k\} \in \mathcal{R}^{N,m}, \quad w_k = A p^{k-1}, \quad k = 1, \ldots, m. \quad (10)$$

This system can be solved if and only if its right-hand side r^m is orthogonal to the kernel of the transposed matrix W^T. In such a unique case $r^m = 0$, and vector $\hat{u}^m$ equals the exact solution u of the original system (1).

Then a new residual is calculated from the original equation $r^m = f - A\hat{u}^m$, rather than from a recurrent relation. We call such a procedure the least squares "restart". In what follows, m iterations of the Chebyshev acceleration with periodical restarts are repeated.

In general, the overdetermined system (10) defines the pseudo solution $c = W^+ r^0$ via the generalized inverse matrix W^+. In [3], in particular, the Moore-Penrose inverse is used. A numerical solution of SLAE (10) can also be obtained by the direct application of the singular value decomposition (SVD) or by the QR-decomposition of the matrix W (description and implementation of these algorithms can be found, for example, in the MKL Intel library [12]).

The considered approach can be interpreted as a variational "implicit" refinement of the spectral Chebyshev iterations in the Krylov subspaces, because the direction vectors $p^0, \ldots, p^{n-1}$ are simultaneously used in formulas (9)–(10). We call such a combined algorithm as the Chebyshev-LSM procedure. It is important to remark that in the proposed approach, the periodic Chebyshev iterations will not be optimal, and the variational LSM improvement compensates a possible stagnation of the initial spectral iterations. In this case, the Chebyshev iterations can be considered to be a tool to compute the basis vectors for the LSM optimization.

3.2   Comparative Analysis of the Chebyshev and Krylov Iterations

Classic restarted versions of different minimal residual methods in the Krylov subspaces used to solve non-symmetric SLAEs, GMRES and semi-conjugate residual (SCR) algorithms are equivalent in terms of the convergence rate to the Chebyshev–LSM approach, described by formulas (9)–(10). As an example, we consider the preconditioned SCR method to solve a non-symmetric SLAE (1). It can be presented as

$$
\begin{aligned}
u^{n+1} &= u^n + \alpha_n p^n = u^0 + \alpha_0 p^0 + \ldots + \alpha_n p^n, \\
r^{n+1} &= r^n - \alpha_n A p^n = r^0 - \alpha_0 A p^0 - \ldots - \alpha_n A p^n.
\end{aligned}
\tag{11}
$$

If the coefficients here are computed as

$$
\alpha_n = \sigma_n / \rho_n, \ \ \rho_n = (Ar^n, \ Ar^n), \ \ \sigma_n = (AB_n^{-1} r^n, \ r^n),
\tag{12}
$$

where the matrices B_n are some preconditioners which can generally vary and will be discussed later, then these relations define the semi-conjugate residual method (see [13–15]) which minimizes the residual norm $||r^{n+1}||$ in the Krylov subspace $\mathcal{K}_{n+1}(r^0, A)$, and which is equivalent, in terms of estimating the number of iterations $n(\varepsilon)$, to the above-considered Chebyshev–LSM algorithm. The residual vectors r^n in these approaches satisfy the orthogonal properties

$$
(AB_k^{-1} r^k, r^n) = \begin{cases} 0, & k < n, \\ \sigma_n \delta_{k,n}, & k = n, \end{cases}
\tag{13}
$$

for all k, n, where $\delta_{k,n}$ is the Kronecker symbol. The direction vectors p^n are defined by the following "long" recursions:

$$
p^{n+1} = B_{n+1}^{-1} r^{n+1} - \sum_{k=0}^{n} \beta_{n,k} p^k, \ \ \beta_{n,k} = (Ap^k, \ AB_{n+1}^{-1} r^{n+1}) / \rho_k, \ \ p^0 = B_0^{-1} r^0.
\tag{14}
$$

Let us remark that formulas (14) implement the orthogonality properties by the Gram–Schmidt procedure. In fact, this procedure should be replaced by a more stable one, namely, a modified Gram-Schmidt orthogonalization.

Let us note that for some *s.p.d.* matrix, from formulas (11), (12) one obtains the conjugate residual method with short recursions in representation (14), namely, $\beta_{n,k} = 0$, $k < n$ and the orthogonal property will take the form $(Ar^n, r^k) = \sigma_n \delta_{k,n}$ instead of (13). If the matrix A is a diagonalizable one, then the convergence rate is defined by inequality (5).

In a general case, to calculate the vectors u^n and r^n via (11)–(14), it is necessary to store all the vectors p^n, p^{n-1}, ..., p^0 and Ap^n, Ap^{n-1}, ..., Ap^0. In practice, these methods are implemented with periodic restarts after a bunch of m iterations. In doing so, the residual vector is computed from the original equation

$$r^{ml} = f - Au^{ml}, \quad p^{ml} = B_n^{-1} r^{m,l} \quad \ell = 0, 1, \ldots,$$

but not from the recurrent form (11), and the calculation of the iterative approximations starts "from the beginning". In doing so, it is necessary to store only the last $m + 1$ vectors p^n, p^{n-1}, ..., p^{n-m}, and Ap^n, Ap^{n-1}, ..., Ap^{n-m}, similar to the Chebyshev-LSM approach (9)–(10). The restarted version of the SCR method, as well as the restarted GMRES one, have a lower convergence rate, but this is the cost one should pay for the less memory usage.

Now, let us compare the computational resources (the volume of arithmetical operations and memory required) which are necessary for the parallel implementation of m iterations in one restart period in the SCR and the Chebyshev–LSM approaches. The most expensive operations in both cases are matrix-vector multiplications and computing scalar vector products. Taking into account these two operation types will suffice for a qualitative comparison of the performances of the algorithms in question because they minimize the same functional in the same Krylov subspace and, consequently, are theoretically equivalent with respect to the convergence rate.

At every m iterations within each restart period the implementation of the restarted SCR method includes the following main stages (for simplicity, we consider the identity preconditioners $B_n = I$ here): m matrix-vector multiplications, $m(m + 3)$ vector-vector operations, computing the $m(m + 2)$ vector dot products. It is important to notice that all these SCR operations are carried out consecutively. Similar computational resources are required for the GMRES implementation (see, e.g., [2]).

In addition, for $m \ll N$, we can neglect the cost of solving SLAE (10) and compute the vector c on all processor units simultaneously. Let us also note that implementation of the Chebyshev acceleration does not require computing the inner vector products, except those for the evaluation of dot product (r^n, r^n) in the stopping criterion. The implementation of the two-term formulas (2), the three-term formulas (7) and the coupled two-term formulas (8) includes calculating $m, 2m, 3m$ vector-vector operations at each restart period, respectively.

After computing the basis vectors and optimal coefficients $c_0, \ldots, c_{m-1}$ at each restart, the corrected least squares iteration can be found by the first formula from (9).

In general, a restarted version of the Chebyshev-LSM procedure can be described as follows. After each l-th restart, the improved iterative solution and the corresponding residual vector are computed by the formulas similar to those from (9):

$$\hat{u}^{ml} = u^{ml} + V^{(l)}c^{(l)}, \quad \hat{r}^{ml} = f - A\hat{u}^{ml}.$$

After that, the vectors $v^k = p^{ml+k}$, $k = 1, \ldots, m$ are calculated by the Chebyshev acceleration in the next restart period, and the new coefficient vector $c^{(l)} = (c_0^{(l)}, \ldots, c_{m-1}^{(l)})^T$ is defined by the LSM procedure (9)–(10), and so on.

3.3 The Second Acceleration Level of the Restarted Chebyshev or SCR Iterations

In this subsection, we describe the construction of an upper acceleration level for the restarted iterative processes, both for the Chebyshev and SCR algorithms. Let us have, after l restarts, some vectors $\hat{u}^0, \hat{u}^m, \ldots, \hat{u}^{lm}$ and the corresponding residuals $\hat{r}^{lm} = f - A\hat{u}^{lm}$. Then we can define the improved vectors

$$\begin{aligned}
\tilde{u}^{lm} &= \hat{u}^{lm} + \hat{c}_1\hat{p}^m + \ldots + \hat{c}_m\hat{p}^{lm}, \\
\tilde{r}^{lm} &= \hat{r}^{lm} - \hat{c}_1 A\hat{p}^m - \ldots - \hat{c}_m A\hat{p}^{lm},
\end{aligned} \tag{15}$$

where the direction vectors are $\hat{p}^{lm} = p^{lm} - p^{(l-1)m}$. The new coefficient vector $\hat{c} = (\hat{c}, \ldots, \hat{c}_m)^T$ is defined by the least squares approach

$$\hat{W}\hat{c}^l = \hat{r}^{lm}, \quad \hat{W} = (A\hat{p}^m, \ldots, A\hat{p}^{lm}) \in \mathcal{R}^{N,l}. \tag{16}$$

This overdetermined system can be solved approximately by means of the left Gauss transformation:

$$\hat{B}_l\hat{c}^l = \hat{W}^T\hat{W}\hat{c}^l = \hat{W}^T\hat{r}^{lm}, \quad \hat{B}_l = \hat{W}^T\hat{W} \in \mathcal{R}^{l,l}, \tag{17}$$

and after that, the improved solution is defined from (15). This procedure can be applied at each restart. One can see that in the case of the Chebyshev acceleration, formulas (9), (10) and (15), (16) correspond to the two-level implementation of the least squares approach. In addition, if one computes the vectors $\hat{p}^{lm}$ in (15) by means of the SCR formulas (11)–(14), then one will have the two-level SCR–LSM iterative approach. The implementation of the LSM procedures (9)–(10) and (15)–(17) for the lower and upper levels are carried out in the same way.

Now we will describe the possibility of some improvement of the upper level of the considered iterative processes. The idea consists in replacing the "implicit" LSM approach in the form of (9)–(10) with an "explicit" recursive procedure in the Krylov subspaces. First, let us consider the conjugate residual (CR) method for a symmetric matrix A.

Flexible, or dynamic, preconditioned CR algorithm is described by relations (11)–(13) where, instead of (14), the direction vectors are defined by the following short recursions:

$$p^0 = B_0^{-1}r^0, \quad p^{n+1} = B_{n+1}r^{n+1} + \beta_n p^n, \quad \beta_n = \sigma_{n+1}/\sigma_n,$$

where B_n are some easily invertible s.p.d. matrices. In our case, we present the preconditioning as an implementation of m Chebyshev iterations in the form

$$u^{n+1} = u^n + \tau_n r^n, \quad r^{n+1} = r^n - \tau_n A r^n = (I - \tau_n A)r^n.$$

One should start with the following m iterations:

$$
\begin{aligned}
r^m &= B_m r^0, \quad B_m = \prod_{k=0}^{m-1} (I - \tau_k A), \\
u^m &= u^0 + B^{-1}r^0, \quad B^{-1} = \tau_0 I + \tau_1 B_1 + \ldots + \tau_m B_{m-1}, \\
r^m &= r^0 - AB^{-1}r^0 = (I - AB^{-1})r^0.
\end{aligned}
\tag{18}
$$

One can see from these formulas that the preconditioning can be written down as

$$B^{-1}r^0 = u^m - u^0, \quad AB^{-1}r^0 = r^m - r^0.$$

Let us define the combined Chebyshev-CR iterative process as a set of the pairs "m Chebyshev iterations + one CR iteration". Formally, we can present the CR algorithm for the restarted steps only:

$$
\begin{aligned}
u^{n_l+1} &= u^{n_l} + \alpha_l p^{n_l}, \\
r^{n_l+1} &= r^{n_l} - \alpha_l A p^{n_l} \\
p^{n_l+1} &= B^{-1}r^{n_l+1} + \beta_l p^{n_l}, \\
\alpha_l &= \sigma_l/\rho_l, \quad \rho_l = (Ar^{n_l}, Ar^{n_l}), \quad \sigma_l = (AB^{-1}r^{n_l}, r^{n_l}).
\end{aligned}
\tag{19}
$$

Here n_l is the total number of iterations from the first up to the l-th restart. The Chebyshev iterations can be presented in the form

$$
\begin{aligned}
u^{n_l} &= u^{n_{l-1}} + B^{-1}r^{n_{l-1}}, \quad n_l = ml, \\
r^{n_l} &= r^{n_{l-1}} - AB^{-1}r^{n_{l-1}} = B_m r^{n_{l-1}}.
\end{aligned}
\tag{20}
$$

A connection between the Chebyshev and CR iterations is provided by the relations $u^{n_l+1} = u^{n_l+1}$, $r^{n_l+1} = r^{n_l+1}$.

In a sense, the considered approaches present a certain special polynomial preconditioning for the iterative processes in the Krylov subspaces. The review of such algorithms is presented in [17]. Let us also mention that a version of the similar two-step Chebyshev and Krylov iterative processes are given in [18,19].

4   Numerical Implementation of the Algorithms

In this section, we describe our numerical implementation of the suggested approaches.

4.1   Model Problem Statement

We consider the Dirichlet problem for the convection-diffusion equation [13]

$$-\frac{\partial^2 u}{\partial x^2} - \frac{\partial^2 u}{\partial y^2} - \frac{\partial^2 u}{\partial z^2} + p\frac{\partial u}{\partial x} + q\frac{\partial u}{\partial y} + r\frac{\partial u}{\partial z} = f(x,y,z),\ (x,y,z) \in \Omega,\quad u|_\Gamma = g(x,y,z),$$

(21)

in the cubic domain $\Omega = (0,1)^3$ with the boundary Γ and the convection coefficients p, q, r which, for simplicity, are assumed to be constant. This boundary value problem is approximated on a cubic grid with the total number of inner nodes $N = L^3$ and the step size $h = 1/(L+1)$:

$$x_i = ih,\ y_j = jh,\ z_k = kh,\ i,j,k = 0,1,\ldots,L+1,$$

by the seven-point finite volume monotone approximation of the exponential type having the second order of accuracy , i.e. $\|u^h - \tilde{u}^h\| = O(h^2)$, $\tilde{u}^h = \{u(x_l, y_l)\} \in \mathcal{R}^{N,N}$ (see [20]) and resulting in the following matrix representation, non-symmetric in general, for a grid node with number l:

$$(Au)_l = a_{l,l}u_l + a_{l,l-1}u_{l-1} + a_{l,l+1}u_{l+1} + a_{l,l-L}u_{l-L} + a_{l,l+L}u_{l+L}$$
$$+ a_{l,l-L^2}u_{l-L^2} + a_{l,l+L^2}u_{l+L^2} = f_l,$$

(22)

Here, l is the "global" number of a grid node in the natural node ordering $l = i + (j-1)L + (j-1)(k-1)L^2$. Generally speaking, the formulas for the coefficients in Eq. (22) may be different and we use the following ones:

$$a_{l,l\pm 1} = -e^{\pm ph/2}/h,\ a_{l,l\pm L} = -e^{\pm qh/2}/h,\ a_{l,l\mp L^2} = -e^{\pm rh/2}/h,$$
$$a_{l,l} = |a_{l,l-1} + a_{l,l-L} + a_{l,l+1} + a_{l,l+L} + a_{l,l+L^2} + a_{l,l-L^2}|.$$

(23)

Equations (22) are written for the inner nodes of the grid, and for the near boundary nodes with the subscripts $i = 1, L$, or $j = 1, L$, or $k = 1, L$ the values of the solution on the boundary should be substituted into the given SLAE and moved to the right-hand side.

The numerical experiments have been carried out using the standard double-precision arithmetic for computing. The right-hand side f and the boundary condition g were taken to correspond the exact solution of the problem (21) $u = x^2 + y^2 + z^2$. Since the convergence rate of iterations depends on the initial error $u - u^0$, its influence has been analyzed on the results for the initial guess $u^0 = 0$. The utilized stopping criterion has been of the form $(r^n, r^n) \le \varepsilon^2(f, f)$, with $\varepsilon = 10^{-7}$. The computations were carried out on the cubic grids with $N = 64^3, 128^3$, and 256^3 nodes and for the restart parameters $m = 8, 16, 32, 64$, and 128. In the tables below, we present the results obtained in solving the problem (21) with zero convection coefficients and $p = q = r = 4$.

The main objective of our experimental research is to demonstrate the high performance of the algorithms presented for very large SLAEs, as well as to study the stability and convergence rate of the LSM approaches with a preliminary usage of the Chebyshev inexpensive iterative processes.

4.2 Numerical Experiments

In each cell of the tables below, we present two values: the upper one is the total number of iterations, and the lower one is the run-time in seconds to obtain the solution. Let us notice that in all of our runs the maximum error $\delta = \max_{i,j,k}\{|1 - u^n_{i,j,k}|\}$ did not exceed the tolerance value ε. All the calculations were carried out on Broadwell processor with Intel©MKL BLAS level 2 functions usage.

In Tables 1, 2, we present the results for the combined Chebyshev-LSM algorithm for symmetric and non-symmetric SLAEs. In all the cases, the boundaries λ_1, λ_N of the matrix spectrum were taken from [20], however the presented results are sufficiently close to each other. These tables correspond to the three-term formulas (6) of the Chebyshev acceleration, and we refer to this version of the algorithm as "CHEB-3-LSM".

Table 1. CHEB-3-LSM, $p = q = r = 0$

$N\backslash m$	8	16	32	64	128	256
64	329	336	352	368	384	384
	0.86	0.52	0.32	0.28	0.15	0.20
128	641	657	672	704	736	768
	14.8	9.4	6.2	6.4	4.9	4.9
256	1801	1281	1313	1344	1408	1472
	527.3	125.7	81.9	65.6	56.7	55.1

Table 2. CHEB-3-LSM, $p = q = r = 4$

$N\backslash m$	8	16	32	64	128	256
64	336	352	352	384	364	384
	0.80	0.45	0.30	0.23	0.22	0.20
128	6141	573	704	704	757	718
	11.7	7.00	5.0	4.1	3.5	3.3
256	1729	1329	13745	1408	1408	1473
	450.0	137.0	83.8	69.9	54.6	52.7

It follows from these tables that in all the cases considered there is an optimal value of m.

Tables 3 and 4 include the similar results but for the two-dimensional case and for another version of the Chebyshev acceleration and non-zero initial guess. Namely, the stable version of the two-step iterations (2) was used. We call such an algorithm as CHEB-2-LSM. It is important to note that without stabilizing the ordering of the parameters τ_n, the iterative process does not converge for large enough values of N and m. In each cell, the low values are the maximum errors δ.

Table 3. CHEB-2-LSM, $u^0 = x^2 + y^2, p, q = 0$.

$N\backslash m$	8	16	32	64	128
31^2	360	220	160	127	160
	$3.0 \cdot 10^{-6}$	$1.8 \cdot 10^{-6}$	$9.9 \cdot 10^{-7}$	$9.8 \cdot 10^{-8}$	$1.1 \cdot 10^{-7}$
63^2	1240	672	416	316	249
	$9.9 \cdot 10^{-6}$	$8.7 \cdot 10^{-6}$	$4.0 \cdot 10^{-6}$	$2.3 \cdot 10^{-6}$	$5.5 \cdot 10^{-7}$
127^2	4415	2271	1241	768	608
	$2.9 \cdot 10^{-5}$	$2.7 \cdot 10^{-5}$	$2.1 \cdot 10^{-5}$	$2.4 \cdot 10^{-5}$	$5.7 \cdot 10^{-6}$

Table 4. CHEB-2-LSM, $u^0 = x^2 + y^2$, $p, q = 4$.

$N\backslash m$	8	16	32	64	128
31^2	272	160	128	128	160
	$2.1 \cdot 10^{-6}$	$1.5 \cdot 10^{-6}$	$1.2 \cdot 10^{-6}$	$7.9 \cdot 10^{-9}$	$3.1 \cdot 10^{-7}$
63^2	680	511	320	256	256
	$7.8 \cdot 10^{-6}$	$5.4 \cdot 10^{-6}$	$3.0 \cdot 10^{-6}$	$6.8 \cdot 10^{-6}$	$2.0 \cdot 10^{-8}$
127^2	2984	1440	928	577	512
	$2.4 \cdot 10^{-5}$	$2.3 \cdot 10^{-5}$	$1.9 \cdot 10^{-5}$	$1.9 \cdot 10^{-5}$	$4.3 \cdot 10^{-6}$

Finally, in Tables 5, 6 the advantage of the two-level LSM approaches (15)–(17) is demonstrated for the symmetric and non-symmetric SLAEs, i.e. for the convection coefficients $p = q = 0$ and $p = q = 4$ respectively. We use the two-step iterations (2), (3) with the stable ordering [5] of the parameters τ_n. And because of another implementation of the least squares approach, this method is named here as "CHEB-D-LSM".

We can see from all the tables that the considered algorithms have the acceptable stability, i.e. the resulting maximum errors (low values in each cell in Tables 1, 2) are comparable with the given stopping criterion for $\varepsilon = 10^{-7}$. Also, the values of the final inner products of the residual vectors (r^n, r^n) (i.e. the lowest values in the cells in Tables 3, 4, 5 and 6) are acceptable as compared with ε.

Table 5. CHEB-D-LSM,$u^0 = x^2 + y^2, p = q = 0$

$N\backslash m$	8	16	32	64	128
31^2	88	112	144	127	160
	$8.1 \cdot 10^{-8}$	$1.3 \cdot 10^{-6}$	$8.0 \cdot 10^{-7}$	$9.8 \cdot 10^{-8}$	$1.1 \cdot 10^{-7}$
63^2	160	160	224	288	249
	$5.9 \cdot 10^{-7}$	$7.5 \cdot 10^{-7}$	$1.1 \cdot 10^{-6}$	$2.5 \cdot 10^{-6}$	$5.5 \cdot 10^{-7}$
127^2	304	304	320	384	512
	$2.2 \cdot 10^{-6}$	$2.3 \cdot 10^{-6}$	$6.9 \cdot 10^{-7}$	$2.9 \cdot 10^{-6}$	$1.2 \cdot 10^{-5}$

Table 6. CHEB-D-LSM,$u^0 = x^2 + y^2, p = q = 4$

$N\backslash m$	8	16	32	64	128
31^2	96	112	158	128	160
	$2.1 \cdot 10^{-7}$	$1.7 \cdot 10^{-7}$	$3.1 \cdot 10^{-7}$	$7.9 \cdot 10^{-9}$	$3.1 \cdot 10^{-7}$
63^2	184	192	224	256	256
	$5.0 \cdot 10^{-7}$	$3.2 \cdot 10^{-7}$	$8.5 \cdot 10^{-8}$	$2.2 \cdot 10^{-6}$	$2.0 \cdot 10^{-8}$
127^2	344	352	384	384	512
	$1.9 \cdot 10^{-6}$	$1.8 \cdot 10^{-6}$	$5.1 \cdot 10^{-7}$	$6.4 \cdot 10^{-6}$	$2.1 \cdot 10^{-6}$

4.3   Remarks on Parallel Implementation

It is important to notice, that the original matrix A is assumed to be stored in some compressed sparse representation (such as CSR format), which is obvious when solving practical algebraic problems arising from approximation of some complicated multi-dimensional boundary value problems on non-structured grids. In general, the algorithms can be implemented on a heterogeneous multiprocessor computer architecture with distributed and hierarchical shared memory (NUMA – Non-Uniform Memory Acess) by means of the hybrid programming, using the Message Passing Interface (MPI) system for communications among cluster nodes, by multi-thread computations (OpenMP-type tools on multi-core CPU), by vectorization of operations with AVX2 (Advanced Vector Extension) instructions as well as by using graphic accelerators (GPGPUs or Intel Phi). The performance of the basic matrix-vector operations in the parallel iterative methods is considered in [21]. As for the presented methods, we can assume that they are applied to a block system of linear equations of the form (1), and the block rows $A_k = \{A_{k,\ell}, \ k = 1,\dots,P\} \in \mathcal{R}^{N_k,N}$, $N_k \cong N/P$, $N_1 + \dots + N_p = N$ of the matrix A are distributed in the memory of the corresponding MPI processes used for the first level of parallelization of the algorithms in the same way as is usually done in the domain decomposition approaches, where every block matrix row corresponds to a computational subdomain (see [2]). It is very important to remark that the convergence rate and the run time of the iterative process substantially depend on the ordering of unknowns and the corresponding block matrix structure. For example, such features arise in solving numerically some system of PDEs with basically different values of the system coefficients.

One of the advantages of the algorithms discussed is provided by a cheap realization of the Chebyshev acceleration due to the absence of the scalar vector multiplications necessary in the conjugate direction methods. In addition, the idea of scalable parallelism as related to the proposed approaches consists in the implicit LSM which is based on simultaneous computations of a large number of inner vector products in contrast to successive computations in the conventional Krylov algorithms. The performance of the approaches on real multi-processor systems with distributed and hierarchical shared memory, as well as application and exploiting of the additional multi-preconditioning, domain decomposition methods, coarse grid corrections and low-rank matrix approximations, which have high mathematical efficiency and rapid execution on a one-processor computer, is a subject matter of further research.

Certainly, the ultimate success of the code optimization depends on minimizing the communication costs which are defined by the big data structure which volume can be measured by terabytes in certain large-scale machine experiments. The communications not only slow down the computational process but are the most energy-consuming operations. So, there is a mathematical problem to find algorithms with a big volume of arithmetic operations but, at the same time, with a small number of communications.

5   Conclusion

We have considered the generalization of the Anderson acceleration for parallel solving non-symmetric large SLAEs with sparse matrices. This generalization is based on the least squares methods applied to some preliminary "cheap" iterative Chebyshev processes which are used not only for the convergence of approximate solutions, but, mainly, for obtaining the basis vectors for an implicit, or a block, implementation of the Krylov type algorithms with using the periodic minimization of the residual vector before restarting the algorithm. The comparative experimental analysis of the combined variational approach and the above-mentioned methods, as well as the spectral Chebyshev acceleration shows a reasonable stability and convergence rate of the iterations of the proposed methods. In particular, the valuable improvement of the iterative processes is achieved by means of the two-level LSM acceleration of the restarted algorithms. The considerable advantage of the Chebyshev spectral algorithms consists in the absence of the vector inner products, which are the obstacles for scalability. Thus, the combination of the Chebyshev and Krylov approaches provides easy-to-get improvements in the efficiency and performance of the resulting method.

References

1. Saad, Y.: Iterative Methods for Sparse Linear Systems. PWS Publishing, Boston (2002)
2. Il'in, V.P.: Problems of parallel solution of large systems of linear algebraic equations. J. Math. Sci. **216**(6), 795–804 (2016). https://doi.org/10.1007/s10958-016-2945-4
3. Pratara, P.P., Suryanarayana, J.E.P.: Anderson acceleration of the Jacobi iterative method. An efficient alternative to Krylov methods for large, sparse linear systems. J. Comput. Phys. **306**, 43–54 (2016). https://doi.org/10.1016/j.jcp.2015.11.018
4. Anderson, D.G.: Iterative procedures for nonlinear integral equations. J. Assoc. Comput. Mach. **12**, 547–560 (1965). https://doi.org/10.1145/321296.321305
5. Samarskii, A.A., Nikolaev, E.S.: Methods for Solving Grid Equations. Nauka, Moscow (1978). (in Russian)
6. Lawson, C.L., Hanson, R.Z.: Solving Least Squares Problems. Prentice-Hall Inc., Englewood Cliffs (1974)
7. Montgomery, P.L.: A block Lanczos algorithm for finding dependencies over GF(2). In: Guillou, L.C., Quisquater, J.-J. (eds.) EUROCRYPT 1995. LNCS, vol. 921, pp. 106–120. Springer, Heidelberg (1995). https://doi.org/10.1007/3-540-49264-X_9
8. Manteufel, T.A.: Adaptive procedure for estimating parameters for the nonsymmetric Tchebychev iteration. Numerische Math. **31**, 183–208 (1978). https://doi.org/10.1007/BF01397475
9. Gutknecht, M.H.: The Chebyshev iteration revisited. Parallel Comput. **28**, 263–283 (2012). https://doi.org/10.1016/S0167-8191(01)00139-9
10. Il'in, V.P.: Finite Difference and Finite Volume Methods for Elliptic Equations. , ICM & MG SBRAS Publisher, Novosibirsk (2001). (in Russian)
11. Zhukov, V.T., Novikova, N.D., Feodoritova, O.B.: Chebyshev iterations based on adaptive update of the lower bound of the spectrum of the matrix. KIAM Preprint N 172, Moscow (2018)
12. Intel®Mathematical Kernel Library. https://software.intel.com/en-us/mkl
13. Il'in, V.P., Itskovich, E.A.: On the semi-conjugate direction methods with dynamic preconditioning. J. Appl. Ind. Math. **3**, 222–233 (2009). https://doi.org/10.1134/S1990478909020082
14. Eisenstat, S.C., Elman, H.C., Schultz, M.H.: Variational iterative methods for nonsymmetric systems of linear equations. SIAM J. Numer. Anal. **20**, 345–357 (1983). https://doi.org/10.1137/0720023
15. Yuan, J.Y., Golub, G.H., Plemmons, R.J., Cecilio, W.A.: Semi-conjugate direction methods for real positive definite systems. BIT Numer. Math. **44**, 189–207 (2004). https://doi.org/10.1023/B:BITN.0000025092.92213.da
16. Liesen, J., Rozloznik, M., Strakos, Z.: Least squares residuals and minimal residual methods. SIAM J. Sci. Comput. **23**, 1503–1525 (2002). https://doi.org/10.1137/S1064827500377988
17. O'Leary, D.P.: Yet another polynomial preconditioner for the conjugate gradient algorithm. Linear Algebra Appl. **154–156**, 377–388 (1991). https://doi.org/10.1016/0024-3795(91)90385-A
18. Elman, H.C., Saad, Y., Saylor, P.E.: Hybrid Chebyshev - Krylov subspace algorithm for solving nonsymmetric systems of linear equations. SIAM J. Sci. Stat. Comput. **7**(3), 840–855 (1986). https://doi.org/10.1137/0907057
19. Baker, A.H., Jessup, E.R., Manteuffel, T.: A technique for accelerating the convergence of restarted GMRES. SIAM J. Matrix Anal. Appl. **26**(4), 962–984 (2005). https://doi.org/10.1137/S0895479803422014

20. Il'in, V.P.: On exponential finite volume approximations. Russ. J. Numer. Anal. Math. Model. **8**, 479–506 (2003). https://doi.org/10.1163/156939803322681158

21. Sidje, R.B.: Alternatives for parallel Krylov subspace basis computation. Numer. Linear Algebra Appl. **4**(4), 305–331 (1997). https://doi.org/10.1002/(SICI)1099-1506(199707/08)4:4⟨305::AID-NLA104⟩3.0.CO;2-D

Supercomputer Simulation

Block Model of Lithosphere Dynamics: New Calibration Method and Numerical Experiments

Valeriy Rozenberg[1,2]([✉])

[1] Krasovskii Institute of Mathematics and Mechanics of the Ural Branch of the Russian Academy of Sciences, 16, Kovalevskoi Str., Ekaterinburg 620990, Russia
`rozen@imm.uran.ru`
[2] Ural Federal University, 19, Mira Str., Ekaterinburg 620002, Russia

Abstract. The article is devoted to the last modification of the spherical model describing lithosphere dynamics and seismicity; the emphasis is on different ways of taking into account random factors that influence the behaviour of block structure elements. The main numerical procedures are effectively parallelized. This provides the possibility to use real data (for example, earthquake catalogues) in the simulation process. The main objective of the presented research is to develop a feasible procedure for automatic tuning of model parameters to obtain the best approximation results concerning the global seismicity. A special quality criterion is introduced based on minimization of the weighted sum of discrepancies between model characteristics and real patterns such as the spatial distribution of epicentres of strongest events and the Gutenberg–Richter law on frequency-magnitude relation. To optimize the time-consuming process of the reduced exhaustive search by specified parameters, the additional parallelization of calculations is applied. The results of some series of numerical experiments are discussed.

Keywords: Block-and-fault models of lithosphere dynamics and seismicity · Parameter calibration procedure · Approximation quality criterion · Parallelization of calculations

1 Introduction: Goals of Seismicity Simulation

The statistical analysis of seismicity as the spatial-temporal sequence of earthquakes in a given area based on real catalogues is heavily hindered by a short history of reliable observation data. The patterns of the earthquake occurrence that are identifiable in real catalogues may be merely apparent and will not necessarily repeat in the future. At the same time, synthetic catalogues obtained from numerical simulations can cover very long time intervals. This allows us to acquire more reliable estimates of seismic flow parameters and to search for premonitory patterns preceding large events. Such a possibility may be needed in expert systems for global/regional seismic risk monitoring [1,2].

L. Sokolinsky and M. Zymbler (Eds.): PCT 2020, CCIS 1263, pp. 181–197, 2020.
https://doi.org/10.1007/978-3-030-55326-5_13

The main result of modelling is a synthetic earthquake catalogue; each of its events is characterized by a time moment, epicentre coordinates, a depth, a magnitude, and, for some models using the region geology, an intensity. The simulation of lithosphere dynamics provides a field of velocities at different depths, acting forces, induced displacements and the character of interaction between structural elements. Due to the absence of adequate seismotectonic process theory, it is common to assume that various features of the lithosphere (e.g., the spatial heterogeneity, hierarchical block structure and different types of nonlinear rheology) are related to the properties of earthquake sequences. The stability of these properties at a quantitative level in different regions allows us to conclude that it is possible to consider the lithosphere as a large dissipative system, which behaviour does not essentially depend on particular details of the specific processes progressing in a geological system.

We can separate a great variety of different approaches to modelling the lithospheric processes (see, for example, [1] and its bibliography) into two main groups. The first (traditional) one relies on a detailed investigation of a specific tectonic fault or a system of faults. Another frequent object of this approach would be a strong earthquake. The investigation is aimed to reproduce certain pre-seismic and/or post-seismic phenomena (relevant to this region or event). In contrast, models of the second type have started to be developed relatively recently. They treat the seismotectonic process in a rather abstract way. The main goal of these simulations is to reproduce such general universal properties of time-spatial series that are inherent in the observed seismicity (primarily, the power-law for earthquake size distribution, i.e., the Gutenberg–Richter law on frequency-magnitude (FM) relation, clustering, migration of events, etc.) However, it seems that an adequate model designed within the second approach should reflect both some universal features of nonlinear systems and the specific geometry of interacting faults.

The block models of lithosphere dynamics and seismicity [2,3] have been developed taking into account both requirements. These models are used to analyze how the basic features of seismicity and block-and-fault dynamics (BAFD) depend on the lithosphere structure and motions. The approach to modelling is based on the concept of the hierarchical block structure of the lithosphere. Tectonic plates are represented as a system of perfectly rigid blocks being in a quasi-static equilibrium state. A model event is a stress drop at a fault separating two blocks; this drop occurs under the action of outer forces. In the model, two main mechanisms are involved in the seismotectonic process: the tectonic loading, with a characteristic rate of a few cm/yr, and the elastic stress accumulation and redistribution, with a characteristic rate of a few km/sec. They are considered over time as a uniform motion and an instantaneous stress drop, respectively. Wave processes are beyond the scope of the existing block models. The plane model, where a structure is restricted by two horizontal planes, has been the most extensively studied. Some approximations to the dynamics of lithosphere blocks for real seismic regions have been built on its basis [2,3]. Simulating the motion of large tectonic plates revealed significant distortions.

Therefore, the spherical geometry was involved [4]. By reason that computational implementation of the spherical modification required much more expenditures of memory and processor time than the plane model, parallel computing technologies were applied.

This paper actually continues the previous studies, both theoretical and numerical, in the spherical block model of lithosphere dynamics and seismicity, see [4–6]. The modification described there contained simple stochastic equations introduced to add uncertainties in the deterministic scheme of block-and-fault interaction. Since this allowed us to improve some properties of synthetic seismicity [6], we consider such a direction of the BAFD model development as a perspective. Now, we suggest using more complicated stochastic differential equations describing the model earthquake mechanism to keep the approximation results improving.

Another aspect of modelling and, in our opinion, the most important one, is related to its main difficulty, namely, to the calibration of the model. Confirmed by previous large numerical experiments, this problem is caused by a huge number of parameters to be fit. The absence of analytical dependencies between the input and output characteristics of block-and-fault interaction along with the necessity for a permanent control over the calibration by an expert, decelerating the process of model development, also make the problem more complicated. Trying to overcome these difficulties, we have added a special block for automatic tuning of the model parameters to the program package. This procedure is based on a formal criterion of approximation quality introduced in the BAFD model for the first time. We use the methodology described in [7] and proposing an empirical procedure for the automatic tuning of the algorithm parameters to get the best approximation results for a specific dynamical inverse problem. The evident similarity of that procedure and the one being developed here is in the fact that both are needed due to vague dependencies between the input and output parameters of approximation schemes. Thus, the novelty of this paper consists in the description of a new calibration procedure decreasing the expert examination and in the discussion of some results of illustrative numerical experiments.

2   Spherical Block Model: A Stochastic Version

Let us briefly describe the last version of the spherical block-and-fault model of lithosphere dynamics and seismicity. For the detailed description of all the modifications of the model, see [4,5]. In this paper, we restrict ourselves to a summary of the basic ideas and principles with an emphasis on the new constructions.

A block structure is a limited and simply connected part of a spherical layer of depth H bounded by two concentric spheres. The outer sphere represents the Earth's surface and the inner one represents the boundary between the lithosphere and the mantle. The partition of the structure into blocks is defined by infinitely thin faults intersecting the layer. Each fault is a part of a cone surface having the same value of the dip angle with the outer sphere at all its points.

The faults intersect along the curves that meet the outer and inner spheres at the points called vertices. A part of such a curve between two respective vertices is called a rib. Fragments of the faults limited by two adjacent ribs are called segments. The common block parts with the limiting spheres are spherical polygons; those on the inner sphere are called bottoms. A block structure may be a part of the spherical shell and be bordered by boundary blocks that are adjacent to boundary segments. Another possibility (impossible in the plane) is to consider the structure covering the whole surface of the Earth without the boundary blocks. The blocks are assumed to be perfectly rigid. All block displacements are considered negligible compared with the block sizes. Therefore, the block structure geometry does not change during the simulation, and the structure does not move as a whole. The gravitation forces remain essentially unchanged by the block displacements and, because the structure is in a quasi-static equilibrium state at the initial time moment, the gravity does not cause block motions.

All blocks (both internal and boundary if specified) have six degrees of freedom and can leave the spherical surface. Each block displacement consists of translation and rotation components. The motions of the boundary blocks, as well as those of the underlying medium, taking into account an external action on the structure, are assumed to be known. As a rule, they are described as rotations on the sphere, i.e. rotation axes and angular velocities are given.

The current modification allows us to set different depths for different blocks (in the range of H) and the changes of fault parameters depending on the depth. This is made to take into account the natural inhomogeneity of the lithosphere with respect to depth. As to random factors, we consider two ways of introducing them into the procedures for calculating the forces acting on the blocks and the resulting displacements causing the model earthquakes. These ways are as follows: adding some noises to the differential equations describing the dynamics of forces and displacements and using random variables when specifying strength thresholds for the medium of tectonic faults.

Since the blocks are perfectly rigid, all the deformations take place at the fault zones and block bottoms; the forces arise at the bottoms due to the relative displacements of the blocks with respect to the underlying medium, and at the faults due to the displacements of the neighbouring blocks or their underlying medium. To calculate different curvilinear integrals, one should discretize (divide into cells) the spherical surfaces of the block bottoms and the cone surfaces of the fault segments. The values of forces and inelastic displacements are assumed to be equal for all points of a cell. Let us present the formulas for the elastic force (f_t, f_l, f_n) acting on a fault per unit area at some point:

$$f_t(\tau) = K_t(\Delta_t(\tau) - \delta_t(\tau)), \quad f_l(\tau) = K_l(\Delta_l(\tau) - \delta_l(\tau)),$$
$$f_n(\tau) = K_n(\Delta_n(\tau) - \delta_n(\tau)). \tag{1}$$

Here τ is the time; (t, l, n) is the rectangular coordinate system with origin at the point of force application (the axes t and l lie in the plane tangent to the fault's surface, the axis n is perpendicular to this plane); $\Delta_t, \Delta_l, \Delta_n$ are the components of the relative displacement in the system (t, l, n) of the structure neighbouring elements; δ_t, δ_l, δ_n are corresponding inelastic displacements at the point; their evolution is described by the quasi-linear stochastic differential equations

$$d\delta_t(\tau) = W_t K_t(\Delta_t(\tau) - \delta_t(\tau))d\tau + \lambda_t \delta_t(\tau)d\xi_t(\tau),$$
$$d\delta_l(\tau) = W_l K_l(\Delta_l(\tau) - \delta_l(\tau))d\tau + \lambda_l \delta_l(\tau)d\xi_l(\tau),$$
$$d\delta_n(\tau) = W_n K_n(\Delta_n(\tau) - \delta_n(\tau))d\tau + \lambda_n \delta_n(\tau)d\xi_n(\tau). \tag{2}$$

Thus, for each displacement, we have the vector equation of dimension that is equal to the number of cells at the segment. At the same time, we assume that standard independent Wiener processes ξ_t, ξ_l and ξ_n (i.e., processes starting from zero with zero mathematical expectation and dispersion equal to τ) are scalar (the same key amplitude characteristics of the outer random noises for the whole segment). Each equation of (2) has a unique solution (in Ito's sense), namely, a normal Markov random process with continuous realizations [8]. The coefficients K_t, K_l, K_n (1)–(2) characterizing the elastic properties of the faults, the coefficients W_t, W_l, W_n (2) characterizing the viscous properties, and the coefficients λ_t, λ_l, λ_n (2) may be different for different faults and may depend on the depth as an option. The formulas for calculating the elastic forces and inelastic displacements at the block bottoms are similar to (1) and (2). At the initial time $\tau = 0$, all the forces and displacements are equal to zero; actually, the source of motion of the model structure is the motion of both the underlying medium and the boundary blocks (if specified) determining the dynamics of relative displacements (like Δ_t, Δ_l, Δ_n) at both the faults and block bottoms.

It seems that the introduction of the quasi-linear equations, where the diffusion term depends on the phase state, into (2) instead of linear ones, as done in [6], allows us to take into account the influence of the displacement values on the level of uncertainties: the higher the value, the greater the possible uncertainty at this time.

The translation vectors of the inner blocks and their rotation angles can be found from the condition that the total force and the total moment of the forces acting on each block are equal to zero. This is the condition of system quasi-static equilibrium and at the same time the condition of energy minimum. The dependence of forces and moments on displacements and rotations of the blocks must be linear (explicit cumbersome formulas are omitted). Therefore, the system of equations for determining these values is also linear:

$$\mathbf{A}\mathbf{w} = \mathbf{b}. \tag{3}$$

Here the components of the unknown vector $\mathbf{w} = (w_1, w_2, ..., w_{6n})$ are the translation vectors of the inner blocks and their rotation angles (n is the number of blocks). The elements of the matrix $\mathbf{A}$ $(6n \times 6n)$ do not depend on the time and can be calculated only once, at the beginning of the process. For realistic values of the model parameters, the matrix $\mathbf{A}$ is non-degenerate; system (3) has a unique solution found at discrete times τ_i.

At every time step τ_i, when computing the force acting on the fault, we find the value of a dimensionless quantity κ (model stress) by the formula

$$\kappa = \frac{\sqrt{f_t^2 + f_l^2}}{P - f_n}. \tag{4}$$

Here P is a parameter, which may be interpreted as the difference between the lithostatic and the hydrostatic pressure. Thus, the κ value is actually the ratio of the magnitude of the force tending to shift the blocks along the fault to the modulus of the force connecting the blocks to each other. For each fault, three strength levels are specified. In general, they depend on time:

$$B > H_f \geq H_s, \quad B = B(\tau_i) = B_0(\tau_i) + \sigma X(\tau_i),$$
$$H_f = H_f(\tau_i) = aB(\tau_i), \quad H_s = H_s(\tau_i) = bB(\tau_i). \tag{5}$$

For each i, we assume that $0 < B_0(\tau_i) < 1$, $0 < \sigma \ll 1$, $X(\tau_i)$ is a normally distributed random value $N(0; 1)$, $0 < a < 1$, $0 < b \leq a$. The initial conditions are such that the inequality $\kappa < B$ is valid for all structure cells.

The interaction between the blocks (between the block and neighbouring underlying medium) is visco-elastic (a normal state) as long as the κ value (4) at the part of the fault separating the structural elements remains below the strength level B. When this level is exceeded at some part of the fault (a critical state), a stress-drop (a failure) occurs in accordance with the dry friction model (such failures represent earthquakes). By failure we mean a slippage by which the inelastic displacements δ_t, δ_l and δ_n in the cells change abruptly to reduce the model stress according to the formulas

$$\delta_t^e = \delta_t + \gamma^e \xi_t f_t, \quad \delta_l^e = \delta_l + \gamma^e f_l, \quad \delta_n^e = \delta_n + \gamma^e \xi_n f_n, \tag{6}$$

where δ_t, f_t, δ_l, f_l, δ_n and f_n are the inelastic displacements and the elastic force per unit area just before the failure. The coefficients $\xi_t = K_l/K_t$ ($\xi_t = 0$ if $K_t = 0$) and $\xi_n = K_l/K_n$ ($\xi_n = 0$ if $K_n = 0$) account for inhomogeneity of the displacements in different directions. The coefficient γ^e is given by

$$\gamma^e = \frac{\sqrt{f_t^2 + f_l^2} - H_f(P - f_n)}{K_l\sqrt{f_t^2 + f_l^2} + K_n H_f \xi_n f_n}. \tag{7}$$

It follows from (1) and (4)–(7) that, after recalculating the new values of the inelastic displacements and elastic forces, the κ value is equal to H_f. Then, the right-hand part of the system (3) is computed, and the translation vectors and angles of rotation for the blocks are found again. If for some cell(s) $\kappa \geq B$, then the entire procedure is repeated. When $\kappa < B$ for all structure cells, the calculations are continued according to the standard scheme. Immediately after the earthquake, it is assumed that the cells in which the failure occurred are in the creep state. This implies that, for these cells, the parameters W_t^s ($W_t^s \gg W_t$), W_l^s ($W_l^s \gg W_l$), and W_n^s ($W_n^s \gg W_n$) are used instead of W_t, W_l and W_n in Eq. (2). Such new values provide a faster growth (compared to the normal state) of the inelastic displacements and, consequently, a decrease of the stress. The cells remain in the creep state so long as $\kappa > H_s$ (we may call this a healing process); when $\kappa \leq H_s$, the cells return to the normal state with the use of W_t, W_l and W_n in (2). Then, the process of stress accumulation starts again.

A synthetic earthquake catalogue is produced as a main result of the numerical simulation. All cells of the same fault, in which the failure occurs at the time τ_i, are considered as a single event. Its epicentre coordinates and depth are the weighted sums of the coordinates and depths of the cells involved in the earthquake. The weighted sum of the vectors $(\gamma^e \xi_t f_t, \gamma^e f_l)$ added to the inelastic displacements δ_t and δ_l computed according to (6) approximates the shift of the blocks along the fault and allows us to determine the mechanism of a synthetic event. Such a mechanism is an important earthquake specific; it informs on the different seismic waves propagation from the earthquake source. Depending on the shift direction and the fault dip angle, the following basic mechanisms are commonly considered: strike-slip, normal faulting and thrust faulting [9]. In the current model version the earthquake magnitude is calculated taking into account its mechanism according to the empirical formulas well-known in seismology: [10]

$$M = D \lg S + E, \tag{8}$$

where S is the total area of the cells (in km^2), $D = 1.02$, $E = 3.98$ for strike-slip, $D = 1.02$, $E = 3.93$ for normal faulting, and $D = 0.90$, $E = 4.33$ for thrust faulting. In addition to the aforesaid, the model informs on the instantaneous kinematics of the blocks and the character of their interaction along the boundaries.

3   Calibration Procedure

The previous rather extensive experience of the BAFD model calibration (primarily consisting in the cumbersome process of fitting numerous parameters that describe the structural element interactions [2,3,5]) gives evidence of the considerable time needed for this procedure (especially concerning the expert contribution in a comparative analysis of the model and real data); this results in the conclusion that the process automation is extremely desirable. A formal criterion of the approximation quality seems to be useful provided that it takes into account the most important seismic characteristics. Then, optimal values of this criterion should be found by the exhaustive search (preferably, optimized), which necessity is explained by lack of any analytical expression for the explicit dependence between input and output parameters of the BAFD models. First, we shall present a brief description of the characteristics, which should be reasonably included in the criterion.

1. *The Spatial Distribution of Epicentres of Strong Events.* To compare real and model distributions, the magnitude threshold for strong events is specified, the real distribution is narrowed to the fault vicinities (since there are no model

events inside the blocks, only at the faults). After that, the Hausdorff distance (in the natural metric space on the Earth's surface) between two sets of epicentres is computed according to the scheme briefly described below. Informally, two sets are close in the Hausdorff distance if every point of either set is close to some point of the other set. The Hausdorff distance is the greatest of all the distances from a point in one set to the closest point in the other set. Thus, for two non-empty sets of epicentres, real E_r and model E_m, the Hausdorff distance is defined as follows:

$$d_{\mathrm{H}}(E_r, E_m) = \max\{ \sup_{e_r \in E_r} \inf_{e_m \in E_m} d(e_r, e_m), \sup_{e_m \in E_m} \inf_{e_r \in E_r} d(e_r, e_m) \}, \quad (9)$$

where $d(e_r, e_m)$ is the distance between points e_r and e_m on the Earth's surface. In addition, we can compute the Hausdorff distance between the sets of epicentres for several strongest events.

2. *The distribution of earthquakes in depth.* To compare real and model distributions in this aspect, we do not need a subtle quantitative analysis since we only need to define the class (shallow, intermediate, and deep) some event belongs to. Therefore, after the magnitude threshold and characteristic depths are determined, we compute a share for every depth interval with respect to the total number of selected events. Then the absolute deviation between real D_r and model D_m data is defined by the following relation:

$$d_{\mathrm{D}}(D_r, D_m) = \sum_{i=1}^{n} |D_r^i - D_m^i|, \quad (10)$$

where n is the chosen number of depth intervals, D_r^i and D_m^i are the shares of the events belonging to the corresponding interval for real and model catalogues, respectively.

3. *The Gutenberg–Richter law.* It characterizes the power distribution of earthquakes in magnitude. The linearity and slop of the FM plot of the dependence of the earthquakes accumulated number N on magnitude M are important. All the FM plots are approximated by the linear least-squares regression $\lg N = c - SM$. The value of S serves as an estimate of the plot slope. The average distance between the plot points and the constructed line is treated as an approximation error A. To compare real $G_r = (S_r, A_r)$ and model $G_m = (S_m, A_m)$ parameters, we apply the following scheme:

$$d_{\mathrm{G}}(G_r, G_m) = \alpha_1 |S_r - S_m| + \alpha_2 |A_r - A_m|, \quad (11)$$

where S_r, S_m are real and model slopes of the corresponding regression lines, A_r, A_m are real and model approximation errors, $\alpha_1, \alpha_2 > 0$, $\alpha_1 + \alpha_2 = 1$ are weights.

Then, we consider the multi-criteria optimization problem, where it is necessary to introduce an appropriate ranking for different calculated variants to find an optimal set of parameters with respect to characteristics 1–3. These variants include the variation of several parameters in some ranges determined as a result of the expertise. An aggregated criterion assumes normalizing the characteristics described above and further summing of the obtained results. In all calculated model variants, the deviations from the etalon values are divided by the maximum (over this parameter) deviation value. Thus, we obtain the normalized values from the segment $[0, 1]$. The minimum weighted sum of these values is treated as the best variant:

$$Er_j = \beta_1 d_{H_j} / \max_i d_{H_i} + \beta_2 d_{D_j} / \max_i d_{D_i} + \beta_3 d_{G_j} / \max_i d_{G_i} \rightarrow \min_j, \qquad (12)$$

where Er_j is the aggregated error of the variant j, $j = 1, \ldots, K$, K is the number of variants under examination, β_1, β_2, $\beta_3 > 0$, $\beta_1 + \beta_2 + \beta_3 = 1$ are weights. Note that $Er = 0$ for the real data and can be equal to 1 in the worst case.

Evidently, there are other possible ways to construct an aggregated approximation quality criterion and a ranking procedure for calculated variants, see [11]. One can use some other parameters, some other weights, for example, some other aggregation rules, etc. For this work, we have chosen the parameters that, in our opinion, are the most appropriate for the comparative analysis of the model and available real data. For example, temporal characteristics of synthetic seismicity are much more difficult to compare with real ones, as the question whether the initial time of simulation corresponds to the real moment generally remains open in such kind of models. We should note that the procedure described in the paper as well as related numerical experiments are the first attempt to automatize the BAFD model calibration process.

4 Numerical Experiments: Conception

The spherical block model is most actively applied to investigating the dynamics and seismicity of the largest tectonic plates covering the whole Earth [4–6]. The geometry of such a structure is assumed to be invariant: 15 plates (North America, South America, Nazca, Africa, Caribbean, Cocos, Pacific, Somalia, Arabia, Eurasia, India, Antarctica, Australia, Philippines and Juan de Fuca), 186 vertices, 199 faults. Boundary blocks are absent since the structure occupies the entire spherical shell; the underlying medium motion is defined as rotations on the sphere according to model HS3-NUVEL1 [12].

To construct a set of K variants to be examined by criterion (9)–(12), we use the variation of the coefficients describing the visco-elastic interaction in formulas (1), (2) and the variation of different noise levels from relations (2), (5) with some reasonable steps.

The previous research [5,6] established that the spherical block model allows us to attain an effective parallelization based on the standard scheme "master-worker" with a unique loading MPI-module [13]. At every moment, the most time-consuming procedures are preserving the information on model events and, substantially, calculating the forces, inelastic displacements and stresses (1), (2), (4)–(6) in all the structure cells (so, in a typical variant for the global system of tectonic plates, we have got more than 250,000 cells at the block bottoms and about 3,500,000 cells at the fault segments). The main calculations are performed independently from each other; therefore, they are uniformly distributed among the processors. Informational exchange between the processors involves only the small dimension values. The detailed description of the parallelization scheme, the procedure for the dynamical loading redistribution and the analysis of some parallelization characteristics are presented in [5].

The recent time calibration experiments showed that the tuning procedure is obviously time-consuming (since it requires a large amount of program runs on the grid with respect to several parameters). Thus, we simultaneously apply two parallel schemes: first, the trivial runs of different variants at different sets of cores and, second, inside every set of chosen cores, the intrinsic parallelization of the BAFD model used and described earlier.

5 Numerical Experiments: Results

This section presents the results of numerical experiments with the block structure outlined in Sect. 4. In the process, the emphasis was made on applying the calibration procedure (9)–(12). We varied two sets of parameters mentioned above in a special way.

As key values, we used the ones tested in [6] on the base of the assumptions common in the BAFD models about the observed seismicity: the coefficients K_t, K_l and K_n are increased, whereas the coefficients W_t, W_l and W_n are decreased for faults with a high level of seismic activity and vice versa. As to the random factors, we assumed that the critical value of ratio κ (4) should be approximately 0.1 and the noise amplitude in the quasi-linear stochastic equations should be at least an order less than the displacement values (the null value of the amplitude corresponds to the noise absence).

As an example, we give the ranges and steps for changing the parameters of the largest subset of the faults: for the coefficients, $K_t, K_l, K_n \in [1, 10]$, $\Delta_K = 1$, $W_t, W_l, W_n \in [0.01, 0.05]$, $\Delta_W = 0.01$, and $W_t^s, W_l^s, W_n^s \in [5, 10]$, $\Delta_s = 5$; for the noise characteristics, $\lambda_t, \lambda_l, \lambda_n \in [0, 0.1]$, $\Delta_\lambda = 0.05$ and $B_0 \in [0.1, 0.2]$, $\Delta_B = 0.1$, $\sigma \in [0, 0.05]$, $\Delta_\sigma = 0.05$. In the variation process, the ratio between coefficients at different subsets of faults is preserved. Thus, we have a grid of dimension $10 \times 5 \times 2 \times 3 \times 2 \times 2$. Let us remind that the main purpose of the calculations is to find (by an exhaustive or optimized search) a variant that minimizes functional (12).

The simulation was performed at the Krasovskii Institute of Mathematics and Mechanics of UB RAS by a hybrid cluster type machine named Uran (1940 Intel Xeon CPUs (used in experiments) and 314 NVIDIA Tesla GPUs; the available operative memory is more than 4 Tb, and the peak performance is about 215 TFlops). For the calculations, we generally used 50 cores per a variant (for the model intrinsic parallelization) and simultaneously run up to 10 variants. For a typical variant till 100 units of dimensionless time, it took about 80 min instead of 68 h for calculations on one core. Thus, due to the effective model parallelization, we managed to examine 1200 variants for 160 h.

Figure 1 shows the plot of the speedup and efficiency (of the intrinsic parallelization) dependence on the number of cores. Here, speedup S_p and efficiency E_p are calculated, respectively, by the formulas $S_p = T_1/T_p$, $E_p = S_p/p$, T_1 is the performance time for the sequential algorithm, T_p is the performance time for the parallel algorithm on p cores. The speedup is equal to the number of engaged cores for an ideal parallel algorithm; in this case, we have the unit efficiency.

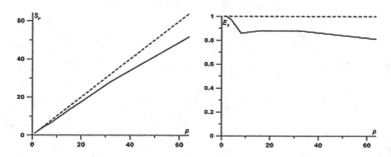

Fig. 1. The dependence of speedup S_p (left) and efficiency E_p (right) on the number of cores p: real plots are marked by a solid line, ideal values, by dashed; horizontal axes show the number p, vertical, the values S_p and E_p

It follows from Fig. 1 that the parallelization efficiency is rather high and does not fall below an acceptable level if the number of engaged processors increases (for example, if $p = 64$ we get $S_p = 51.84$ and $E_p = 0.81$).

During the examination, we detected the variant minimizing functional (12) (the optimal one according to the procedure suggested; hereinafter, variant **A**). Its key parameters are as follows: $K_t = K_l = K_n = 8$, $W_t = W_l = W_n = 0.02$, $W_t^s = W_l^s = W_n^s = 10$, $\lambda_t = \lambda_l = \lambda_n = 0.1$, $B_0 = 0.1$, $\sigma = 0$ (the last value means that the noises in level B are absent).

The simulation results for variant **A** comparing with the real data can be found below. The comparative analysis of spatial distributions between the epicentres of strong events recorded in the real catalogue [14], including the events for the period 01.01.1900–31.12.2018 without any restrictions on depth and area (Fig. 2), and those in the model catalogue (Fig. 3), shows that a number of common qualitative features revealed earlier [5,6] is preserved. In particular, the most important patterns of the global seismicity should be noted: two main seismic belts, the Circum-Pacific and Alpine-Himalayan, where most of the strong earthquakes occur, and the increased seismic activity associated with triple junctions of plate boundaries. It seems very promising that the strongest events in the model occur approximately at the same places as in reality. The distinctions between the synthetic and real seismicity are also obvious: the absence of model events inside the plates (in the model, earthquakes only occur at faults) and at some boundaries (for example, the north and east of African plate).

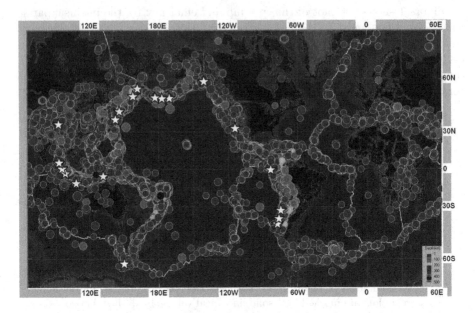

Fig. 2. Registered seismicity: epicentres of strong earthquakes with $M \geq 6.0$, NEIC, 01.01.1900–31.12.2019 [14], 20 strongest events are marked with stars

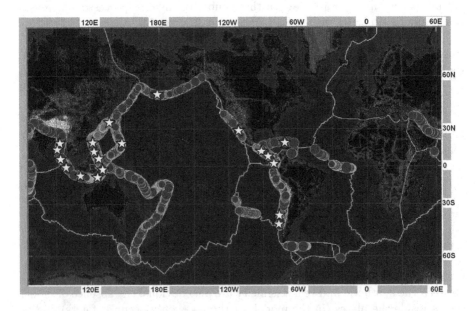

Fig. 3. Synthetic seismicity: epicentres of strong earthquakes with $M \geq 6.0$, 100 units of model time, variant **A**, 20 strongest events are marked with stars

To compare the distributions in Fig. 2 and Fig. 3 quantitatively, we restricted the real one by removing all events inside the plates. Then we found the Hausdorff distance between the two sets according to (9). Here, $d_H(E_r, E_m) = 3450$ km. In Fig. 4, one can see the points providing this value. The possible explanation of a rather large value of this criterion (even for the best variant) is in the fact that, as a rule, there are no model earthquakes at some boundaries since real events there, maybe, are not stipulated by reasons taken into account in the BAFD models as, for example, the geometric incompatibility in a local fault system, see [15].

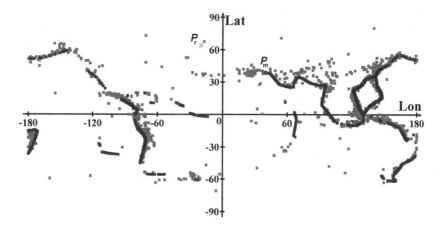

Fig. 4. Two sets of epicentres, real (red) and model (blue), for calculating the Hausdorff distance. The points providing d_H are marked by P_r and P_m (Color figure online)

As to the distribution in depth (see Table 1), we observe both the noticeable similarity of the shallow earthquake share in the real and synthetic seismicity and the essential increase in the shares of model events for average depths. The last fact is mainly explained by the specified values of block depths (most of them range from 30 to 50 km). Here, we have $d_D(D_r, D_m) = 0.36$. It is evident that to redistribute the events, we need to get more specific information on how the model parameters change depending on the depth. On the other side, the block model is mainly intended to model the events just in the surface layer of the Earth's crust.

We analysed the parameters of the Gutenberg–Richter law characterizing the earthquake power distribution in magnitude. Figure 5 shows the FM plots on a logarithmic scale for the real and synthetic seismicity, while Table 2 contains some quantitative results. The real and model FM plots are approximated by the linear least-squares regressions $\lg N = c_r - S_r M$ and $\lg N = c_m - S_m M$, respectively. The values of S_r and S_m serve as the slope estimates, whereas c_r and c_m are constants. The average distances between the plot points and the constructed lines are treated as approximation errors A_r and A_m. For all the

Table 1. Distribution of earthquakes in depth (in shares with respect to the total number of events with $M \geq 4.0$): NEIC and variant **A** catalogues for 100 time units

Depth	NEIC	Variant **A**
up to 10 km	0.16	0.14
[10, 40 km]	0.47	0.65
over 40 km	0.37	0.21

model variants tested, the magnitude interval $[5.0, 8.5]$ of a sufficient linearity is considered; for the NEIC catalogue, the whole range is $[4.0, 9.0]$. The FM plot for the registered global seismicity is almost linear and its slope is very close to 1. The slope estimates and approximation errors are presented in Table 2. Here, we have $d_G(G_r, G_m) = 0.125$ for $\alpha_1 = \alpha_2 = 0.5$.

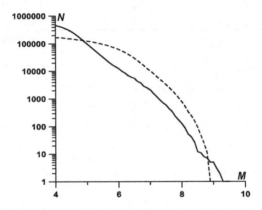

Fig. 5. The FM plots constructed for real (NEIC, events with $M \geq 4.0$, 01.01.1900–31.12.2018, *solid line*) and model (variant **A**, events with $M \geq 4.0$, 100 time units, *dashed line*) catalogues; N is the accumulated number of earthquakes, M is the magnitude (color figure online)

Table 2. Simulation results: NEIC and variant **A** catalogues

Parameters	NEIC	Variant **A**
Number of events	463811	167763
Slope estimate S	1.0	0.86
Approximation error A	0.16	0.27

We should note that the linearity of a model plot even in some magnitude range is a positive sign since it testifies to the features of the distribution law for model events, which corresponds to the real distribution and, thereby, provides the possibility of studying real patterns using a synthetic catalogue, see [16]. This quality index has been essentially improved compared to the previous versions of the model [4–6], although an insufficient (with respect to the real seismicity) relative number of small events (with magnitude 4.0–4.5) remains the same causing almost horizontal part of the model plot in the range of small magnitudes. The experiments confirmed that the share of small events increased when disturbing the Eq. (2). The less total number of events in variant **A** comparing with NEIC is explained by the fact that there are no model earthquakes inside the plates. As to the events nearby the faults, the real and model numbers are similar. It seems that a key factor for increasing the interval of linearity of an FM plot and changing its slope is the model calibration (with respect to the coefficients in Eqs. (1), (2) for faults and block bottoms, the noise levels in relations (4)–(6), the steps of temporal and spatial discretization, etc.) We hope that the procedure suggested in this paper is useful for this objective. As to the resulting value of criterion (12), we obtained the maximum distances in the sense of (9)–(11) $d_H = 5360$ km, $d_D = 0.66$ and $d_G = 1.1$; so we get $Er \approx 0.43$ for $\beta_1 = \beta_2 = \beta_3 = 1/3$. It is interesting to note that variant **A** is not the best by each of three criteria (9)–(11) but the best according to their sum (12).

Summarizing the brief analysis of the conducted experiments, we conclude that the application of the suggested procedure for automatic tuning of the model parameters in order to obtain the best approximation results when simulating the global seismicity can be considered as a perspective. Obviously, the quality criterion (9)–(12) used in the calculations should be treated as an example of possible relations of such kind. In addition, the new stochastic quasi-linear term in Eq. (2) and the random temporal fluctuations of strength levels (5), which reflect the fault medium strength variability (not amenable to a precise analytical description and actually non-predictable) were tested during the experiments. According to the parameter values for variant **A**, we can preliminarily assume that quasi-linear equations are more suitable than linear ones (since the properties under investigation were improved in comparison with [6]), whereas the necessity to disturb the strength levels is rather disputable.

6 Conclusive Remarks

For the last modification of the spherical BAFD model, different ways of taking into account random factors that influence the behaviour of block structure elements were tested and some results in favour of the quasi-linear equations describing the displacement evolution were obtained. The main objective of the presented research to develop a feasible procedure enabling automatic tuning of the model parameters to obtain the best approximation results concerning the global seismicity. A special quality criterion was introduced based on the minimization of the weighted sum of discrepancies between the model characteristics and the real patterns such as the spatial/depth distribution of epicenrers and the Gutenberg–Richter law on frequency-magnitude relation. During the numerical experiments, we found the best variant according to this criterion. Its parameters are in agreement with our assumptions. This fact testifies to sufficient adequacy of the current spherical BAFD model modification and to the necessity of advanced research in this field with the use of the elaborated procedure.

References

1. Gabrielov, A.M., Newman, W.I.: Seismicity modeling and earthquake prediction: a review. Geophys. Monograph 83, **18**, 7–13 (1994). IUGG, Washington
2. Keilis-Borok, V.I., Soloviev, A.A. (eds.): Nonlinear Dynamics of the Lithosphere and Earthquake Prediction. Springer, Heidelberg (2003). https://doi.org/10.1007/978-3-662-05298
3. Ismail-Zadeh, A.T., Soloviev, A.A., Sokolov, V.A., Vorobieva, I.A., Muller, B., Schilling, F.: Quantitative modeling of the lithosphere dynamics, earthquakes seismic hazard. Tectonophysics **746**, 624–647 (2018)
4. Rozenberg, V.L., Sobolev, P.O., Soloviev, A.A., Melnikova, L.A.: The spherical block model: dynamics of the global system of tectonic plates and seismicity. Pure. Appl. Geophys. **162**, 145–164 (2005). https://doi.org/10.1007/s00024-004-2584-4
5. Melnikova, L.A., Rozenberg, V.L.: A stochastic modification of the spherical block-and-fault model of lithosphere dynamics and seismicity. Numer. Meth. Program. **16**, 112–122 (2015). (in Russian)
6. Melnikova, L., Mikhailov, I., Rozenberg, V.: Simulation of global seismicity: new computing experiments with the use of scientific visualization Software. In: Sokolinsky, L., Zymbler, M. (eds.) PCT 2017. CCIS, vol. 753, pp. 215–232. Springer, Cham (2017). https://doi.org/10.1007/978-3-319-67035-5_16
7. Melnikova, L.A., Rozenberg, V.L.: Algorithm of dynamical input reconstruction for a stochastic differential equation: tuning of parameters and numerical experiments. Bull. South Ural State Univ. Ser. Comput. Math. Software Eng. **8**(4), 15–29 (2019). (in Russian) https://doi.org/10.14529/cmse190402
8. Oksendal, B.: Stochastic Differential Equations: an Introduction with Application. Springer, New York (1998). https://doi.org/10.1007/978-3-662-03620-4. Mir, Moscow (2003)
9. Aki, K., Richards, P.G.: Quantitative Seismology: Theory and Methods. Freeman, San Francisco (1980). https://doi.org/10.1002/gj.3350160110. Mir, Moscow (1983)
10. Wells, D.L., Coppersmith, K.L.: New empirical relationships among magnitude, rupture length, rupture width, rupture area, and surface displacement. Bull. Seism. Soc. Am. **84**(4), 974–1002 (1994)
11. Gilboa, I.: Theory of Decision under Uncertainty. Cambridge University Press, Cambridge (2009). https://doi.org/10.1017/CBO9780511840203
12. Gripp, A.E., Gordon, R.G.: Young tracks of hotspots and current plate velocities. Geophys. J. Int. **150**, 321–361 (2002). https://doi.org/10.1046/j.1365-246x.2002.01627.x
13. Gergel, V.P.: Theory and Practice of Parallel Computing. Binom, Moscow (2007). (in Russian)
14. Global Hypocenters Data Base, NEIC/USGS, Denver, CO. http://earthquake.usgs.gov/regional/neic/

15. Gabrielov, A.M., Keilis-Borok, V.I., Jackson, D.D.: Geometric incompatibility in a fault system. Proc. Natl. Acad. Sci. U.S.A. **93**(9), 3838–3842 (1996). https://doi.org/10.1073/pnas.93.9.3838

16. Soloviev, A.A.: Transformation of frequency-magnitude relation prior to large events in the model of block structure dynamics. Nonlin. Processes Geophys. **15**, 209–220 (2008). https://doi.org/10.5194/npg-15-209-2008

Parallel Computational Algorithm for the Simulation of Flows with Detonation Waves on Fully Unstructured Grids

Alexandr I. Lopato[1]($\boxtimes$) (iD), Artem G. Eremenko[2]($\boxtimes$) (iD), Pavel S. Utkin[1]($\boxtimes$) (iD), and Dmitry A. Gavrilov[2]($\boxtimes$) (iD)

[1] Institute for Computer Aided Design, Russian Academy of Sciences, Moscow, Russia
lopato2008@mail.ru, pavel_utk@mail.ru
[2] Moscow Institute of Physics and Technology, Dolgoprudnyi, Russia
artem-temoff@mail.ru, gavrilov.da@mipt.ru

Abstract. Numerical simulation of flows with detonation waves (supersonic regime of combustion) in the computational domains, corresponding to the realistic technical systems, does not allow using classical structured grids and numerical schemes. One of the solutions is using fully unstructured grids. In relation to gas dynamics problems of chemically inert media, the apparatus of high approximation order schemes on fully unstructured computational grids has been developed and the shock waves problems have been formulated. The analysis of previous publications about the application of such approaches to the study of high-speed flows with chemical reactions revealed several papers only. This work is dedicated to the development of the parallel computational algorithm for the simulation of two-dimensional flows with detonation waves on fully unstructured grids with the examples of its application to the solution of the cellular detonation wave propagation problem.

Keywords: Detonation wave · Numerical simulation · Unstructured grid · Parallel computations · CFD General Notation System (CGNS) · ParMETIS programs for parallel graph partitioning

1 Introduction

Mathematical modeling of detonation processes is an important tool for clarifying the detonation waves' (DW) propagation mechanisms, reducing the risks and costs of full-scale experiments. As noted in the review [1], the majority of the works on numerical studies of gas detonation problems are carried out on structured computational grids. However, the usage of structured grids is limited to very simple computational domains. The complex shaped computational domains enhance the role of unstructured grids. In addition, programming of adaptive mesh refinements (AMR) approach for unstructured grids is sometimes relatively easier than that for structured grids, as discussed in [1–3]. On the

© Springer Nature Switzerland AG 2020
L. Sokolinsky and M. Zymbler (Eds.): PCT 2020, CCIS 1263, pp. 198–208, 2020.
https://doi.org/10.1007/978-3-030-55326-5_14

other hand, the methods based on the unstructured grids usage are generally more diffusive than those based on the structured grids (see [1], for example). The accuracy of the grid function gradient reconstruction on the unstructured grid is less than for on the structured one, especially for the rapidly changing physical values. As a result, to solve the strong shock waves (SW) problems with detonation one may need a finer resolution than that achievable for the structured grids.

The question of the required unstructured grid resolution and comparison with the results obtained from simulation of detonation in the H_2/Air mixture using a structured grid was considered in [1]. The formation of cellular detonation in the channel in a two-dimensional (2D) case was studied. The series of calculations on unstructured grids with the average cell size 1, 2.5, 3, 5 μm and calculation on the structured grid with the cell size 5 μm were carried out. The results showed that the unstructured grid simulation carried out with the same grid resolution as for the structured grid simulation could not capture the same DW features. Some elements of the solution, such as triple point, were proven not to have been fully and clearly captured by using the unstructured grid with the cell size 5 μm. The effect of the grid resolution on the transverse waves depended not only on the resolution around the detonation front, but also on the resolution around the shear layer behind the detonation front. The simulations carried out with the cell size of 2.5 μm and more could not capture the vortices in the shear layer area. On the other hand, the calculation with the cell size 1 μm captured the vortices like the 5 μm grid. Thus, an insufficient grid resolution can cause errors in the DW structure description. The authors noted that the most important advantage of the unstructured grids is the easier AMR application that allows adding/removing grid nodes without big code modifications.

The paper [4] presents a mathematical model, the numerical method and a parallelization technique for the problems of detonation initiation and propagation in complex-shaped three-dimensional tubes. Based on the three-dimensional computational domain decomposition, the quantitative characteristics of parallelization quality are presented. The authors used unstructured computational grids (sets of structured parts) and the second approximation order numerical algorithm, but the details of grid function interpolation on the unstructured grids need to be explained.

We proposed a second order numerical algorithm for the simulation of flows with DW in [5]. The algorithm was applied to the simulation of DW initiation and propagation in a variable cross-section axisymmetric channel filled with the hydrogen-air mixture. The channel modeled a full-scale device for the worn-out tire utilization. In [6], the mechanisms of detonation initiation in multi-focused systems were investigated using the proposed algorithm. The objective of this work is parallelization of the numerical algorithm proposed in [5].

2   Mathematical Model and Numerical Method

The mathematical model is based on 2D Euler equations written in the Cartesian frame supplemented by one-stage chemical kinetics of hydrogen combustion in the air [7]:

$$\frac{\partial \mathbf{U}}{\partial t} + \frac{\partial \mathbf{F}}{\partial x} + \frac{\partial \mathbf{G}}{\partial y} = \mathbf{S},$$

$$\mathbf{U} = \begin{bmatrix} \rho \\ \rho u \\ \rho v \\ e \\ \rho Z \end{bmatrix}, \ \mathbf{F} = \begin{bmatrix} \rho u \\ \rho u^2 + p \\ \rho u v \\ u(e+p) \\ \rho Z u \end{bmatrix}, \ \mathbf{G} = \begin{bmatrix} \rho v \\ \rho u v \\ \rho v^2 + p \\ v(e+p) \\ \rho Z v \end{bmatrix}, \ \mathbf{S} = \begin{bmatrix} 0 \\ 0 \\ 0 \\ 0 \\ \rho \omega \end{bmatrix}, \quad (1)$$

$$e = \rho \epsilon + \frac{\rho}{2}(u^2 + v^2), \ \epsilon = \frac{p}{\rho(\gamma-1)} + ZQ, \ p = \frac{\rho}{\mu}RT, \ \omega = -A\rho Z \exp(-\frac{E}{RT}).$$

Here ρ is the gas density, u and v are the velocity components, p is the pressure, ϵ is the specific internal energy of gas, e is total energy per unit of volume, Q is the heat released in the chemical reaction, Z is the mass fraction of the reactive mixture component, ω is the chemical reaction rate, A is the preexponent factor, E is the activation energy, μ is the molar mass, R is the universal gas constant, and T is the temperature. The gas is assumed to be ideal with the specific heat ratio γ. The following parameters of the mixture are used [7]:

$$\gamma = 1.17, \ \mu = 21 \ \frac{g}{mol}, \ Q = 5.02 \ \frac{MJ}{kg}, \ E = 113 \ \frac{kJ}{mol}, \ A = 6.85 \cdot 10^9 \ \frac{m^3}{kg \cdot s}.$$

The distinctive feature of the computational technique is numerical solution of the governing equations on completely unstructured computational grids with triangular cells. The computational algorithm is based on the Strang splitting principle in terms of physical processes, namely, convection and chemical reactions. The spatial part of (1) is discretized using the finite volume method. The numerical flux is calculated using Advection Upstream Splitting Method (AUSM) [8] extended for the a two-component mixture case. It should be noted that choosing AUSM for flux calculation is not a compulsory requirement. Thus, the simulations of the flows with DW in [9] were successfully carried out with the Courant-Isaacson-Rees numerical flux. To increase the approximation order, the grid functions were reconstructured [10] and a minmod limiter was applied. Time integration is carried out with the second-order Runge-Kutta method [11]. The time step is chosen dynamically based on the stability criterium. At the second stage of the algorithm, a system of ordinary differential chemical kinetics equations for Z variable and temperature in each computational cell of the grid is solved.

The detailed numerical algorithm description with the verification results is presented in [5].

Numerical experiments consist of following steps:

- computational grid construction;
- computational domain and computational grid decomposition for parallel computation;

- parallel supercomputer computation;
- visualization and processing of the results.

3   Computational Grid Construction

In this work, the computational grid was built with the open-source SALOME software [12] which provides ample opportunities for the computational domain geometry construction and filling it with cells of various types. The software can be downloaded and installed on Linux platforms and has been experimentally modified for Windows.

The computational domain was built in the GEOM module. MESH grid construction module of SALOME provides a wide range of algorithms, particularly suited for finite-element and finite-volume methods. Group naming function provides for the local boundary identification. Various files formats facilitate the result output visualization or other post-processing operations. SALOME provides several ways for the triangular computation grid generation. In particular, there are three main triangulation algorithms: Projection 1D–2D, Delaunay, and Frontal. In this study, the computation was carried out with the Projection 1D–2D grid generation algorithm.

The algorithm is based on the advancing front method. The computational grid generation starts from the boundary of the computational domain. Nodes on the domain boundaries are searched by recursive splitting of the region in halves to the fineness, limited by the algorithm parameters (reference size of the triangle edge). Then the iterative process of searching for nodes connected to the boundary edges takes place. The nodes form a new layer of triangles and a new border, separating the area with the generated grid from the area without it. As a result, the current algorithm status is always represented by the advancing boundary front. The coordinates of the new boundary nodes are determined by the area in front of the advancing boundary, in accordance with the rules and criteria presented in [13].

The result of using this and other two algorithms is shown in Fig. 1. The triangulation region is assumed to be a rectangle of $0.1 \times 0.02\,\mathrm{m}^2$. In SALOME, the edge length of $50\,\mu\mathrm{m}$ is specified as the reference length of the triangle edges. Delaunay triangulation distributes of the maximum triangles edge lengths similarly to the normal Gauss distribution. For the Frontal algorithm, all the constructed triangular cells have the same maximum edge length value. In Projection 1D-2D triangulation, a part of the cells has the maximum edge length equal to the selected reference value in SALOME, with the length of the remaining cells being higher and satisfying a certain distribution, and the dominating number of triangles (about 75%) having the maximum edge length of about $87\,\mu\mathrm{m}$. Thus, the selected algorithms produce qualitatively different triangle distribution patterns in terms of geometrical parameters. Selection of the optimal algorithm for the computations depends on many factors, including triangulation time, computational grid quality, computational domain geometry, and computational algorithm features when dealing with unstructured grids.

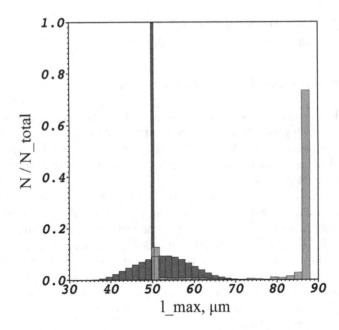

Fig. 1. The percentage distribution of the maximum triangle lengths for Delaunay algorithm (red color), Frontal algorithm (green color), Projection 1D-2D algorithm (blue color). The horizontal axis corresponds to the maximum triangle edge length in μm. The vertical axis corresponds to the percentage of triangles with the selected maximum edge length (Color figure online)

The MESH module in SALOME software provides an option of creating a group of grid entities and edges in particular. Create a group called a "wall" for the edges that form the boundary of the computation domain – rectangle. For these edges, the slip-conditions are implemented. The groups are exported along with grid objects to a small number of file formats, one of which is Computational Fluid Dynamics (CFD) General Notation System (CGNS) [14].

4   CGNS

CGNS provides a standard for recording and recovering the computer data associated with the numerical solution of CFD problems [14]. The intent is to facilitate the exchange of CFD data between sites, applications codes, and across computing platforms, and to stabilize the archiving of CFD data. CGNS Application Program Interface (API) is an independent platform including embedded C/C++ application. The data are stored in a compact binary format and are accessible through a complete and extensible library of functions.

A CGNS file containing a grid is a hierarchical data type with the following structure. This entity is organized into a set of tree-structured nodes according to certain rules that allow users to easily access the necessary information. The rules are described in Standard Interface Data Structures (SIDS) [14]. The topmost node is called the "root node". Each node can be a "parent" for one or more "child" nodes. A node can also have a child node referenced to a node elsewhere in the file or to a single node. The links are transparent to the user: the user "sees" the associated child nodes as if they really exist in the current tree. The structure of the file with the studied grid is shown in Fig. 2.

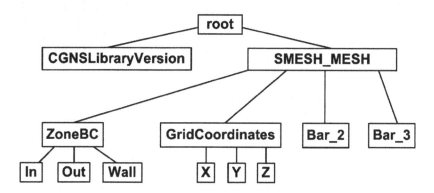

Fig. 2. Hierarchical data type of the CGNS file format

The first level of the CGNS file hierarchical structure determines its version – section CGNSLibraryVersion – and dimension of the problem – section SMESH_MESH. CGNS file version defines API for the subsequent dealing with the file structure and possible node types. In this work, version 3.4.0 [15] is used. The next level characterizes the geometrical model, namely, the number of the boundary vertices (section Bar_2), the number of cells (section Bar_3), the coordinates of nodes (section GridCoordinates) and the boundary conditions (section ZoneBC). For various problems and geometries, three types of boundary conditions are introduced: "in" corresponds to the inflow conditions, "out" corresponds to the extrapolation conditions and "wall" corresponds to the slip-conditions.

5   Parallelization

For parallelization, the computational grid as an unstructured graph is divided into subdomains by the number of computational cores using ParMETIS open source software [16]. It should be noted that ParMETIS is quite demanding to the parameters of the working environment, and choosing a suitable operation system and a compiler configuration may present a serious challenge. So, in this study the stable work of ParMETIS was achieved by the configuration with RedHat linux and gcc version 4.1.2. Attempts to run ParMETIS on Ubuntu or another system with gcc of another version were unsuccessful. The partitioning procedure would constantly fail due to the memory leaks. The analysis of different configurations showed that MacOSx systems are suitable for partitioning domains using ParMETIS, but the systems are not widespread for clusters.

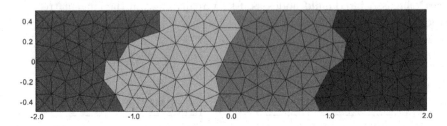

Fig. 3. The example of the computational domain partition into four subdomains using ParMETIS. Axis are in centimeters

The objective of the graph partitioning problem is to compute a k-way partitioning that would ensure the minimized number of edges that straddle different partitions. This objective is commonly referred to as the edge-cut. When partitioning is used to distribute a graph or a grid among the computational cores of a parallel computer, the objective of minimizing the edge-cut is only an approximation of the true communication cost resulting from the partitioning. Based on this concept, ParMETIS provides the mpmetis program for partitioning grids arising in the finite element or finite volume methods. As an input, the program uses the element-node array of the grid and computes a k-way partitioning for both of its elements and nodes. The program first converts the grid into either a dual graph or a nodal graph and then uses the graph partitioning API routines for graph partitioning. ParMETIS returns the file with the grid partition. According to the file, the grid parts are sent to the appropriate cores. The example of partition of the rectangular-shaped computational domain $0.04 \times 0.01 \, \mathrm{m}^2$ into four subdomains using ParMETIS is shown in Fig. 3.

Due to the explicit integration scheme with respect to time, the data exchange of conservative vectors are required in the cells with common edges for different cores and in the neighboring cells because of the second approximation order of the method. The smaller the number of boundary edges, the smaller is the amount of the boundary triangles data to be sent to the computational cores. For the given example, the groups of cells exchanged between the neighboring cores are connected with dashed lines in Fig. 4. The cells involved in the calculation of the flux through the edge highlighted by the bold line are marked in dots. The programming solution of parallelization, including computational cores data exchanges for the correct calculation of fluxes at each stage of the Runge-Kutta method, is carried out using Message Passing Interface (MPI) library. Note that a similar approach in terms of the combination of "CGNS + METIS + MPI" was used in [4] for the DW simulation.

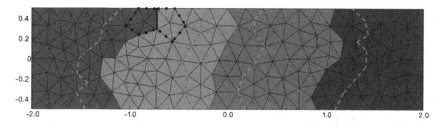

Fig. 4. The example of the computational domain partition into four subdomains using ParMETIS with marked cells for the computational cores data exchanges. Axis are in centimeters

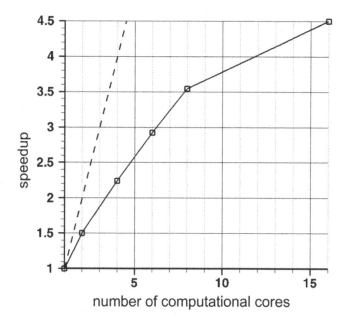

Fig. 5. Speedup estimation. Dashed line shows ideal speedup. Solid line presents simulations

In this work, the simulations were carried out using MVS-100K supercomputer at the Joint Supercomputer Center of RAS (JSCC RAS). This supercomputer supports MPI and contains about two thousand computational cores that allows user to build grids with a large number of cells. More details about the characteristics of the supercomputer can be found in [17]. To estimate the speedup of the parallel algorithm, several simulations on different number of computational cores on the supercomputer MVS-100K we carried out. The minimum number of cores throughout the test was 1. The maximum number of cores was 16. The problem from the Sect. 6 with the number of computational cells of about 1 million was considered. Figure 5 illustrates the estimation of the algorithm speedup. Apparently, not very good results on the speedup are caused by imperfect programming solutions and need to be improved.

To visualize the results obtained, the Visit software [18] was used. Earlier versions of this software had problems with CGNS files, but with the introduction of version 2.13.3, these problems have disappeared.

6   Numerical Experiments

Let us consider the initiation and propagation of DW in a plane channel with the width $H = 1$ cm and the length $L = 10$ cm, filled with a quiescent reactive hydrogen-air mixture. Detonation is initiated as a result of instantaneous energy release in a short region with the length $l = 2$ mm and the entire width of the channel adjacent to the left wall of the channel. In this region, the higher values of $T_l = 1500$ K and pressure $p_l = 40$ atm are set at the initial time moment with respect to the rest of the channel. The computations are performed on the grid containing about 4 million triangular cells. The energy in the initiation region is enough to form an overdriven planar DW, which begins to propagate in the channel. To visualize the flow pattern we use the "numerical soot footprints" (see Fig. 6). To construct it, the maximum pressure in each cell for the entire time is plotted. Such visualization corresponds to the experimental method of DW multidimensional structures recording using smoked foil or plates installed on the wall of the channel used for the DW propagation. At the distance of approximately 6 cm from the initiation region, the DW front changes its planar nature and the system of transverse waves typical of the two-dimensional detonation appears. In the computations describing the appearance of this structure, we do not use any disturbing factors. The trigger to the emergence of the so-called DW cellular structure is the unstructured type of computational grid that creates the small vertical components of gas velocity. At the same time, it is known that even in a uniform computational grid with rectangular cells the cellular structure is formed in the same manner and the triggering mechanism in this case is a set of errors of machine calculations (for example, see [19]). Despite the fact that the appearance of detonation cells is caused by the arbitrary factors, the characteristic scale of the transverse size cells is already determined by the macroscopic parameters of the problem that is typical of the hydrodynamic instability development problems.

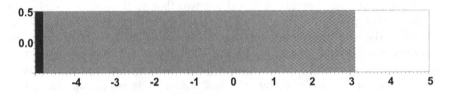

Fig. 6. "Numerical soot footprints" in the initiation and propagation of the DW in the plane channel. Axis are in centimeters

The direct initiation of detonation leads to the overdrive of the DW propagation. Then, a typical pattern of DW decrease to Chapman-Jouguet parameters takes place. The velocity of the DW front gets close to the theoretical value:

$$D_{CJ} = \sqrt{\gamma + \frac{1}{2}(\gamma^2 - 1)Q} + \sqrt{\frac{1}{2}(\gamma^2 - 1)Q} \approx 1925\,\frac{m}{s}.$$

The problem of the DW propagation in the plane channel is the standard one in the detonation theory and can be considered using the general finite volume approach and the Cartesian grid. In [7], the channel with the height of 2 cm and the length about 60 cm was considered. It was a Cartesian grid with adaptive refinement. The comparison of the obtained simulation results with [7] will be discussed in the further study.

7   Concluding Remarks

The parallel computational algorithm was developed for the study of two-dimensional flows with detonation waves on completely unstructured computational grids. The main stages of the computational experiment, such as construction of the computational grid, decomposition of the computational domain and the computational grid for parallel computation, parallel computation on the supercomputer, visualization and processing of the results are considered. Construction of the computational grid is carried out using the SALOME software. The MESH grid construction module of the SALOME software provides a wide range of algorithms. Based on the test problem example, three triangulation algorithms were compared. The structure of the CGNS file format relevant for our computations is described. The parallelization of the algorithm is carried out by decomposition of the computational domain. The computational grid is divided into subdomains using the ParMETIS open source software. The important features of ParMETIS usage including the requirements to the configuration of the operating system are considered. For the correct execution of the computational algorithm the data exchange between groups of cells at the boundaries of subdomains is provided using MPI library. The parallel simulations were performed using MVS-100K supercomputer at the JSCC RAS.

The work demonstrates the possibility of numerical modeling of the initiation and propagation of a gaseous detonation wave in the two-dimensional statement on completely unstructured computational grid with triangular cells.

Acknowledgments. The work of A.I. Lopato and A.G. Eremenko is supported by the Council for Grants of the President of the Russian Federation for the State Support of Young Russian Scientists (project no. MK-244.2020.1). The work of P.S. Utkin is carried out under the state task of the ICAD RAS.

References

1. Togashi, F., Lohner, R., Tsuboi, N.: Numerical simulation of H_2/air detonation using unstructured mesh. Shock Waves **19**, 151–162 (2009). https://doi.org/10.1007/s00193-009-0197-7

2. Loth, E., Sivier, S., Baum, J.: Adaptive unstructured finite element method for two-dimensional detonation simulations. Shock Waves **8**(1), 47–53 (1998). https://doi.org/10.1007/s001930050097

3. Figueira-da-Silva, L.F., Azevedo, J.L.F., Korzenowski, H.: Unstructured adaptive grid flow simulations of inert and reactive gas mixtures. J. Comput. Phys. **160**(2), 522–540 (1998). https://doi.org/10.1006/jcph.2000.6470

4. Semenov, I.V., Akhmedyanov, I.F., Lebedeva, A.Y., Utkin, P.S.: Three-dimensional numerical simulation of shock and detonation waves propagation in tubes with curved walls. Sci. Technol. Energetic Mater. **72**, 116–122 (2010)

5. Lopato, A.I., Utkin, P.S.: Numerical study of detonation wave propagation in the variable cross-section channel using unstructured computational grids. J. Combust. **3635797**, 1–8 (2018). https://doi.org/10.1016/0021-9991(89)90222-2

6. Utkin, P.S., Lopato, A.I., Vasil'ev, A.A.: Numerical and experimental investigation of detonation initiation in multifocused systems. In: Proceedings of 27th International Colloquium on the Dynamics of Explosions and Reactive Systems (ICDERS), Paper 105. Beijing, China (2019)

7. Gamezo, V., Ogawa, T., Oran, E.: Flame acceleration and DDT in channels with obstacles: effect of obstacle spacing. Combust. Flame **155**, 302–315 (2008). https://doi.org/10.1016/j.combustflame.2008.06.004

8. Liou, M.-S., Steffen Jr., C.J.: A new flux splitting scheme. J. Comput. Phys. **107**, 23–39 (1993). https://doi.org/10.1006/jcph.1993.1122

9. Lopato, A.I., Utkin, P.S.: The usage of grid-characteristic method for the simulation of flows with detonation waves. In: Petrov, I.B., Favorskaya, A.V., Favorskaya, M.N., Simakov, S.S., Jain, L.C. (eds.) GCM50 2018. SIST, vol. 133, pp. 281–290. Springer, Cham (2019). https://doi.org/10.1007/978-3-030-06228-6_23

10. Chen, G., Tang, H., Zhang, P.: Second-order accurate Godunov scheme for multi-component flows on moving triangular meshes. J. Sci. Comput. **34**, 64–86 (2008). https://doi.org/10.1007/s10915-007-9162-8

11. Shu, C.W., Osher, S.: Efficient implementation of essentially non-oscillatory shock-capturing schemes. J. Comput. Phys. **77**, 439–471 (1988). https://doi.org/10.1016/0021-9991(89)90222-2

12. Salome. https://www.salome-platform.org/

13. Schoberl, J.: An advancing front 2D/3D-mesh generator based on abstract rules. Comput. Vis. Sci. **1**, 41–52 (1997). https://doi.org/10.1007/s007910050004

14. CGNS. https://cgns.github.io/

15. CGNS version 3.4.0. https://github.com/CGNS/CGNS/releases/tag/v3.4.0/

16. ParMETIS. http://glaros.dtc.umn.edu/gkhome/metis/parmetis/overview/

17. JSCC resources. https://www.jscc.ru/resources/hpc/

18. Visit. https://wci.llnl.gov/simulation/computer-codes/visit/

19. Gamezo, V.N., Desbordes, D., Oran, E.S.: Two-dimensional reactive flow dynamics in cellular detonation waves. Shock Waves **9**, 11–17 (1999). https://doi.org/10.1007/s001930050

Using Supercomputer Technologies to Research the Influence of Abiotic Factors on the Biogeochemical Cycle Variability in the Azov Sea

Alexander I. Sukhinov[1], Yulia V. Belova[1] (ID), AlexanderE. Chistyakov[1,5] (ID), Alena A. Filina[2,5] (ID), Vladimir N. Litvinov[1,3] (ID), Alla V. Nikitina[2,4,5(✉)] (ID), and Anton L. Leontyev[2,4,5] (ID)

[1] Don State Technical University, Rostov-on-Don, Russia
sukhinov@gmail.com, cheese_05@mail.ru
[2] Supercomputers and Neurocomputers Research Center, Taganrog, Russia
j.a.s.s.y@mail.ru, leontyev_anton@mail.ru
[3] Azov-Black Sea Engineering Institute of Don State Agrarian University,
Zernograd, Russia
litvinovvn@rambler.ru
[4] Southern Federal University, Rostov-on-Don, Russia
nikitina.vm@gmail.com
[5] Science and Technology University "Sirius", Sochi, Russia

Abstract. The paper covers the mathematical modelling of the main biogenic matter transformations in the production-destruction processes of phytoplankton populations in the Azov Sea, taking into account the influence of external factors, including the salinity and temperature. A multi-species mathematical model of the phytoplankton population dynamics taking into account the transport and transformation of nutrients was developed and researched. A numerical algorithm of salinity and temperature field restoration in shallow water for the case of the Azov Sea was offered to simulate biogeochemical cycles of the main pollutants, including nitrogen, phosphorus and silicon. A discrete analogue of the proposed water ecology model problem was developed based on the schemes of the increased accuracy taking into account the partial filling of the computational cells. A parallel algorithm adapted for hybrid computing systems using the NVIDIA CUDA architecture was developed for the numerical implementation of the proposed interrelated mathematical models of biological kinetics.

Keywords: Abiotic factors · Salinity · Temperature · Biogeochemical cycle · Azov sea · Parallel algorithm · GPU

1 Introduction

Many scientists have been engaged in mathematical modelling of hydrobiological processes that have a significant influence on the ecological situation and

© Springer Nature Switzerland AG 2020
L. Sokolinsky and M. Zymbler (Eds.): PCT 2020, CCIS 1263, pp. 209–223, 2020.
https://doi.org/10.1007/978-3-030-55326-5_15

water reproduction processes. Note that the fundamental works in the field of creating the mathematical models, methods for diagnosis and change prognosis for water ecosystems were performed by such authors as Lotka A.J., Volterra V., Logofet D.O., Hutchinson G.E., Monod J., Mitscherlich E.A., Odum H.T. [1], Gause G.F., Vinberg G.G. [2], Abakumov A.I. [3], Menshutkin V.V., Rukhovets L.A., Vorovich I.I. [4], Gorstko A.B., Marchuk G.I. [5], Boulion V., Imberger A., Jorgensen S.E. [6], Vollenweider R.A. Fleishman B.S., who developed and researched the potential ecosystem effectiveness of stochastic models. The model of biological kinetics was described by Krestin S.V. and Rosenberg G.S. [7], where a possible explanation of the blue-green algae outbreak phenomenon and a more complex process of "water blooming" is given in the framework of the system of interacting and competing species and the "predator – prey" model.

The spatial heterogeneity of aquatic ecosystems was researched in papers by Leonov A.V. [8], Wiens J.J. [9]. Organogenic elements of different nature (also called biophilic, biogenic or organic elements) are distinguished among the multiple naturally occurring forms of a substance. To name a few, they are represented by carbon, nitrogen, phosphorus, sulphur and oxygen, the aggregated fractions of which are interconnected and classified as suspended and dissolved, biological and chemical types [10].

An analysis of the existing mathematical models of water hydrobiology showed that many of these models are oriented towards the use of hydrodynamic models that do not take into account the complex geometry of the coastline and bottom, surge phenomena, bottom friction and wind tensions, turbulent exchange, Coriolis force, river flows, evaporation, etc. Wind currents prevail in shallow waters, such as the Azov Sea. Since the researched waters are shallow, the bottom surface has a significant effect on their current pattern. Some of currently developed three-dimensional hydrophysical models are implemented in well-known packages, for example, MARS 3D, POM (Princeton Ocean Models), CHTDM (Climatic Hydro Termo Dynamic Model), NEMO (Nucleus for European Modelling of the Ocean). The oceanological models have proven to be well suited for modelling the hydrodynamic processes in the waters as deep as 50–100 m and more. But they cannot be used to calculate the flow fields in shallow water if the water depth is comparable with the surface wavelengths. Therefore, the developed models are complex, spatially heterogeneous 3D models, the numerical implementation of which on the basis of the finite-difference approach requires the processing of millions of nodes of the computational domain.

In this regard, it is necessary to develop parallel algorithms for the numerical solution of the defined problems inherent in the biological kinetic model, focused on the high-performance computing systems.

2   Problem Statement

The model is based on the system of convection-diffusion-reaction equations [11], the form of which for each F_i is:

$$\frac{\partial q_i}{\partial t} + u\frac{\partial q_i}{\partial x} + v\frac{\partial q_i}{\partial y} + w\frac{\partial q_i}{\partial z} = div(k\ grad\ q_i) + R_{q_i}, \tag{1}$$

where q_i is a concentration of the i-th component, [mg/l]; $i \in M$, $M = \{F_1, F_2, F_3, PO_4, POP, DOP, NO_3, NO_2, NH_4, Si\}$; $\{u, v, w\}$ are components of water flow velocity vector, [m/s]; k is a turbulent exchange coefficient, [m²/s]; R_{q_i} is a chemical-biological source, [mg/(l · s)]. In (1), the index i indicates the type of substance (see Table 1).

Table 1. Nutrients in the model of phytoplankton dynamics

No	Designation	Title
1	F_1	*Chlorella vulgaris* (green algae)
2	F_2	*Aphanizomenon flos-aquae* (blue-green algae)
3	F_3	*Sceletonema costatum* (diatom algae)
4	PO_4	Phosphates
5	POP	Suspended organic phosphorus
6	DOP	Dissolved organic phosphorus
7	NO_3	Nitrates
8	NO_2	Nitrites
9	NH_4	Ammonium
10	Si	Dissolved inorganic silicon (silicic acids)

Chemical-biological sources are described by the following equations [12]:

$$R_{F_i} = \varphi_{F_i}(1 - K_{F_i R})q_{F_i} - K_{F_i D}q_{F_i} - K_{F_i E}q_{F_i}, i = \overline{1, 3},$$

$$R_{POP} = \sum_{i=1}^{3} s_P K_{F_i D} q_{F_i} - K_{PD}q_{POP} - K_{PN}q_{POP},$$

$$R_{DOP} = \sum_{i=1}^{3} s_P K_{F_i E} q_{F_i} + K_{PD}q_{POP} - K_{DN}q_{DOP},$$

$$R_{PO_4} = \sum_{i=1}^{3} s_P \varphi_{F_i}(K_{F_i R} - 1)q_{F_i} + K_{PN}q_{POP} + K_{DN}q_{DOP},$$

$$R_{NH_4} = \sum_{i=1}^{3} s_N \varphi_{F_i}(K_{F_i R} - 1)\frac{f_N^{(2)}(q_{NH_4})}{f_N(q_{NO_3}, q_{NO_2}, q_{NH_4})}q_{F_i}$$

$$+ \sum_{i=1}^{3} s_N(K_{F_i D} + K_{F_i E})q_{F_i} - K_{42}q_{NH_4},$$

$$R_{NO_2} = \sum_{i=1}^{3} s_N \varphi_{F_i} (K_{F_i R} - 1) \frac{f_N^{(1)}(q_{NO_3}, q_{NO_2}, q_{NH_4})}{f_N(q_{NO_3}, q_{NO_2}, q_{NH_4})} \cdot \frac{q_{NO_2}}{q_{NO_2} + q_{NO_3}} q_{F_i}$$

$$+ K_{42} q_{NH_4} - K_{23} q_{NO_2},$$

$$R_{NO_3} = \sum_{i=1}^{3} s_N \varphi_{F_i} (K_{F_i R} - 1) \frac{f_N^{(1)}(q_{NO_3}, q_{NO_2}, q_{NH_4}) \cdot q_{NO_3} \cdot q_{F_i}}{f_N(q_{NO_3}, q_{NO_2}, q_{NH_4}) \cdot (q_{NO_2} + q_{NO_3})} + K_{23} q_{NO_2},$$

$$R_{Si} = s_{Si} \varphi_{F_3} (K_{F_3 R} - 1) q_{F_3} + s_{Si} K_{F_3 D} q_{F_3},$$

where $K_{F_i R}$ is a specific respiration rate of phytoplankton; $K_{F_i D}$ is a specific rate of phytoplankton death; $K_{F_i E}$ is a specific rate of phytoplankton excretion; K_{PD} is a specific rate of autolysis POP; K_{PN} is a phosphatification coefficient POP; K_{DN} is a phosphatification coefficient DOP; K_{42} is a specific rate of oxidation of ammonium to nitrites in the nitrification process; K_{23} is a specific rate of oxidation of nitrites to nitrates in the nitrification process; s_P, s_N, s_{Si} are normalization coefficients between the content of N, P, Si in organic matter.

The growth rate of phytoplankton is determined by the expressions:
$$\phi_{F_{1,2}} = K_{NF_{1,2}} f_T(T) f_S(S)) \min\{f_p(q_{PO_4}), f_N(q_{NO_3}, q_{NO_2}, q_{NH_4})\},$$
$$C_{F_3} = K_{NF_3} f_T(T) f_C(C)) \min\{f_p(q_{PO_4}), f_N(q_{NO_3}, q_{NO_2}, q_{NH_4}), f_{Si}(q_{Si})\},$$
where K_{NF} is the maximum specific growth rate of phytoplankton.

The dependences of temperature and salinity have the forms:

$$f_T(T) = \exp\left(-\alpha \left(T - T_{opt}\right)^2\right) / T_{opt}^2,$$

where T_{opt} is an optimal temperature for this type of phytoplankton; α is a coefficient that determines the interval width of phytoplankton survival depending on the temperature.

$$f_C(C) = \exp\left(-\beta \left(C - C_{opt}\right)^2\right) / C_{opt}^2,$$

where C_{opt} is an optimal salinity for this type of phytoplankton; β is a coefficient that determines the interval width of phytoplankton survival depending on the salinity.

Figure 1 shows the model scheme of a biogeochemical transformation of phosphorus, nitrogen and silicon forms.

The functions, describing biogenic content [13]:

– for phosphorus $f(p)(q_{PO_4}) = \dfrac{q_{PO_4}}{q_{PO_4} + K_{PO_4}}$, K_{PO_4} is a half-saturation constant of phosphates;

– for silicon $f(Si)(q_{Si}) = \dfrac{q_{Si}}{q_{Si} + K_{Si}}$, K_{Si} is a half-saturation constant of silicon;

– for nitrogen $f_N(q_{NO_3}, q_{NO_2}, q_{NH_4}) = f_N^{(1)}(q_{NO_3}, q_{NO_2}, q_{NH_4}) + f_N^{(2)}(q_{NH_4})$,
$$f_N^{(1)}(q_{NO_3}, q_{NO_2}, q_{NH_4}) = \frac{(q_{NO_3} + q_{NO_2}) \exp(-K_{psi} q_{NH_4})}{K_{NO_3} + q_{NO_3} + q_{NO_2}},$$

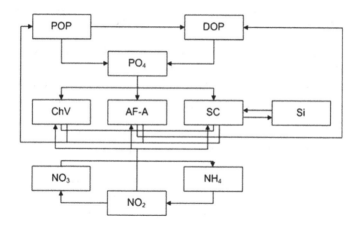

Fig. 1. The biogeochemical transformation scheme

$$f_N^{(2)}(q_{NH_4}) = \frac{(q_{NH_4})}{K_{NH_4} + q_{NH_4}}, \; K_{NO_3} \text{ is a half-saturation constant of nitrates;}$$
K_{NH_4} is a half-saturation constant of ammonium; K_{psi} is a coefficient of ammonium inhibition.

For system (1), it is necessary to specify the vector field of the water flow velocities, as well as the initial values of concentration functions q_i [14]:

$$q_i(x, y, z, 0) = q_i^0(x, y, z), (x, y, z) \in \overline{G}, t = 0, i \in M. \qquad (2)$$

Let the boundary Σ of a cylindrical domain G to be sectionally smooth, and $\Sigma = \Sigma_H \cup \Sigma_o \cup \sigma$, where Σ_H is water bottom surface, Σ_o is an unperturbed surface of the water environment, σ is the lateral (cylindrical) surface. Let u_n be the normal component of the water flow velocity vector with respect to the Σ surface, $\mathbf{n}$ be the outer normal vector with respect to the Σ. Let us assume the concentrations q_i in the form:

$$q_i = 0 \text{ on } \sigma, \text{ if } u_n < 0, \; i \in M; \frac{\partial q_i}{\partial n} = 0 \text{ on } \sigma, \text{ if } u_n \geq 0, \; i \in M; \qquad (3)$$

$$\frac{\partial q_i}{\partial z} = 0, \; i \in M \text{ on } \Sigma_o; k\frac{\partial q_i}{\partial z} = \varepsilon_{1,i}q_i, \; i \in \{F_1, F_2, F_3\}, k\frac{\partial q_i}{\partial z} = \varepsilon_{2,i}q_i \text{ on } \Sigma_H,$$

$i \in \{PO_4, POP, DOP, NO_3, NO_2, NH_4, Si\}$, $\varepsilon_{1,i}, \varepsilon_{2,i}$ are sedimentation rate of algae and nutrients to the bottom.

Note that the obtained values will be further used to simulate the spatially inhomogeneous distribution of substances taking into account the aquatic environment movement.

3   Restoring the Salinity and Temperature Fields

It is necessary to take into account the salinity and temperature, which are usually set on the maps at separate points or level contours for modeling the dynamics of phytoplankton population development. It is undesirable to use such maps due to a calculation error. This reveals the problem of hydrological information processing. Using smoother functions to approximate the functional dependences describing the salinity and temperature fields makes it possible to increase the accuracy of hydrodynamic calculations.

To obtain the salinity function, one can solve the diffusion equation, which for a long time is reduced to the Laplace equation [15,16]. However, the salinity function obtained in this way cannot have sufficient smoothness at the field value set points. Therefore, we used the equation to obtain the schemes of a high order of accuracy for the Laplace equation:

$$\Delta C - \frac{h^2}{12}\Delta^2 C = 0, \qquad (4)$$

where C is a concentration of the researched water salinity.

The isolines of salinity and temperature were obtained for solving the problem of hydrological information processing. For this, a recognition algorithm was used. Using the interpolation algorithm described above and by superimposing the boundaries of the region, we obtained the maps of salinity and temperature for the Azov Sea (Fig. 2, 3). The initial image of salinity isolines (isotherms) for the Azov Sea (July 2019) (a) and the restored salinity field of the Azov Sea (b) are given in Fig. 2. The initial image of temperature isolines (isotherms) of the Azov Sea (July 2019) (a) and the restored temperature field of the Azov Sea (b) are given in Fig. 3.

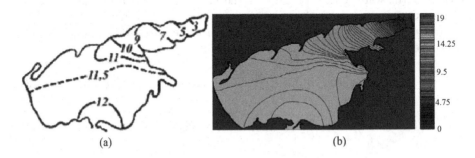

(a)   (b)

Fig. 2. The results of the algorithm for restoring the salinity field

The resulting salinity and temperature fields for summer will be further used in modeling the dynamics of phytoplankton populations.

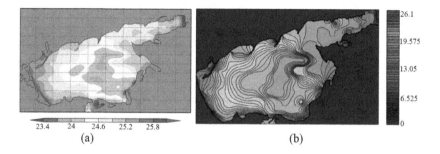

Fig. 3. The results of the algorithm for restoring the temperature field

4   The Parallel Algorithm for Numerical Implementation of the Problem of the Pollutant Transport Process

Each equation of the system (1)–(3) can be represented by the diffusion-convection-reaction equation in the three-dimensional case [17].

The discrete analogue of the proposed water ecology model was developed on the basis of schemes of increased order of accuracy taking into account the partial filling of computational cells [18,19]. The modified alternating triangular method (MATM) was used for solving the grid equations [20,21]. The methods solving such problems with a better convergence rate are described and compared with the known methods [22].

4.1   Parallel Algorithm for MCS

We described the parallel algorithm for solving the problems (1)–(3). The parallel implementation of the developed algorithm is based on the Message Passing Interface (MPI) technologies. The peak performance of a multiprocessor computer system (MCS) is 18.8 TFlops. As computing nodes, 128 HP ProLiant BL685c homogeneous 16-core Blade servers of the same type were used, each of which is equipped with four quad-core AMD Opteron 8356 2.3 GHz processors and 32 GB RAM.

In the parallel implementation of the developed numerical algorithms, we used the decomposition methods of computational domains for various computational labor hydrophysics problems and took into account the architecture and parameters of the MCS (Table 2). In Table 2: A, E are acceleration and efficiency of the parallel algorithm; p is a number of processors. Note that the efficiency is the ratio of acceleration to the number of quad-core processors of the computing system.

Figure 4 shows the dependence of the acceleration (a) and efficiency (b) on the number of processors for the developed parallel algorithm of MATM for solving the problem of biogeochemical cycle variability.

It was established that the maximum acceleration was 63 times on 512 processors (see Table 2, Fig. 4).

Table 2. The acceleration and efficiency of the parallel MATM algorithm

p	Time, s	A (practical)	E (practical)	A_t (theoretical)
1	3.700073	1	1	1
2	1.880677	1.967	0.984	1.803
4	1.265500	2.924	0.944	3.241
8	0.489768	7.555	0.731	7.841
16	0.472151	7.837	0.490	9.837
32	0.318709	11.610	0.378	14.252
64	0.182296	20.297	0.363	26.894
128	0.076545	48.338	0.317	55.458
256	0.063180	58.563	0.229	65.563
512	0.058805	62.921	0.123	72.921

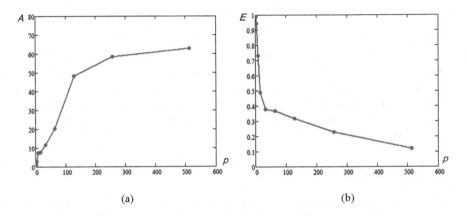

(a)   (b)

Fig. 4. The graphs of the acceleration and efficiency dependence on the number of processors

4.2   Parallel Algorithm for NVIDIA Tesla K80

For the numerical implementation of the proposed interrelated mathematical water hydrobiological models, we developed parallel algorithms. It will be adapted for the hybrid computer systems using the NVIDIA CUDA architecture which will be used for the mathematical modeling of the phytoplankton production and destruction process (Fig. 5).

The NVIDIA Tesla K80 computing accelerator has high computing performance and supports all modern closed (CUDA) and open technologies (OpenCL, DirectCompute). The NVIDIA Tesla K80 specifications: the GPU frequency is 560 MHz; the GDDR5 video memory is 24 GB; the video memory frequency is 5000 MHz; the video memory bus digit capacity is 768 bits. The NVIDIA CUDA platform characteristics: Windows 10 (x64) operating system; CUDA

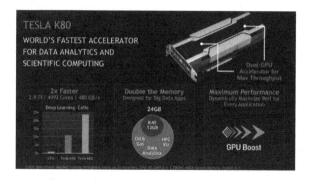

Fig. 5. The hybrid computer systems using the NVIDIA CUDA architecture

Toolkit v10.0.130; Intel Core i5-6600 3.3 GHz processor; DDR4 of 32 GB RAM; the NVIDIA GeForce GTX 750 Ti video card of 2 GB; 640 CUDA cores.

Using the GPU with CUDA technology is required to address the effective resource distribution at solving the system of linear algebraic equations (SLAE), occurring at discretization of the developed model (1)–(3). The dependence of the SLAE solution time on the matrix dimension and the number of nonzero diagonals was obtained to implement the corresponding algorithm (see Fig. 6). Due to this, in particular, we can choose the grid size and determine the solution time of SLAE based on the number of nonzero matrix diagonals.

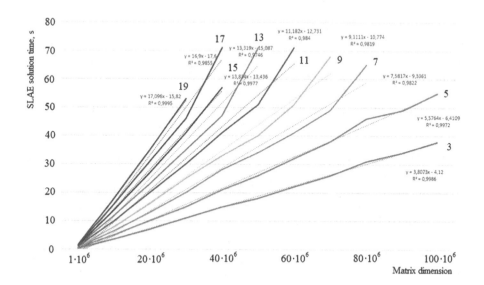

Fig. 6. The graph of SLAE solution time dependence on the order of a square matrix with a given number of nonzero diagonals

The integration process based on parallel algorithms slows down the computational procedure with the GPU using the previously developed software. So, it is necessary to estimate the slowing time depending on the dimension of the tape slows for a different number of non-zero elements in the main matrix diagonal. As a result of the conducted research, a number of regression lines with corresponding determination coefficients were obtained, indicating an acceptable degree of approximation.

As a result of the developed parallel algorithm for solving the biological kinetics problem, it is necessary to modify the element storage format of a sparse matrix from a system of grid equations. The elements of this matrix for internal nodes are a repeating sequence. In the case of high-dimensional problems, the standard storage leads to inefficient memory usage.

Using the Compressed Sparse Rows (CSR) matrix storage format excludes the need to store null elements. However, all nonzero elements, including many repeating ones, are stored in the corresponding array. This drawback is not critical when using computing systems with shared memory. However, in heterogeneous and distributed computing systems, it can adversely affect the performance when data transferring between the nodes.

The paper proposes a modification of the CSR format to increase the efficiency of data storage with a repeating sequence of CSR1S elements. The proposed format is effectively used for modeling continuous hydrobiological processes by the finite difference method. In addition, to change a differential operator instead of repeatedly searching and replacing values in an array of nonzero elements, it is enough to just change them in an array that preserves a repeating sequence.

Notations in Table 3: R is a number of rows in the matrix; R_{seq} is a number of rows in the matrix with the repeating sequence of elements; C is a number of columns in the matrix; N_{nz} is a number of nonzero matrix elements; N_{seq} is a number of the repeating sequence elements; B_{nz} is memory capacity for storing one nonzero element; B_{idx} is a memory capacity for storing one index.

We estimated the required memory capacity in CSR and CSR1S formats:
$P_{csr} = N_{nz}B_{nz} + N_{nz}B_{idx} + (R + 1)B_{idx} = N_{nz}B_{nz} + (N_{nz} + R + 1)B_{idx}$;
$P_{csr1s} = (N_{nz} - N_{seq}(R_{seq} + 1))B_{nz} + (N_{nz} - R_{seq}(N_{seq} + 1) + R + 1)B_{idx}$.

We shall denote the ratio of the lines containing the repeating sequence to the total number of lines: $k_r = R_{seq}/R$.

Then, $P_{csr1s} = (N_{nz} - N_{seq}(k_rR+1))B_{nz} + (N_{nz} - k_rR(N_{seq}+1) + R + 1)B_{idx}$.

The effective function libraries have been developed to solve SLAEs in CSR format with GPUs using the CUDA technology. The proposed approach is to use the CSR1S format in the model processing with further conversion to the CSR format to solve the resulting SLAE on GPU using NVIDIA CUDA technology. This led to the problem of developing an algorithm for matrix conversion from CSR1S to CSR format in the minimum time.

The computational experiments were performed with fivefold repetition and recording of the average calculation time. We performed an experimental research to obtain the dependence of the execution time of the conversion

Table 3. Characteristics of the data storage formats

Storage format characteristic	CSR format	CSR1S format
Number of arrays	3	5
The array size of nonzero elements	$N_{nz}B_{nz}$	$(N_{nz} - N_{seq}R_{seq})B_{nz}$
The array size of column indices where nonzero elements are located	$N_{nz}B_{idx}$	$(N_{nz} - N_{seq}R_{seq})B_{idx}$
The array size of the first nonzero row elements indices	$(R+1)B_{idx}$	$(R - R_{seq} + 1)B_{idx}$
The array size to store a repeating sequence	-	$N_{seq}B_{nz}$
The array size for storage the indices of columns in which the first elements of repeating sequence are located	-	$R_{seq}B_{idx}$

algorithm on N_{seq} and the k_r coefficient for a sequential implementation, as well as the parallel implementation using the TPL library, and the parallel implementation using the NVIDIA CUDA platform. N_{seq} ranged from 3 to 19 in increments of 2; k_r ranged from 0 to 1 in increments of 0.1; the matrix dimension is 10^6.

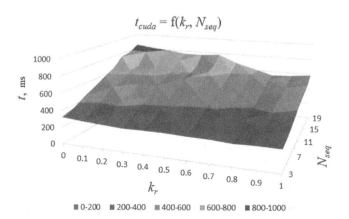

Fig. 7. The time dependence of the matrix conversion from CSR1S to CSR format

According to the performed experiments, the algorithm using the parallel computing on CPU is more efficient than the sequential algorithm and the algorithm using NVIDIA CUDA. The plane inclination (Fig. 7) shows a decreasing calculation time with an increasing k_r and N_{seq} that allows it to be used asynchronously at working with matrices with high k_r and N_{seq}, thereby unloading the central processor to perform other operations.

The analysis of CUDA architecture characteristics showed the algorithm applicability for the numerical implementation of the developed mathematical models of hydrobiological process to design high-performance information systems.

5   Modeling the Nutrients Distribution and Dynamics of Phytoplankton Populations in the Azov Sea

The experimental software complex (SC) was designed for the mathematical modeling of shallow water ecosystems for the Azov Sea case using MCS and GPU that allowed increasing the efficiency of algorithms for solving hydrophysics and biological kinetic problems. Using the "Azov3d" SC we can construct the operational flow forecast turbulence of the water environment – the velocity field on grids with high resolution [23, 24]. The SC was used to calculate the three-dimensional velocity vector of the Azov Sea. The SC takes into account such physical parameters as the Coriolis force, the turbulent exchange, the complex geometry of bottom surface and coastline, the evaporation, river flows, wind-surges phenomena, wind currents, and friction bottom. In addition, the SC is capable of calculating a velocity field without pressure, hydrostatic pressure (used as an initial approximation for the hydrodynamic pressure), hydrodynamic pressure and three-dimensional velocity field of the water flow. The SC involves the numerical implementation of model problems related to the hydrodynamics and biological kinetics, including the model (1)–(3).

When solving the problems of biological kinetics, the following physical dimensions of the simulated region (the Azov Sea) were taken into account: the surface area under research was $37605 \, \text{km}^2$; the length was $343 \, \text{km}$; the width was $231 \, \text{km}$. The numerical experiment was performed on a sequence of condensing grids: computational nodes.

Figures 8, 9, 10, 11 and 12 show the concentration dynamics of three types of phytoplankton and nutrients in the Azov Sea over time for a uniform initial distribution. The time interval was 30 days. The figures reflect the influence of the abiotic factors (salinity and temperature) on the development of three types of phytoplankton, the absorption of phosphates and nitrogen forms by phytoplankton, the transition of phosphorus and nitrogen forms from one to another, as well as the absorption of silicon by diatoms. The results of the software implementation when simulating the change of *Aphanizomenon flos-aquae* blue-green algae (a) and *Sceletonema costatum* diatom (b) concentration distributions are given in Fig. 8.

Figures 9, 9, 10, 11 and 12 represent the influence of such abiotic factors as salinity, temperature and biogenic regime on production and destruction processes of the phytoplankton ensembles.

At the salinity level $(C : 0 \div 9‰)$, green and blue-green algae prevail in the Taganrog Bay, and diatoms prevail in the main part of the Azov Sea $(C : 9 \div 12‰)$ (Fig. 8).

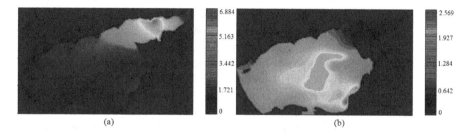

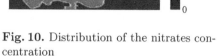

Fig. 8. The simulation of phytoplankton population dynamics (Color figure online)

Fig. 9. Distribution of the dissolved organic phosphorus concentration

Fig. 10. Distribution of the nitrates concentration

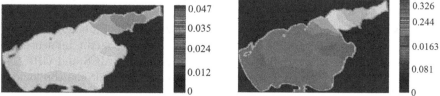

Fig. 11. Distribution of the ammonium concentration

Fig. 12. Distribution of the dissolved inorganic silicon (silicic acids) concentration

6 Conclusion

A multi-species mathematical model of phytoplankton population dynamics was developed and researched taking into account the transport and transformation of nutrients, the convective and diffusion transfers, the absorption and release of nutrients by phytoplankton, as well as the transformation cycles of phosphorus, nitrogen and silicon forms.

The numerical algorithm based on the modified Laplace equation was proposed to reconstruct the salinity and temperature fields in shallow water for the Azov Sea case. A discrete analogue of the proposed model problem of the aquatic ecology was calculated using the high order of accuracy schemes taking into account the partial filling of computational cells. The distributions of three phytoplankton species concentrations (green, blue-green, and diatoms) were obtained taking into account the influence of external and internal factors on their development dynamics.

The numerical implementation of the proposed interrelated mathematical models of biological kinetics was based on the developed parallel algorithm oriented on MCS. It was found that the maximum acceleration was 63 times on 512 processors of MCS. The parallel algorithm adapted for the hybrid computing systems using the NVIDIA CUDA architecture was performed using the CSR1S element storage modified format. It is possible to improve the performance of the computer system with the memory shared for data transfer between the nodes. The computational experiments were made to find the execution time of the algorithm using the developed conversion of the data storage format. As a result, it was established that the algorithm using parallel computing on CPU is more efficient than the sequential algorithm and the algorithm using NVIDIA CUDA.

The developed complex of interrelated mathematical hydrobiology models enables performing the predictive modeling of the abiotic factor influence on the biogeochemical cycle variability in the Azov Sea using MCS and GPU, as well as performing the numerical experiments for various hydrometeorological situations.

Acknowledgment. The reported study was funded by RFBR, project number 19-31-51017.

References

1. Odum, H.T. : System Ecology. Wiley, New York (1983). 644 p.
2. Vinberg, G.G.: Some results of the practice of production-hydrobiological methods. In: Production of Populations and Communities of Aquatic Organisms and Methods of Its Research, pp. 3–18. UNC AN USSR, Sverdlovsk (1985)
3. Abakumov, A.I.: Signs of of water ecosystems stability in mathematical models. In: Proceedings of the Institute of System Analysis of RAS. System Analysis of the Problem of Sustainable Development, vol. 54, pp. 49–60. ISA RAS, Moscow (2010)
4. Vorovich, I.I. et al.: Rational use of water resources of the Azov Sea basin: mathematical models. Science (1981). 360 p.
5. Marchuk, G.I., Sarkisyan, A.S.: Mathematical modelling of ocean circulation. Science (1988). 297 p.
6. Jorgensen, S.E., Mejer, H., Friis, M.: Examination of a lake model. Ecolog. Model. 4(2–3), 253–278 (1978)
7. Rozenberg, G.S., et al.: Some thoughts on the Federal law on the Volga River. Probl. Reg. Global Ecol. 28(1), 9–17 (2019). (in Russian). Samarskaya Luka
8. Leonov, A.V.: Matematicheskoe modelirovanie processov biotransformacii veshhestv v prirodnyh vodah. Vod. Resursy. 26(5), 624–630 (1999). (in Russian)
9. Wiens, J.J.: Faster diversification on land than sea helps explain global biodiversity patterns among habitats and animal phyla. Ecol. Lett. 18, 1234–1241 (2015)
10. Menshutkin, V.V., Rukhovets, L.A., Filatov, N.N.: Modeling of freshwater lake ecosystems (review). 2. Models of freshwater lake ecosystems. Water Resour. 41(1), 24–38 (2014)

11. Yakushev, E.V., Mikhailovsky, G.E.: Mathematical modeling of the influence of marine biota on the carbon dioxide ocean-atmosphere exchange in high latitudes. In: Jaehne, B., Monahan, E.C. (eds.) Air-Water Gas Transfer, AEON Verlag & Studio, pp. 37–48. Heidelberg University, Hanau (1995)
12. Sukhinov, A., Isayev, A., Nikitina, A., Chistyakov, A., Sumbaev, V., Semenyakina, A.: Complex of models, high-resolution schemes and programs for the predictive modeling of suffocation in shallow waters. In: Sokolinsky, L., Zymbler, M. (eds.) PCT 2017. CCIS, vol. 753, pp. 169–185. Springer, Cham (2017). https://doi.org/10.1007/978-3-319-67035-5_13
13. Gushchin, V.A., Sukhinov, A.I., Nikitina, A.V., Chistyakov, A.E., Semenyakina, A.A.: A model of transport and transformation of biogenic elements in the coastal system and its numerical implementation. Comput. Math. Math. Phys. **58**(8), 1316–1333 (2018). https://doi.org/10.1134/S0965542518080092
14. Sukhinov, A.I., Chistyakov, A.E., Nikitina, A.V., Belova, Y.V., Sumbaev, V.V., Semenyakina, A.A.: Supercomputer modeling of hydrochemical condition of shallow waters in summer taking into account the Influence of the Environment. In: Sokolinsky, L., Zymbler, M. (eds.) PCT 2018. CCIS, vol. 910, pp. 336–351. Springer, Cham (2018). https://doi.org/10.1007/978-3-319-99673-8_24
15. Vladimirov, V.S., Zharinov, V.V.: Uravnenija matematicheskoj fiziki. 2-e izdanie, stereotipnoe. Fizmatlit, Moskva (2004). ISBN 5-9221-0310-5, 400 p. (in Russian)
16. Samarskij, A.A., Vabishhevich, P.N.: Chislennye metody reshenija zadach konvekcii-diffuzii. Jeditorial URSS, Moskva (1999). (in Russian)
17. Sukhinov, A.I., et al.: Game-theoretic regulations for control mechanisms of sustainable development for shallow water ecosystems. Autom. Remote Control **78**(6), 1059–1071 (2017). https://doi.org/10.1134/S0005117917060078
18. Sukhinov, A., Nikitina, A., Chistyakov, A., Sumbaev, V., Abramov, M., Semenyakina, A.: Predictive modeling of suffocation in shallow waters on a multiprocessor computer system. In: Malyshkin, V. (ed.) PaCT 2017. LNCS, vol. 10421, pp. 172–180. Springer, Cham (2017). https://doi.org/10.1007/978-3-319-62932-2_16
19. Sukhinov, A.I., Chistyakov, A.E., Semenyakina, A.A., Nikitina, A.V.: Parallel realization of the tasks of the transport of substances and recovery of the bottom surface on the basis of schemes of high order of accuracy. Comput. Meth. Program. New Comput. Technol. **16**(2), 256–267 (2015)
20. Belocerkovskij, O.M., Gushhin, V.A., Shhennikov, V.V.: Use of the splitting method to solve problems of the dynamics of a viscous incompressible fluid. Comput. Math. Math. Phys. **15**(1), 190–200 (1975)
21. Konovalov, A.N.: The method of steepest descent with adaptive alternately-triangular preamplification. Differ. Eqn. **40**(7) (2004). 953 p.
22. Sukhinov, A.I., Chistyakov, A.E.: Adaptive modified alternating triangular iterative method for solving grid equations with a non-self-adjoint operator. Math. Models Comput. Simul. **4**(4), 398–409 (2012). https://doi.org/10.1134/S2070048212040084
23. Nikitina, A.V., Kravchenko, L.V., Semenov, I.S., Belova, Y.V., Semenyakina, A.A.: Modeling of production and destruction processes in coastal systems on a super-computer. MATEC Web Conf. **226**(2), 172–180 (2017). https://doi.org/10.1051/matecconf/201822604025
24. Nikitina, A.V., et al.: Optimal control of sustainable development in the biological rehabilitation of the Azov Sea. Math. Models Comput. Simul. **9**(1), 101–107 (2017). https://doi.org/10.1134/S2070048217010112

Simulating Relativistic Jet on the NKS-1P Supercomputer with Intel Broadwell Computing Nodes

Igor Kulikov$^{(\boxtimes)}$ [ORCID], Igor Chernykh, Dmitry Karavaev, Ekaterina Genrikh,
Anna Sapetina, Victor Protasov, Alexander Serenko, Vladislav Nenashev,
Vladimir Prigarin, Ivan Ulyanichev, and Sergey Lomakin

Institute of Computational Mathematics and Mathematical Geophysics SB RAS,
Novosibirsk, Russia
kulikov@ssd.sscc.ru, chernykh@parbz.sscc.ru, kda@opg.sscc.ru,
mesyats@gmail.com, afsapetina@gmail.com, inc_13@mail.ru, fafnur@yandex.ru,
arni.12@mail.ru, vovkaprigarin@gmail.com, wmzonacomvn@mail.ru, soj@sscc.ru

Abstract. This paper presents the results of modeling the interaction of a relativistic jet with an inhomogeneous galactic medium. To numerically solve the equations of special relativistic hydrodynamics, a new numerical method was developed combining the Godunov method, flow calculation by the Rusanov method, and piecewise linear representation of the solution. The numerical method was implemented in program code for supercomputer architectures with distributed memory using MPI. The results of studying the code performance on the Intel Broadwell computing nodes of the NKS-1P cluster located in the Siberian Supercomputer Center are presented.

Keywords: Relativistic hydrodynamics · Parallel computing · Intel broadwell · Computational astrophysics · Numerical methods

1 Introduction

Many astrophysical phenomena are associated with gas motion at relativistic velocities. The source of such currents are relativistic jets [1] in active galactic nuclei [2,3], microquasars [4], blazars [5,6], gamma bursts [7], extragalactic jets [8], core collapse [9] and stellar collapse in massive stars [10], black holes [11–14] and binary black holes [15], neutron stars [16–19], pulsars [20], and gravitational waves [21]. A number of frameworks and codes have been developed for simulation of the relativistic flows. We will list the most famous.

The **Einstein Toolkit** [22] framework is developed to solve the equations of relativistic hydrodynamics in a special and general form, with and without magnetic field and other effects. The framework consists of several blocks:

© Springer Nature Switzerland AG 2020
L. Sokolinsky and M. Zymbler (Eds.): PCT 2020, CCIS 1263, pp. 224–236, 2020.
https://doi.org/10.1007/978-3-030-55326-5_16

- a cactus parallel computing infrastructure [23], which provides meta-computing, parallel output to HDF5 format, web-based interface and visualization tools;
- two approaches to organize computations using adaptive nested grids. The PUGH approach, which is based on local refinement of a regular grid and provides scalability up to 130,000 processes. The Carpet approach, which implements a multi-level nesting of the grids [24,25] (note that both approaches fit the requirements for accuracy and stability [26]);
- the computing factory concept that is used to implement the framework [27]. It offers a set of abstractions to organize a supercomputer modeling such as the remote access, configuration and assembly, sending and modeling management (these abstractions hide a part of the low-level structures, allowing one to focus on solving the applied problems);
- the Kranc [28] package, which is used to translate tensor systems of partial differential equations, written in the Wolfram Mathematica package for computer algebra, into C or Fortran code.

The numerical methods are based on the reconstruction of primitive variables using total variation diminishing methods (TVD), piecewise-parabolic method (PPM), and essentially non-oscillatory methods (ENO). The Harten-Lax-van Leer (HLL) and the Roe and Marquin [29] methods are used as solvers of the Riemann problem.

The code **CAFE-R** [30] is developed to simulate special relativistic hydrodynamics processes taking into account radiation. The purpose of the code is to model high-energy astrophysics, in which matter and radiation are strongly coupled, while magnetic fields and the gravitational field barely influence the process: relativistic jets from the active nuclei of galaxies [2], microquasars [4], blazars [5], and gamma-ray bursts [7]. The radiation model is described in the code description in details. In addition, a state equation determining the temperature is explicitly formulated. It is important to reduce this state equation, as, even in the case of relativistic interaction velocities, there is a need to fully take into account the matter thermodynamics. The equations of special relativistic hydrodynamics with radiation are solved using the operator separation method, similar to the approach from [31] and [32]. To solve the Riemann problem, a classic HLL solver is used. The code was verified on the problem of a shock wave in a weak, medium, and strong formulation. Similar shock wave regimes were considered taking into account radiation. In a multidimensional formulation, the code was verified on the problem of the radiation pulse, the dynamics of a single beam, the collision of pulses, the problem of the radiation effect on the shadow. In a three-dimensional formulation, the model problem of jet dynamics was solved.

The problems of relativistic jet simulation are a popular theme. Therefore, for such tasks, the code **GENESIS** [33] was developed. The code is adapted to use an arbitrary equation of state, which allows one to simulate gas chemical kinetics, as well as to use stellar equations of state [34]. The code description in detail presents a parallel implementation of the calculations, data distribution scheme, and memory usage. Separately, it provides the procedure for restoring primitive (physical) variables in details. Discontinuity decay with different values of the Lorentz factor is solved as a test problem. It should also be noted that the code detailed description allows developing your own implementation based on it.

The code **PLUTO** [35] is intended for all astrophysical flows, including relativistic ones. The solver is based on a combination of HLL and PPM methods. The paper thoroughly describes the code testing. The code takes the magnetic field and relativistic Lorentz factor-factor into account. The code is adapted to an arbitrary coordinate system. It is worth noting that the code is used quite a lot to solve astrophysical problems.

The code **RAM** [36] is intended for solving the problems of special relativistic hydrodynamics. The code is based on ENO and PPM adaptive mesh (AMR) schemes. The paper describes a detailed testing of the code and presents the equations of special relativistic hydrodynamics in cylindrical and spherical coordinates. The code also focuses on modeling a relativistic jet.

The code **TESS** enables solving special relativistic hydrodynamics equations on Voronoi moving grids [37]. The code implements the ability to calculate using both a stationary Eulerian grid and a fully Lagrangian grid. In this sense, the code implements a full Arbitrary Lagrangian Eulerian (ALE) approach. The code describes the details of the numerical method, where the term responsible for the grid movement and the velocities sum determination taking into account the relativistic Lorentz factor are of particular importance. The code was subjected to shock tube tests, including the Brio-Wu magneto-hydrodynamic test in the relativistic mode. The Rayleigh-Taylor and the Kelvin-Helmholtz instabilities evolution is considered separately.

Software to solve the equations of special relativistic hydrodynamics for modeling the relativistic extragalactic jet [8], the collapse problem of nucleus [9] and jets from the active galactic nucleus and gamma radiation sources [1] were also developed. In [8], the method was verified on the shock tube problem and the hydrodynamics of an extragalactic jet at various values of the Lorentz factor was studied in a sufficient way. The paper [9] describes the neutrino transport implementation since the nucleus collapse and the subsequent supernova explosion are among the sources of such a particle flux. It also presents an algorithm for constructing the collision integral. The paper [1] describes the approach of using adaptive meshes and their adjustment to obtain the best resolution of instabilities' development in jets of different nature. In general, three approaches, described above with their capabilities, correspond to the program codes that we discussed earlier.

The paper describes a new numerical method for solving the relativistic hydrodynamics equations, implemented in a parallel code for classical supercomputer architectures. To reduce the method computational complexity, a piecewise linear method of reconstructing the solution and using the Rusanov scheme on regular grids was used. This allowed us to maintain small solution dissipation with a significant increase in the calculation speed compared to the piecewise parabolic representation and the Roe scheme. The second section describes the new parallel code for simulating a relativistic jet evolution process. The third section is devoted to a brief description of the program code structure and the study of its parallel implementation on the computing nodes of the NKS-1P cluster located in the Siberian Supercomputer Center, based on Intel Broadwell processors. The fourth section describes the results of computational experiments on the interaction of a relativistic jet with an inhomogeneous galactic medium. The fifth section presents the conclusions.

2   The Numerical Method

To describe the relativistic gas dynamics, we will use physical (primitive) variables: ρ is density, v is velocity vector and p is pressure. Let us introduce the concept of special enthalpy h, which is determined by the equation:

$$h = 1 + \frac{\gamma}{\gamma - 1} \frac{p}{\rho}, \tag{1}$$

where γ is adiabatic index. The speed of sound c_s is determined by the equation:

$$c_s^2 = \gamma \frac{p}{\rho h}. \tag{2}$$

In this paper, we take the speed of light $c \equiv 1$. In this case, the Lorentz factor Γ will be determined by the equation:

$$\Gamma = \frac{1}{\sqrt{1 - (v/c)^2}} = \frac{1}{\sqrt{1 - v^2}}. \tag{3}$$

Thus, the speed modulus is limited to one.

We will introduce the concepts of conservative variables: the relativistic density $D = \Gamma \rho$, the relativistic momentum $M_j = \Gamma^2 \rho h v_j$, where v_j are the components of the velocity vector v for $j = 1, 2, 3$ and the total relativistic energy $E = \Gamma^2 \rho h - p$. In this case, the relativistic hydrodynamics equations in the form of conservation laws are written in the form:

$$\frac{\partial D}{\partial t} + \frac{\partial (D v_k)}{\partial x_k} = 0, \tag{4}$$

$$\frac{\partial M_j}{\partial t} + \frac{\partial (M_j v_k + p \delta_{jk})}{\partial x_k} = 0, \tag{5}$$

$$\frac{\partial E}{\partial t} + \frac{\partial (E + p) v_k}{\partial x_k} = 0. \tag{6}$$

Or in a compact vector form in three-dimensional space:

$$\frac{\partial U}{\partial t} + \sum_{k=1}^{3} \frac{\partial F_k}{\partial x_k} = 0 \tag{7}$$

where the vector of conservative variables U and the vector of flows F_k are given in the form:

$$U = \begin{pmatrix} D \\ M_j \\ E \end{pmatrix} = \begin{pmatrix} \Gamma\rho \\ \Gamma^2\rho h v_j \\ \Gamma^2\rho h - p \end{pmatrix}, \tag{8}$$

$$F_k = \begin{pmatrix} \rho\Gamma v_k \\ \rho h \Gamma^2 v_j v_k + p\delta_{jk} \\ \rho h \Gamma^2 v_k \end{pmatrix},$$

where δ_{jk} is the Kronecker symbol. Note that in the numerical method, calculations are performed in conservative variables D, M_j, E, taking into account a nonlinear dependence between the conservative variables and primitive ρ, v, p. To restore the primitive variables, we use the procedure described in [38].

We consider a region in which a uniform grid with $\triangle x$ steps is introduced. The time step τ is selected from the Courant condition, based on the wave propagation velocities. Their analytical form is given below. Integer indexes will be used for numbering the cells. Fractional indexes will be used for numbering the nodes. Let us write a one-dimensional analog of Godunov's scheme for Eq. (7) for an arbitrary cell with the number i:

$$\frac{U_i^{n+1} - U_i^n}{\tau} + \frac{F_{i+1/2}^{n+1/2} - F_{i-1/2}^{n+1/2}}{\triangle x} = 0, \tag{9}$$

where $F_{i\pm1/2}^{n+1/2}$ is the solution of the linearized Riemann problem for the equations of special relativistic hydrodynamics. To solve the linearized Riemann problem, we will use the Rusanov scheme, which we consider for two neighboring cells defined as left (L) and right (R):

$$F = \frac{1}{2}\left(F^L + F^R\right) + \frac{\lambda_{max}}{2}\left(U^L - U^R\right), \tag{10}$$

where λ_{max} is the maximum speed of relativistic characteristics [39]:

$$\lambda_1 = v_1\frac{\left(1 - c_s^2\right) - c_s\Gamma^{-1}\sqrt{1 - v_1^2 - c_s\left(v_2^2 + v_3^2\right)}}{1 - c_s^2 v^2},$$

$$\lambda_{2,3,4} = v_1, \tag{11}$$

$$\lambda_5 = v_1\frac{\left(1 - c_s^2\right) + c_s\Gamma^{-1}\sqrt{1 - v_1^2 - c_s\left(v_2^2 + v_3^2\right)}}{1 - c_s^2 v^2}.$$

To calculate the vector of conservative variables U and flow vectors F_k, a piecewise linear reconstruction of the physical variables q is used. Which for the i cell can be written as:

$$q^L = q_i - \frac{\delta q_i}{2}\left(1 - \frac{\lambda_{max}\tau}{\triangle x}\right),\tag{12}$$

$$q^R = q_i + \frac{\delta q_i}{2}\left(1 - \frac{\lambda_{max}\tau}{\triangle x}\right).$$

Any known limiter can be used to determine the value of δq_i. The three-dimensional Godunov's scheme for Eq. (7) can be written in the following form:

$$\frac{U_{i,k,l}^{n+1} - U_{i,k,l}^{n}}{\tau} + \frac{F_{x,i+1/2,k,l}^{n+1/2} - F_{x,i-1/2,k,l}^{n+1/2}}{\triangle x} + \tag{13}$$

$$+\frac{F_{y,i,k+1/2,l}^{n+1/2} - F_{y,i,k-1/2,l}^{n+1/2}}{\triangle y} + \frac{F_{z,i,k,l+1/2}^{n+1/2} - F_{z,i,k,l-1/2}^{n+1/2}}{\triangle z} = 0.$$

The obtained formulas correspond to the solution integration along the characteristics [31].

3   The Parallel Code

The developed numerical method for solving the equations of special relativistic hydrodynamics allows for using a geometric decomposition of the computational domain with one layer of overlapping subdomains. Using a uniform grid in Cartesian coordinates to solve the hydrodynamics equations allows one to use an arbitrary Cartesian topology to decompose the computational domain. This organization of the computations has potentially infinite scalability. The program code implements one-dimensional decomposition by MPI technology Fig. 1.

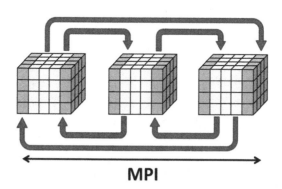

Fig. 1. Geometrical decomposition of computations

Various program codes on Intel Broadwell architectures were studied earlier [40]. Due to the linear algorithmic complexity of the program code basis, we will study the acceleration and scalability of the software implementation. Cluster NKS-1P is produced by RSC Technologies and includes 20 computing nodes RSC Tornado over Intel Omni-Path Fabric, two 16-core Intel Xeon E5-2697A v4 2.6 GHz processors with 128 GiB DDR4 RAM per compute node. Features of the Intel Broadwell node architecture of the NKS-1P supercomputer. In such a case,

Table 1. The speed up research.

Nodes	Processors	Cores	Threads	MPI-Processes	Speed Up
1	1	1	1	1	1.000
1	1	2	1	2	1.981
1	1	1	2	2	2.133
1	2	1	1	2	1.706
2	1	1	1	2	2.142
1	1	4	1	4	3.813
1	1	2	2	4	4.095
1	2	2	1	4	3.817
1	2	1	2	4	3.778
2	2	1	1	4	4.059
2	1	2	1	4	3.833
4	1	1	1	4	3.812
1	1	8	1	8	5.064
1	1	4	2	8	6.159
1	2	2	2	8	6.196
1	2	4	1	8	6.098
2	2	1	2	8	7.085
4	1	1	2	8	6.706
4	2	1	1	8	6.818
1	1	16	1	16	6.948
1	1	8	2	16	6.801
1	2	4	2	16	6.879
2	2	2	2	16	8.097
4	2	1	2	16	9.015
4	2	2	1	16	9.009
1	1	16	2	32	4.956
1	2	8	2	32	4.934
1	2	16	1	32	4.962
2	2	4	2	32	6.294
4	2	2	2	32	7.253

we will consider the number of MPI processes for a given number of nodes, processors, cores and the number of threads on the core, taking into account Hyper-Threading technology.

The program code acceleration was studied on a computational grid of size 256^3. For this, the computation time was measured for a different number of processes used. The acceleration (Q) was calculated by the Eq. (14):

$$Q = \frac{Time_1}{Time_K} \tag{14}$$

where $Time_1$ is the computation time when one MPI process is used. $Time_K$ is the computation time when using K processes. The acceleration study results are shown in Table 1.

Due to a comparatively small size of the computational grid, the acceleration starts to decline quite quickly. At the same time, its maximum values are achieved when using two threads per core, i.e. when using HyperThreading. Apparently, this is due to more efficient use of the first and second-level cache, where the data used by both processes is located. The greatest acceleration is achieved when using the process of pure communications. That is when using the maximum possible number of nodes, with exclusive use of the socket and thread running on one of the cores.

The program code scalability was studied on a computational grid of size $256K \times 256 \times 256$. Thus, the subdomain size was 256^3 for each MPI process. For this, the computation time was measured for a various number of processes used. The scalability (M) was calculated using the Eq. (15):

$$M = \frac{Time_1}{Time_K} \tag{15}$$

where $Time_1$ is the computation time when one MPI process is used. $Time_K$ is the computation time when K processes are used. The scalability study results are shown in the Table 2.

Ultra-scalability with a small number of processes is achieved using Hyper-Threading technology. The reasons for this are the same as in the acceleration analysis. Note that the transition to 32 processes gives a significant fall in scalability. In this case, more than one core is used per processor, a similar but less significant drawdown occurs with fewer processes. Apparently, the Intel Broadwell nodes and the software infrastructure used are focused on using the MPI + OpenMP model when a process uses one processor with a large number of cores. This situation is more preferable than the situation when it was more efficient to use MPI technology for work distribution between the cores.

Table 2. The scalability reserach.

Nodes	Processors	Cores	Threads	MPI-Processes	Scalability
1	1	1	1	1	1.000
1	1	1	2	2	0.996
1	1	2	1	2	0.993
1	2	1	1	2	0.994
2	1	1	1	2	1.031
1	1	2	2	4	0.965
1	2	2	1	4	1.036
2	2	1	1	4	1.027
2	1	1	2	4	1.067
1	2	2	2	8	0.829
2	2	1	2	8	1.021
4	1	1	2	8	1.039
4	2	1	1	8	0.969
2	2	2	2	16	0.816
4	2	1	2	16	0.964
4	2	2	1	16	0.943
4	2	2	2	32	0.734
4	2	4	1	32	0.769
4	1	4	2	32	0.765

4   The Numerical Simulation

We will simulate a galactic jet with density $\rho_J = 10^{-3}$ cm^{-3} and of radius $R_J = 200$ parsec moving with the Lorentz factor $\Gamma = 5$ and with the relativistic Mach number of $\mathcal{M} = 8$. The galaxy atmosphere has a temperature $T_A = 10^7$ Kelvin and an irregular density of $\rho_A = 10^{-2} \pm 10^{-3}$ cm^{-2}. An adiabatic exponent is selected as $\gamma = 5/3$. Figure 2 shows the results of simulating the galactic jet evolution. From the simulation results it is seen that the shock wave goes forward; the shock wave propagation velocity corresponds to the speed of light. Behind the shock front, there is a shell region separating the shock front and hotspot, where the maximum temperature is reached. The inner part of the cocoon flow is limited by the contact surface, and the cocoon, in its turn, contains a jet. The backflow streams are propagating on the outside of the cocoon and closer to the base, which interacts with the jet stream. This leads to the evolution of the Kelvin–Helmholtz instability, which characteristic evolution time corresponds to the results of a computational experiment. Using a high-order accuracy numerical scheme ensures low dissipation of the numerical solution. However, we do not get a high order in the shock wave regions. This experimental fact is described in details in the papers of S.K. Godunov [41] and V.V. Ostapenko [42].

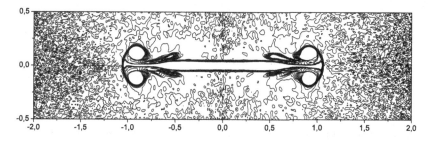

Fig. 2. Density isolines in the equatorial density of a galactic jet

5 Conclusion

The new program code enables simulating special relativistic hydrodynamics flows on supercomputer architectures with distributed memory. The code is based on the new numerical method combining the Godunov method, flow calculation by the Rusanov method, and piecewise linear representation of the solution. Such a combination allows us to preserve conservatism, correctly reproduce the shock waves, and ensure a high order of accuracy on smooth solutions and small dissipation at discontinuities. The simulation results of the relativistic jet interaction with an inhomogeneous galactic medium are presented. The results of studying the program code performance on the computing nodes (based on Intel Broadwell processors) of the NKS-1P cluster located in the Siberian Supercomputer Center are presented. Ultra-scalability with a small number of processes is achieved using HyperThreading technology.

Acknowledgements. The research work was supported by the Grant of the Russian Science Foundation (project 18-11-00044).

References

1. Wang, P., Abel, T., Zhang, W.: Relativistic hydrodynamic flows using spatial and temporal adaptive structured mesh refinement. Astrophys. J. Suppl. Ser. **176**, 467–483 (2008). https://doi.org/10.1086/529434
2. Karen Yang, H.-Y., Reynolds, C.S.: How AGN jets heat the intracluster medium-insights from hydrodynamic simulations. Astrophys. J. **829** (2016). article number 90, https://doi.org/10.3847/0004-637X/829/2/90
3. Khoperskov, S., Venichenko, Y., Khrapov, S., Vasiliev, E.: High performance computing of magnetized galactic disks. Supercomput. Front. Innov. **5**, 103–106 (2018). https://doi.org/10.14529/jsfi180412
4. Bosch-Ramon, V., Khangulyan, D.: Understanding the very-high-energy emission from microquasars. Int. J. Mod. Phys. D. **18**, 347–387 (2009). https://doi.org/10.1142/S0218271809014601
5. Fromm, C.M., Perucho, M., Mimica, P., Ros, E.: Spectral evolution of flaring blazars from numerical simulations. Astron. Astrophys. **588** (2016). article number A101. https://doi.org/10.1051/0004-6361/201527139

6. Janiuk, A., Sapountzis, K., Mortier, J., Janiuk, I.: Numerical simulations of black hole accretion flows. Supercomput. Front. Innov. **5**, 86–102 (2018). https://doi.org/10.14529/jsfi180208

7. Nagakura, H., Ito, H., Kiuchi, K., Yamada, S.: Jet propagations, breakouts, and photospheric emissions in collapsing massive progenitors of long-duration gamma-ray bursts. Astrophysi. J. **731** (2011). article number 80. https://doi.org/10.1088/0004-637X/731/2/80

8. Hughes, P., Miller, M., Duncan, G.: Three-dimensional hydrodynamic simulations of relativistic extragalactic jets. Astrophys. J. **572**, 713–728 (2002). https://doi.org/10.1086/340382

9. Nagakura, H., Sumiyoshi, K., Yamada, S. Three-dimensional Boltzmann hydro code for core collapse in massive stars. I. Special relativistic treatments. Astrophys. J. Suppl. Ser. **214** (2014). article number 16. https://doi.org/10.1088/0067-0049/214/2/16

10. O'Connor, E., Ott, C.: A new open-source code for spherically symmetric stellar collapse to neutron stars and black holes. Class. Quantum Gravity **27** (2010). article number 114103. https://doi.org/10.1088/0264-9381/27/11/114103

11. Komissarov, S.: Electrodynamics of black hole magnetospheres. Mon. Not. R. Astron. Soc. **350**, 427–448 (2004). https://doi.org/10.1111/j.1365-2966.2004.07598.x

12. Komissarov, S.: Observations of the Blandford-Znajek process and the magnetohydrodynamic Penrose process in computer simulations of black hole magnetospheres. Mon. Not. R. Astron. Soc. **359**, 801–808 (2005). https://doi.org/10.1111/j.1365-2966.2005.08974.x

13. Palenzuela, C., Garrett, T., Lehner, L., Liebling, S.: Magnetospheres of black hole systems in force-free plasma. Phys. Rev. D. **82** (2010). article number 044045. https://doi.org/10.1103/PhysRevD.82.044045

14. Palenzuela, C., Bona, C., Lehner, L., Reula, O.: Robustness of the Blanford-Znajek Mechanism. Class. Quantum Gravity **28** (2011). article number 4007. https://doi.org/10.1088/0264-9381/28/13/134007

15. Palenzuela, C., Lehner, L., Liebling, S.: Dual jets from binary black holes. Science **329**, 927–930 (2010). https://doi.org/10.1126/science.1191766

16. Komissarov, S.: Simulations of the axisymmetric magnetospheres of neutron stars. Mon. Not. R. Astron. Soc. **367**, 19–31 (2006). https://doi.org/10.1111/j.1365-2966.2005.09932.x

17. Duez, M.D., Liu, Y.T., Shapiro, S.L., Shibata, M., Stephens, B.C.: Collapse of magnetized hypermassive neutron stars in general relativity. Phys. Rev. Lett. **96** (2006). article number 031101. https://doi.org/10.1103/PhysRevD.77.044001

18. Duez, M.D., Liu, Y.T., Shapiro, S.L., Shibata, M., Stephens, B.C.: Evolution of magnetized, differentially rotating neutron stars: simulations in full general relativity. Phys. Rev. D. **73** (2006). article number 104015. https://doi.org/10.1103/PhysRevD.73.104015

19. Shibata, M., Duez, M.D., Liu, Y.T., Shapiro, S.L., Stephens, B.C.: Magnetized hypermassive neutron-star collapse: a central engine for short gamma-ray bursts. Phys. Rev. Lett. **96** (2006). article number 031102. https://doi.org/10.1103/PhysRevLett.96.031102

20. Marsh, T., et al.: A radio-pulsing white dwarf binary star. Nature **537**, 374–377 (2016). https://doi.org/10.1038/nature18620

21. Coleman Miller, M., Yunes, N.: The new frontier of gravitational waves. Nature **568**, 469–476 (2019). https://doi.org/10.1038/s41586-019-1129-z

22. Löffler, F., et al.: The Einstein Toolkit: a community computational infrastructure for relativistic astrophysics. Class. Quant. Grav. **29** (2012). article number 115001. https://doi.org/10.1088/0264-9381/29/11/115001

23. Goodale, T., et al.: The cactus framework and toolkit: design and applications. In: Palma, J.M.L.M., Sousa, A.A., Dongarra, J., Hernández, V. (eds.) VECPAR 2002. LNCS, vol. 2565, pp. 197–227. Springer, Heidelberg (2003). https://doi.org/10.1007/3-540-36569-9_13

24. Schnetter, E., Hawley, S., Hawke, I.: Evolutions in 3D numerical relativity using fixed mesh refinement. Class. Quantum Gravity **21**, 1465–1488 (2004). https://doi.org/10.1088/0264-9381/21/6/014

25. Schnetter, E., Diener, P., Dorband, E.N., Tiglio, M.: A multi-block infrastructure for three-dimensional time-dependent numerical relativity. Class. Quantum Gravity **23**, S553–S578 (2006). https://doi.org/10.1088/0264-9381/23/16/S14

26. Lehner, L., Liebling, S., Reula, O.: AMR, stability and higher accuracy. Class. Quantum Gravity **23**, S421–S446 (2006). https://doi.org/10.1088/0264-9381/23/16/S08

27. Thomas, M., Schnetter, E.: Simulation factory: taming application configuration and workflow on high-end resources. In: 2010 11th IEEE/ACM International Conference on Grid Computing, pp. 369–378 (2010). https://doi.org/10.1109/GRID.2010.5698010

28. Husa, S., Hinder, I., Lechner, C.: Kranc: a Mathematica package to generate numerical codes for tensorial evolution equations. Comput. Phys. Commun. **174**, 983–1004 (2006). https://doi.org/10.1016/j.cpc.2006.02.002

29. Donat, R., Font, J.A., Ibanez, J.M., Marquina, A.: A flux-split algorithm applied to relativistic flows. J. Comput. Phys. **146**, 58–81 (1998). https://doi.org/10.1006/jcph.1998.5955

30. Rivera-Paleo, F.J., Guzman, F.S.: CAFE-R: a code that solves the special relativistic radiation hydrodynamics equations. Astrophys. J. Suppl. Ser. **241** (2019). article number 28. https://doi.org/10.3847/1538-4365/ab0d8c

31. Kulikov, I., Vorobyov, E.: Using the PPML approach for constructing a low-dissipation, operator-splitting scheme for numerical simulations of hydrodynamic flows. J. Comput. Phys. **317**, 318–346 (2016). https://doi.org/10.1016/j.jcp.2016.04.057

32. Roedig, C., Zanotti, O., Alic, D.: General relativistic radiation hydrodynamics of accretion flows - II. Treating stiff source terms and exploring physical limitations. Mon. Not. R. Astron. Soc. **426**, 1613–1631 (2012). https://doi.org/10.1111/j.1365-2966.2012.21821.x

33. Aloy, M., Ibanez, J., Marti, J., Muller, E.: GENESIS: a high-resolution code for three-dimensional relativistic hydrodynamics. Astrophys. J. Suppl. Ser. **122**, 122–151 (1999). https://doi.org/10.1086/313214

34. Timmes, F.X., Arnett, D.: The accuracy, consistency, and speed of five equations of state for stellar hydrodynamics. Astrophys. J. Suppl. Ser. **125**, 277–294 (1999). https://doi.org/10.1086/313271

35. Mignone, A., et al.: PLUTO: a numerical code for computational astrophysics. Astrophys. J. Suppl. Ser. **170**, 228–242 (2007). https://doi.org/10.1086/513316

36. Zhang, W., MacFayden, A.: RAM: a relativistic adaptive mesh refinement hydrodynamics code. Astrophys. J. Suppl. Ser. **164**, 255–279 (2006). https://doi.org/10.1086/500792

37. Duffel, P., MacFayden, A.: TESS: a relativistic hydrodynamics code on a moving Voronoi mesh. Astrophys. J. Suppl. Ser. **197** (2011). article number 15. https://doi.org/10.1088/0067-0049/197/2/15

38. Nunez-de la Rosa, J., Munz, C.-D.: XTROEM-FV: a new code for computational astrophysics based on very high order finite-volume methods - II. Relativistic hydro- and magnetohydrodynamics. Monthly Not. Roy. Astron. Soc. **460**, 535–559 (2016). https://doi.org/10.1093/mnras/stw999

39. Lamberts, A., Fromang, S., Dubus, G., Teyssier, R.: Simulating gamma-ray binaries with a relativistic extension of RAMSES. Astron. Astrophys. **560** (2013). article number A79. https://doi.org/10.1051/0004-6361/201322266

40. Saini, S., Hood, R.: Performance evaluation of intel broadwell nodes based super-computer using computational fluid dynamics and climate applications. In: 2017 IEEE 19th International Conference on High Performance Computing and Communications Workshops (HPCCWS), Bangkok, pp. 58–65 (2017). https://doi.org/10.1109/HPCCWS.2017.00015

41. Godunov, S.K., Manuzina, Y.D., Nazareva, M.A.: Experimental analysis of convergence of the numerical solution to a generalized solution in fluid dynamics. Comput. Math. Math. Phys. **51**, 88–95 (2011). https://doi.org/10.1134/S0965542511010088

42. Ladonkina, M.E., Neklyudova, O.A., Ostapenko, V.V., Tishkin, V.F.: On the accuracy of the discontinuous galerkin method in calculation of shock waves. Comput. Math. Mat. Phys. **58**(8), 1344–1353 (2018). https://doi.org/10.1134/S0965542518080122

Parallel Technologies in Unsteady Problems of Soil Dynamics

Yury V. Perepechko$^{(\boxtimes)}$ ⓘ, Sergey E. Kireev ⓘ, Konstantin E. Sorokin ⓘ,
Andrey S. Kondratenko ⓘ, and Sherzad Kh. Imomnazarov ⓘ

N. A. Chinakal Institute of Mining, SB RAS, Novosibirsk, Russia
perep@igm.nsc.ru, kireev@ssd.sscc.ru, konst_sorokin_85@ngs.ru,
kondratenko@misd.ru, imom@omzg.sscc.ru

Abstract. The paper focuses on parallel technologies in numerical modeling of unsteady soil dynamics for the trenchless technology of well construction. The consolidated cores being formed at the pulse pressing of a hollow pipe into the soil is a feature of the trenchless technology used for underground pipeline laying. In contrast to the soil that is simulated as a granular medium, the consolidated core is simulated as a porous medium due to its shear stresses. The equations of granular medium dynamics are approximated by a completely implicit scheme following the finite volume method, while the equations of porous medium dynamics are approximated by an explicit scheme based on the WENO-Runge-Kutta method. Very fast processes of forming and breaking a consolidated plug require using an explicit scheme. Nonlinear models of porous core medium and models of granular soil medium involve thermal processes and are thermodynamically compatible. In this work, we examine the patterns and efficiency of a parallel algorithm in reference to the used differential approximation methods of the equations describing the continuous heterophase media mechanics.

Keywords: Mathematical modeling · Parallel technologies ·
Trenchless technologies · Granular media · Porous media ·
Two-velocity hydrodynamics · Conservation law method · Control
volume method · Decomposition method

1 Introduction

The development of various technologies based on the heterophase media properties leads to the needs to enhance both mathematical models of such continuous media and numerical methods of their solution. Considering various dissipative and surface phenomena, the mathematical models of the heterophase media have complex and non-linear nature of the differential equation systems, which describe them. This is particularly true for the problems of thermal effects, compressibility, and rheology of the heterophase medium components. For such systems, it is important to develop parallelizing methods in the application of

© Springer Nature Switzerland AG 2020
L. Sokolinsky and M. Zymbler (Eds.): PCT 2020, CCIS 1263, pp. 237–250, 2020.
https://doi.org/10.1007/978-3-030-55326-5_17

hydrodynamic system equations. In this paper, these methods are discussed in the application to the trenchless well-drilling technology with a batch removal of a soil core in the light of urban infrastructure [1–7]. This technology includes the widespread models of continuous heterophase media. Such processes as the air or water-saturated soil entering into a pipe, the soil core consolidation with a low-permeable plug being formed, and the plug transfer along the pipe with its subsequent destruction generate the need for the heterophase media models simulating the granular and porous media saturated with gas, liquid or their mixture. To obtain governing equations of such media dynamics, methods that guarantee the thermodynamic compatibility of the model equations and, consequently, the physical consistency of the simulation results are used. The numerical methods that we have applied in this work were also selected based on the criterion of being able to keep the solution consistency. This approach is ubiquitous and enables applying the mathematical and numerical methods created to a wide class of technological issues from different research fields of industrial and natural systems [8–11]. A numerical analysis of the saturated granular medium dynamics was performed by the method of control volume, which guarantees the integral conservation of the main model parameters [12,13]. A completely implicit scheme is used, and due to the fact that typical times of the soil motion process in a pipe are hydrodynamic one needs to ensure the possible calculations for sufficiently long periods of time. In this problem, the processes occurring in a consolidated soil plug are characterized by rapid sonic times. A numerical analysis of the saturated medium dynamics is therefore conducted using the WENO-Runge-Kutta method showing its efficiency for the hyperbolic equation systems [14–16]. In both models, the parallel implementation is based on the domain decomposition techniques.

2   The Mathematical Model

The simulated system is represented by an internal cavity of a pipe, which periodically supplies water-saturated soil or water-air mixture. The pipe discharges these media from its right end (see Fig. 1). When a soil core has been formed, it is pushed by the ejecting medium of the water and air mixture with a minor content of soil particles through the right opening of the pipe. During the ejection, the core gets consolidated with a porous soil plug being formed, which is subsequently destroyed [1–3,7]. Thus, it is necessary to simulate two types of continuous media during the system operation – a saturated granular media and a saturated deformable porous media. It should be noted that the soil plug consolidation and destruction is characterized by small times, which are lesser than the typical times of soil injection into the pipe.

Granular Medium Dynamics (GMD) Model. A compressible granular medium consisting of soil particles and a saturating medium (water, air or gas-water mixture) was considered as a soil and soil core model. A single volume of such a two-phase medium is characterized by densities and velocities of phases, entropy, and specific content of the soil granules. Thermodynamic properties of such

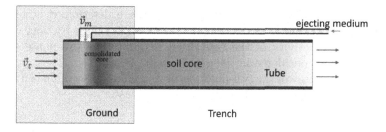

Fig. 1. Pipe intrusion into the soil under the impact action at a periodic injection of an ejecting medium near the downhole face

a two-phase medium are determined by the functional dependence of its bulk density on the internal energy $E_0 = E_0(\rho, J, \mathbf{j}_0, S)$:

$$dE_0 = TdS + \mu d\rho + \varsigma\sigma dJ + (\mathbf{w}, d\mathbf{j}_0),$$

where $\rho = \rho_1 + \rho_2$, $\mathbf{j} = \rho_1\mathbf{u}_1 + \rho_2\mathbf{u}_2$, ρ_1, ρ_2 are partial densities of the soil particles and saturating medium: $\rho_n = \phi_n\rho_n^{ph}$, ρ_n^{ph} are physical densities, ϕ_n is volume contents of the phases; $\mathbf{u}_n$ is a corresponding velocity of the phases; $\mathbf{w} = \mathbf{u}_1 - \mathbf{u}_2$ is a relative velocity of the phase movement; μ is a two-phase medium chemical potential; T, S are the temperature and entropy density of the two-phase medium; ς is specific surface area of soil granules; σ is a surface tension tensor, which can depend on temperature: $\sigma = \sigma_0(T_c - T)/(T_c - T_{ref})$; a specific content of the soil granules J determines a partial density of the granular phase ρ_1; $\mathbf{j}_0 = \mathbf{j} - \rho\mathbf{u}_2$ is density of the phase relative momentums.

A thermodynamically compatible system of governing equations, which describes the saturated granular medium hydrodynamics can be obtained within the conservation laws [8,10,11,17]. Taking into account the dissipative effects, harmonization of the thermodynamics principles, conservation laws of mass, energy, momentum, and Galilean invariance of the equations defines the following kind of governing equations of the saturated granular media dynamics [1,2]

$$\frac{\partial \rho}{\partial t} + \operatorname{div} \mathbf{j} = 0, \quad \frac{\partial \rho_1}{\partial t} + \operatorname{div}(\rho_1\mathbf{u}_1) = 0, \tag{1}$$

$$\frac{\partial j_i}{\partial t} + \partial_k\left(\rho_1 u_{1i}u_{1k} + \rho_2 u_{2i}u_{2k} + p\delta_{ik}\right) = \delta_k\left(\eta_1 u_{1ik} + \eta_2 u_{2ik}\right) + \rho g_i, \tag{2}$$

$$\frac{\partial u_{2i}}{\partial t} + u_{2k}\partial_k u_{2i} = -\frac{1}{\rho}\partial_i p + \varsigma\frac{\rho_1}{\rho}\partial_i\sigma + \frac{\rho_1}{2\rho}\partial_i\mathbf{w}^2 + bw_i + \frac{1}{\rho_2}\partial_k\left(\eta_2 u_{2ik}\right) + g_i, \tag{3}$$

$$\frac{\partial \rho s}{\partial t} + \operatorname{div}(s\mathbf{u}) = \frac{\rho_2}{T}b\mathbf{w}^2 + \frac{1}{T}\kappa\Delta T + \frac{1}{2T}\eta_1 u_{1ik}u_{1ik} + \frac{1}{2T}\eta_2 u_{2ik}u_{2ik}. \tag{4}$$

p is pressure; $\mathbf{u} = \mathbf{j}/\rho$ is a mean velocity of the two-phase medium; $u_{nik} = \partial_k u_{ni} + \partial_i u_{nk} - \frac{2}{3}\delta_{ik}\operatorname{div}\mathbf{u}_n$ are viscous deformation velocities; $\mathbf{g}$ is the gravitational acceleration. Kinetic coefficients of the interphase friction b, phase shear viscosity η_n, thermal conductivity κ of the two-phase medium are the thermodynamic parameter functions. The dissipative phenomena are determined according to Onsager's reciprocal relations – this model includes interphase friction, viscosity, and thermal conductivity of the two-phase medium. The two-phase granular medium motion is considered in the gravity field.

The internal energy of the two-phase saturated granular medium can be given as an additive function of the internal energy of phases

$$e_0 = e_1\left(\rho_1, s\right) + x\left(e_2\left(\rho_2, s\right) - e_1\left(\rho_1, s\right)\right) + \frac{1}{2}x\left(1 - x\right)w_k w_k.$$

Here $e_0 = E_0/\rho$, $s = S/\rho$ are mass densities of the internal energy and two-phase medium entropy, respectively; $x = x_2$ is a mass content of saturating phase "2": $x_n = \phi_n \rho_n/\rho$. The specific densities of internal energies of phases e_n are taken into account by the following relations

$$e_n = \frac{1}{\rho_n^2}p_{n0}\delta\rho_n + T_0\delta s + \frac{1}{2}\kappa_n\delta s\delta s + \frac{1}{\rho_n^2}\pi_n\delta s\delta\rho_n + \frac{1}{2\rho_n^3}K_n\delta\rho_n\delta\rho_n. \qquad (5)$$

In relations (5) the following notions are used $\delta s = s - s_0$, $\delta\rho_n = \rho_n - \rho_{n0}$. The coefficients κ_n, π_n are related to the coefficients of thermal expansion β_n, coefficients of volumetric expansion K_n and heat capacities of phases c_{pn} by ratios

$$\pi_n = T_0\frac{K_n\beta_n}{c_{pn}}, \quad \kappa_n = \frac{T_0}{c_{pn}}\left(1 + T_0\frac{K_n\beta_n^2}{\rho_{n0}c_{pn}}\right). \qquad (6)$$

Determining pressure and temperature in the phases

$$p_n = \rho_n^2\frac{\partial e_n}{\partial\rho_n}, \quad T_n = \frac{\partial e_n}{\partial\rho_n} \qquad (7)$$

determine hydrodynamic pressure and temperature of the two-phase medium

$$p = p_1 + x\left(p_2 - p_1\right), \quad T = T_1 + x\left(T_2 - T_1\right), \qquad (8)$$

Consequently, the equation of state has a form of

$$\delta\rho = \rho\alpha\delta p - \rho\beta\delta T, \quad \delta s = c_p\delta T/T - \beta\delta p/\rho.$$

Here the coefficients of volume compression α, heat expansion β, and specific heat capacity of the two-phase medium c_p are taken as additive in the phases.

Porous Medium Dynamics (PMD) Model. A deformable porous medium consisting of a porous soil matrix and a saturating medium (water, air or gas-water mixture) was considered as a model of the consolidated core. Shear stresses and corresponding deformations are additional properties distinguishing a porous and granular media. In this case, the internal energy bulk density of the saturated porous medium will also depend on the strain tensor $E_0 = E_0\left(\rho, \rho_1, \phi, \mathbf{j}_0, \epsilon_{ik}, S\right)$:

$$dE_0 = TdS + \mu d\rho + \mu_1 d\rho_1 + (\mathbf{w}, d\mathbf{j}_0) + \tilde{\sigma}_{ik}d\epsilon_{ik}.$$

Here ϵ_{ik} is a strain tensor of the saturated porous medium, $\tilde{\sigma}_{ik}$ is a coupled stress tensor. The generalized internal energy in this model depends on the overall entropy of the two-phase system S due to the assumption of phase equilibrium by temperature.

A thermodynamically compatible system of the governing equations describing the saturated porous medium hydrodynamics can be obtained by the method of thermodynamically compatible system of the conservation laws [9]. The governing equations of the compressible fluid medium flowing through the deformable porous matrix, which consider the dissipative phase interaction in the form of the bulk phase concentrations relaxation to equilibrium values and interphase friction, can be represented as follows according to [15, 18]

$$\frac{\partial \rho}{\partial t} + \operatorname{div} \mathbf{j} = 0, \quad \frac{\partial \rho_1}{\partial t} + \operatorname{div}(\rho_1 \mathbf{u}_1) = 0, \quad \frac{\partial \phi \rho}{\partial t} + \operatorname{div}(\phi \mathbf{j}) = -\lambda (p_1 - p_2), \quad (9)$$

$$\frac{\partial j_i}{\partial t} + \partial_k (\rho_1 u_{1i} u_{1k} + \rho_2 u_{2i} u_{2k} + p \delta_{ik} - \phi \sigma_{ik}) = \rho g_i, \quad (10)$$

$$\frac{\partial w_i}{\partial t} + \partial_i \left(\frac{1}{2} u_{1i}^2 - \frac{1}{2} u_{2i}^2 + e_1 + \frac{\phi}{\rho_1} p_1 - e_2 - \frac{1-\phi}{\rho_2} p_2 \right) = u_i (\partial_k w_i - \partial_i w_k) - b w_i, \quad (11)$$

$$\frac{\partial \epsilon_{ij}}{\partial t} + u_k \partial_k \epsilon_{ij} + \epsilon_{ik} \partial_j u_k + \epsilon_{kj} \partial_i u_k - \frac{1}{2} (\partial_j u_i + \partial_i u_j) = 0, \quad (12)$$

$$\frac{\partial \rho s}{\partial t} + \partial_k (s u_k) = \frac{\rho_2}{T} b \mathbf{w}^2 + \frac{1}{\rho_2 T} \lambda (p_1 - p_2)^2. \quad (13)$$

The given equations represent the conservation laws of the two-phase medium density, solid phase mass and volumetric concentrations, total momentum, and relative velocities of the liquid phase movement in relation to the porous matrix, elastic deformation, and specific entropy density of the heterophase medium.

The internal energy of the saturated porous medium is also defined as an average of the mass concentration of the phase internal energies, which is supplemented by the kinetic energy of the phase relative motions:

$$e_0 = e_1 (\rho_1, \epsilon_{ij}, s) + x (e_2 (\rho_n, s) - e_1 (\rho_1, \epsilon_{ij}, s)) + \frac{1}{2} (1 - x) x w_i w_i.$$

The equation of state is determined by the ratios for phase pressure $p_n = \rho_n^2 e_{n,\rho_n}$, shear stresses $\sigma_{ij} = \rho_1 (\delta_{jm} - 2\epsilon_{jm}) e_{1,\epsilon_{im}}$, and temperature $T = x e_{1,s} + (1 - x) e_{2,s}$. In contrast to the expression for granular medium, the internal energy of the porous medium is further dependent on the strain tensor

$$e_1 = \frac{1}{\rho_1^2} p_{10} \delta \rho_1 + T_0 \delta s + \frac{1}{2} \kappa_1 \delta s \delta s + \frac{1}{\rho_1^2} \pi_1 \delta s \delta \rho_1 + \frac{1}{2\rho_1^3} K_1 \delta \rho_1 \delta \rho_1$$
$$+ \frac{1}{\rho_1} \mu \left(\epsilon_{ij} \epsilon_{ij} - \frac{1}{3} \epsilon_{ll} \epsilon_{ll} \right).$$

Here, μ is a shear elastic modulus, K_1 is a bulk elastic modulus of the porous solid phase. Other parameters are determined by relations (5)–(8).

3   Numerical Algorithms

Control Volume Method. In order to solve the problem of a granulated core movement in a pipe, the control volume method is used to ensure compliance with the conservation laws on any finite volume [12,13]. A numerical implementation of the model on a uniform rectangular grid in combination with an implicit time scheme [1,2] was used. To calculate the flows on the faces of control volumes, a non-linear second-order HLPA scheme [19] is utilized. To approximate the diffusion terms, a central difference scheme is applied. The model equations are nonlinear. To solve this problem, we used the IPSA [20] iteration procedure for a computational algorithm, which is based on a simple iteration method and is closed by a global iterative procedure to calculate the pressure consistent with the phase flow velocity fields. The basic computational complexity of the algorithm is to solve the systems of linear algebraic equations (SLAE) resulting from digitalization of the system equations. To solve the motion and transport equations, the alternating direction method (ADM) modification is used. To solve the pressure correction equation, an implementation of the BiCGStab method from the PETSc library is used.

WENO-Runge-Kutta Method. To numerically approximate the equations of porous consolidated core dynamics, a method combining the WENO reconstruction of the initial data with the Runge-Kutta method of high order time and space integration accuracy method [16,18] was used. The choice of this method is determined by the divergent type and hyperbolicity of the model equations. These properties are determined by the possibility to represent the model equations in terms of generating a thermodynamic potential [9]. This method was successfully employed in the continuum mechanics problems [15,18] and in the analysis of acoustic fields in a well and porous fluid-saturated rock [21]. The utilized method is explicit, which constrains a time step due to the Courant criterion determined by the calculated maximum velocity of the acoustic vibrations. A uniform rectangular calculation grid is used as in the first problem.

4   Parallel Implementations

Parallel implementations for both GMD and PMD problems are based on a domain decomposition method. A 2D simulation domain is divided into subdomains along both directions. All arrays of the mesh values are appropriately cut and distributed among the parallel processes. The number of subdomains in each direction is an implementation parameter. Shadow edges of the subdomains have a width of two cells for the GMD problem, and three cells for the PMD problem, which corresponds to the size of the difference schemes used. The same parameter determines the minimum allowable size for each subdomain dimension. The MPI library was used to parallelize both problems. In addition, computations in each MPI process were parallelized using the OpenMP threads.

The parallel PMD code was implemented using C language. All stages of the PMD problem solution by the WENO-Runge-Kutta method belong to the class of stencil computations. For the parallel solution of such problems, an independent recalculation of the mesh values in each subdomain and exchange of boundary values are sufficient. The parallelization efficiency of stencil computations is usually high. Particular attention during the program implementation was given to the vectorization, which is necessary to achieve the highest possible performance. In this case, the vectorization was performed by means of the Intel vectorizing compiler by specifying the corresponding directives in the code.

The parallel GMD code was implemented using the Fortran language. The following SLAE solvers are used in the program:

- our own implementation of the ADM adapted to the considered class of problems and optimized for specific conditions of use,
- the implementation of BiCGStab method from the PETSc library [22],
- the PARDISO solver from the Intel MKL library [23].

The parallel ADM solver is built upon the original distributed 2D arrays and only requires the border elements transfer during the computation. The parallel BiCGStab and PARDISO solvers require a distributed sparse SLAE matrix to be built. To minimize the overhead, the distribution of the sparse matrix elements used in the program is consistent with the 2D decomposition of the solution vector. No special attention was paid to the GDM code vectorization. But the use of Intel MKL library in PARDISO and BiCGStab solvers to some extent guarantees certain low-level optimizations, such as vectorization. More details of the GMD code implementation may be found in [1, 2].

5   Simulation Results

The problems of motion of the water or water-gas mixture through the saturated granulated soil core inside a pipe and deformation of the consolidated core in the field of gravity under the periodic action of a pneumatic hammer and the injection of ejecting medium near the downhole face were solved in the conditions close to natural ones (see Fig. 1). The processes occurring in such a system have been studied within two problems based on (1)–(4) and (9)–(13) system equations. The core dynamics and the distribution evolution of the granular and liquid phases in this system have been investigated as part of the GMD problem. The obtained values of the GMD problem parameters for the flow regimes from the experimental observations were used to set a PMD problem to calculate the distribution deformation fields and stresses in a porous core assuming its consolidation. Introducing the core consolidation and plug destruction criterion and obtaining the integrated model equations are subjects of further researches.

We performed computations for the problem characterized by a periodic injection of saturated granular soil through the left pipe end and periodic ejecting medium supply through the service opening. The calculations were performed for different values of the pipe wall friction, relative water and air content in the ejecting medium, interphase friction (determining permeability of the granular soil), velocity and frequency of soil introduction into the pipe, and injection of

Fig. 2. The regime of the pneumatic hammer's action on the soil core (red color) and ejecting medium (blue color); $\mathbf{v}_t$ is the velocity of pipe introduction into the soil at the downhole pipe end, $\mathbf{v}_m$ is the velocity of ejecting medium at the boundary with the pipe (Color figure online)

the ejecting medium. The main parameters for analyzing the obtained result are the distribution of specific soil particle content at different moments of the soil core motion and stress distributions in the consolidated core. All calculations were made for the pipe horizontal location, and pipe deformation was not taken into account.

The computational domain size was 1.5–2 m, and the cross-section was set as 0.22 m. The opening for the ejecting media injection on the top wall was located at a distance of an order of the diameter of a pipe of 0.22 m from downhole edge of a pipe. The diameter of the opening was equal to of 0.01 m (see Fig. 1).

The soil core dynamics in the pipe was generated by the pneumatic hammer and ejecting medium action, which was governed by the compressor operation regime. There are two regimes according to the experiment (see Fig. 2): (I) under the action of a pneumatic hammer, the pipe is incorporated into the soil at an average velocity of 0.5 cm/s, the ejecting mixture is not supplied through the service opening; (II) pneumatic hammer does not operate, and the pipe does not move; the ejecting mixture is fed through the opening. In the computations, the time of regime (I) was specified from 8 to 45 s, the time of regime (II) – from 3 to 5 s. Such operating regimes define the following boundary conditions.

- Regime (I) – soil saturated with water or water-air mixture is pushed into the pipe at the inlet boundary at a rate 0.005 m/s under the pulse action with a frequency of 1.5 Hz. The service opening is taken as impermeable.
- Regime (II) – the inlet boundary is assumed to be impermeable, no mass flows through it. The flow rate of water or water-air mixture is set from 1 m/s to 40 m/s through the service opening.

At the inner pipe walls, a condition of partial adhesion was set with the friction coefficient varying at the boundary corresponding from full adhesion to a complete slip of the ground plug. The pipe output boundary was considered to be open.

The properties of sand were selected to act as of the soil particle physical properties in the GMD problem. The soil granule size was set to 1 mm. A sand volume content in the saturated soil was taken as $\phi = 0.8$. The ejecting medium physical properties were assumed to be the properties of water or an air-water mixture having a water content of 100% to 10%. The initial ground temperature was taken to be 20 °C.

Figure 3 shows the computation results of the pipe filling with soil, the core formation, and its removal during the injection through the service opening. The core boundary heterogeneity is determined by the specific selection of physical parameters of the heterophase granular medium corresponding to the sand saturated with the water-air mixture. The presence of clay particles in the natural ground also contributes to the soil particle consolidation.

Fig. 3. (a) shows the removal of the soil core from the pipe under the influence of the ejecting mixture; (c) illustrates the formation of a ground core at the introduction of a hollow pipe into the soil. The distribution of soil particles content is shown. Light color corresponds to a high concentration of soil granules, dark color is a water-air ejecting mixture. There is practically no friction on the pipe walls (Color figure online)

Numerical experiments with varying the model parameters also showed occurance of different regimes of the ejecting mixture movement through the core. In some cases, the core was partially captured along the pipe cross-section (Fig. 4a) and even an air/water-air channel was formed. The ejecting mixture was pushed out from the pipe through this channel and did not remove the core. Figure 4b displays how the channel is formed at the pipe downhole, while Fig. 4c illustrates how it is formed at the pipe front end.

The parameters of the study system corresponding to the case in Fig. 3 were used in the PMD problem, in which the core is assumed to be consolidated and is simulated by a deformable porous medium. The consolidated core physical properties (density, elastic modulus, and thermal conductivity) correspond to compacted sand or loose sandstone. The initial porosity of a soil plug was taken as $\phi = 0.2$. The ejecting medium physical properties were assumed to be similar to the GMD problem.

Figures 5b and 5c demonstrate the stress field propagation patterns in the consolidated core after turning off the compressor, which was injecting the ejecting mixture through the service opening in the pipe top wall.

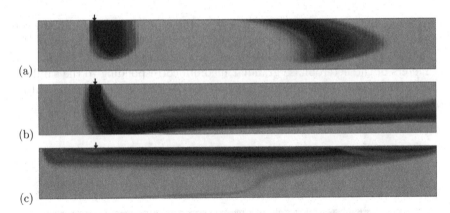

Fig. 4. The core soil partially removed from the pipe by the ejecting mixture (a); an air channel formation at the pipe bottom wall (b); an air channel being formed along the pipe upper wall (c). In (b) and (c) cases, no core removal occurs. The colors correspond to Fig. 3 (Color figure online)

Fig. 5. Distribution of core density ρ_1 (a) and σ_{xx} (b), σ_{xy} (c) components of the stress tensor in the consolidated core for the same time interval

6   Performance Evaluation

Studying the parallelizing efficiency of differential approximations for the two hydrodynamic equation systems presented above is particularly interesting. This section presents the performance evaluation results for the two problems under consideration. All computations were performed on cluster MVS-10P OP [24], "Skylake" partition with nodes containing 2 × 18-core Intel Xeon Gold 6154 3.0 GHz (36 cores per node, 2 threads per core), 192 GB RAM. The following software was used: Intel C/C++ Compiler v.17.0.4, Intel Fortran Compiler v.17.0.4, Intel MKL v.2017.0 update 3, PETSc v.3.10.2 (built with Intel MKL), Intel MPI library v.2017 update 3.

To investigate the correctness and performance of the parallel programs, we used the problem in the statement corresponding to the real system under study. The system had a rectangular geometry with a size of 1.0 m × 0.2 m. The injection border was on the left; the open border – on the right. On the upper wall closer to the left side, there was an opening for injecting a two-three-phase medium. As an execution time estimate, the average duration of one time step obtained over the first few steps is used. In both problems, grids of the same size were used.

To assess the performance of the developed codes, first of all, it is necessary to determine the launch parameters that give the least execution time. Such parameters include the number of processes and threads used in the cluster node, the aspect ratio in the process grid, as well as special implementation parameters of the problem, such as choosing the SLAE solver for solving various equations in the GMD problem.

In [2], such an analysis was performed for the GMD code. In particular, it was found out that the ADM solver gives the best results for solving the motion and transport equations, and the BiCGStab from the PETSc library showed the best time for solving the pressure correction equation. Further research showed that when using this combination of solvers, the smallest execution time of the entire program is obtained when parallelizing with MPI processes only, without OpenMP. In this case, the number of MPI processes in the node should be equal to the number of cores. It turned out that the use of Hyperthreading technology, which enables executing two threads by a single core, does not accelerate the execution. In addition, the best topology of a 2D grid of MPI processes turned out to be the closest to the square one, which is set by default by the MPI library. Completely similar results were obtained from the analysis of the PMD code. These parameters were later used to evaluate the performance of both codes.

Figure 6 shows the parallelization characteristics of the GMD code. The runs were performed on 1, 2, 4, 8, 16, 32, and 36 cores within the cluster node, as well as on 1, 2, 4, 8, 16, and 32 nodes, using 36 cores in each node. It can be seen that with an increase in the number of cores, the parallelization efficiency noticeably decreases.

When analyzing the parallelization efficiency, a general rule is known: the larger the problem, the better it is scaled. So, for this task, the smallest efficiency is observed on the smallest 500 × 100 mesh. But the next largest 1000 × 200 mesh shows the best parallelization efficiency within one cluster node (up to 36 cores). Larger tasks within the node show less efficiency, probably due to the fact that their data no longer fit in the cache of the node processors. As a result, more frequent accesses of several processes to the relatively slow RAM of the node slowed down the execution. As the number of nodes increased, the amount of cache available to the task began to increase, and efficiency began to grow. Starting from 8 nodes (288 cores), the performance graphs began to satisfy the general rule. The greatest acceleration achieved on the GMD problem is 387 times on a 4000 × 800 grid using 32 cluster nodes (1152 cores).

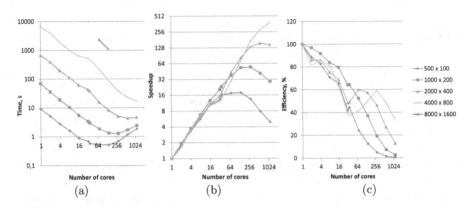

Fig. 6. The GMD problem. Execution time (a), speedup (b) and parallelization efficiency (c) for various mesh sizes, depending on the number of processor cores

Figure 7 shows the parallelization characteristics of the PMD code. Although the parallelization efficiency decreases with an increase in the number of cores, the acceleration graphs on available resources have not reached the saturation point and continue to grow. As usual, a larger tasks scale is better, and on the largest grid of 8000 × 1600, the parallelization efficiency does not fall below 80%. There are no noticeable effects associated with the cache in this task due to the simple data access pattern, which is a sequential traversal of several arrays. The greatest acceleration on the PMD problem is 976 times on a grid of 8000 × 1600 size using 32 cluster nodes (1152 cores). With the parameters similar to the best acceleration on the GMD problem (387 times), the acceleration on the PMD problem was 950 times. Thus, it is obvious that the PMD code scales better.

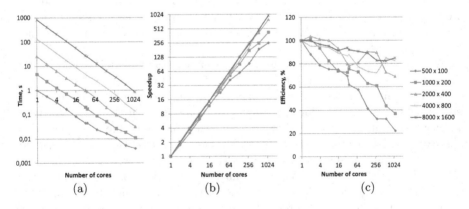

Fig. 7. The PMD problem. Execution time (a), speedup (b) and parallelization efficiency (c) for various mesh sizes, depending on the number of processor cores

7   Conclusion

We presented some approaches to apply high-performance computing systems to solving the nonstationary problems of the hydrodynamic equation-based modeling of the multivelocity heterophase media dynamics. Parallel implementations of the granular and porous media numerical models have proven to be efficient. The developed parallel models have been successfully applied in the numerical research (modeling) of the trenchless well drilling with batch removal of soil during the urban infrastructure development. Further development of the presented models is connected with the plug formation mechanism and the plug destruction at a stress field variation as the core moves in the pipe.

This work was performed at the Chinakal Institute of Mining, SB RAS, and supported by the Russian Science Foundation (grant No. 17-77-20049).

References

1. Perepechko, Y., Kireev, S., Sorokin, K., Imomnazarov, S.: Modeling of nonstationary two-phase flows in channels using parallel technologies. In: Sokolinsky, L., Zymbler, M. (eds.) PCT 2018. CCIS, vol. 910, pp. 266–279. Springer, Cham (2018). https://doi.org/10.1007/978-3-319-99673-8_19

2. Perepechko, Y., Kireev, S., Sorokin, K., Imomnazarov, S.: Use of parallel technologies for numerical simulations of unsteady soil dynamics in trenchless borehole Drilling. In: Sokolinsky, L., Zymbler, M. (eds.) PCT 2019. CCIS, vol. 1063, pp. 197–210. Springer, Cham (2019). https://doi.org/10.1007/978-3-030-28163-2_14

3. Smolyanitsky, B.N., et al.: Modern technologies for the construction of extended wells in soil massifs and technical means for controlling their trajectories. Publishing House of the Siberian Branch of the Russian Academy of Sciences, Novosibirsk (2016)

4. Kondratenko, A.S., Petreev, A.M.: Features of the earth core removal from a pipe under combined vibro-impact and static action. J. Min. Sci. 44(6), 559–568 (2008). https://doi.org/10.1007/s10913-008-0063-5

5. Grabe, J., Pucker, T.: Improvement of bearing capacity of vibratory driven open-ended tubular piles. In: Meyer, V. (ed.) Proceedings of 3rd International Symposium on Frontiers in Offshore Geotechnics 2015 in Oslo (Norway), vol. 1, pp. 551–556. Taylor & Francis Group, London (2015)

6. Labenski, J., Moormann, C., Ashrafi, J., Bienen, B.: Simulation of the plug inside open steel pipe piles with regards to different installation methods. In: Kuliešius, V., Bondars, K., Ilves, P. (eds.) Proceedings of 13th Baltic Sea Geotechnical Conference, pp. 223–230, Vilnius Gediminas Technical University (2016). https://doi.org/10.3846/13bsgc.2016.034

7. Danilov, B.B., Kondratenko, A.S., Smolyanitsky, B.N., Smolentsev, A.S.: Improving the technology of drilling wells in the soil by the method of forcing. Phys. Techn. Probl. Devel. Mineral Resour. 3, 57–64 (2017)

8. Khalatnikov, I.M.: An Introduction to the Theory of Superfluidity. W.A. Benjamin, New York (1965)

9. Godunov, S.K., Romenskii, E.I.: Elements of Continuum Mechanics and Conservation Laws. Springer, Boston (2003). https://doi.org/10.1007/978-1-4757-5117-8

10. Dorovsky, V.N.: Mathematical models of two-velocity media. I. Math. Comput. Model. **21**(7), 17–28 (1995). https://doi.org/10.1016/0895-7177(95)00028-Z

11. Dorovsky, V.N., Perepechko, Y.V.: Mathematical models of two-velocity media. II. Math. Comput. Model. **24**(10), 69–80 (1996). https://doi.org/10.1016/S0895-7177(96)00165-3

12. Arun Manohar, G., Vasu, V., Srikanth, K.: Modeling and simulation of high redundancy linear electromechanical actuator for fault tolerance. In: Srinivasacharya, D., Reddy, K.S. (eds.) Numerical Heat Transfer and Fluid Flow. LNME, pp. 65–71. Springer, Singapore (2019). https://doi.org/10.1007/978-981-13-1903-7_9

13. Date, A.W.: Introduction to Computational Fluid Dynamic. Cambridge University Press, New York (2005)

14. Romenski, E., Drikakis, D., Toro, E.F.: Conservative models and numerical methods for compressible two-phase flow. J. Sci. Comput. **42**(1), 68–95 (2010). https://doi.org/10.1007/s10915-009-9316-y

15. Perepechko, Y.V., Romenski, E.I., Reshetova, G.V.: Modeling of compressible multiphase flow through porous elastic medium. Seismic Technol. **4**, 78–84 (2014)

16. Shu, C.W.: Essentially non-oscillatory and weighted essentially non-oscillatory schemes for hyperbolic conservation laws. NASA/CR-97-206253, ICASE. report No. 97-65 (1997)

17. Dorovsky, V.N., Perepechko, Y.V.: Hydrodynamic model of the solution in fractured porous media. Russ. Geol. Geophys. **9**, 123–134 (1996)

18. Romenski, E.I., Perepechko, Y.V., Reshetova, G.V.: Modeling of multiphase flow in elastic porous media based on thermodynamically compatible system theory. In: Proceedings of the 14th European Conference on the Mathematics of Oil Recovery (Catania, Italy, 8–11 September). European Association of Geoscientists and Engineers (2014)

19. Wang, J.P., Zhang, J.F., Qu, Z.G., He, Y.L., Tao, W.Q.: Comparison of robustness and efficiency for SIMPLE and CLEAR algorithms with 13 high-resolution convection schemes in compressible flows. Numer. Heat Transfer, Part B **66**, 133–161 (2014). https://doi.org/10.1080/10407790.2014.894451

20. Yeoh, G.H., Tu, J.: Computational Techniques for Multi-phase Flows. Butterworth-Heinemann, Oxford (2010). https://doi.org/10.1016/B978-0-08-046733-7.00003-5

21. Dorovsky, V.N., Romenski, E.I., Fedorov, A.I., Perepechko, Y.V.: Resonance method for measuring the permeability of rocks. Russ. Geol. Geophys. **7**, 950–961 (2011)

22. PETSc: Portable, Extensible Toolkit for Scientific Computation. https://www.mcs.anl.gov/petsc

23. MKL, Intel® Math Kernel Library. http://software.intel.com/en-us/intel-mkl

24. MVS-10P cluster, JSCC RAS. http://www.jscc.ru

Parallel Implementation of Stochastic Simulation Algorithm for Nonlinear Systems of Electron-Hole Transport Equations in a Semiconductor

Karl K. Sabelfeld[ID] and Anastasiya Kireeva[(✉)][ID]

Institute of Computational Mathematics and Mathematical Geophysics,
Novosibirsk, Russia
karl@osmf.sscc.ru, kireeva@ssd.sscc.ru

Abstract. The paper presents parallel implementation of a stochastic model of electron and hole transport in a semiconductor. The transfer process is described by a nonlinear system of drift-diffusion-Poisson equations, which is solved by combining different stochastic global algorithms. The nonlinear system includes drift-diffusion equations for electrons and holes, and a Poisson equation for the potential, which gradient enters the drift-diffusion equations as drift velocity. We consider a two-dimensional problem and introduce a cloud of nodes that in particular can be chosen as a regular mesh on the domain. The electron and hole drift and diffusion are simulated by the stochastic cellular automata (CA) algorithm. The algorithm calculates the electron and hole concentrations in all lattice points. The drift velocity is calculated as a gradient of the solution to the Poisson equation with a right-hand side depending on the electron and hole densities. A global Monte Carlo (GMC) random walk on spheres algorithm enables calculating the solution of the Poisson equation and its derivatives in all lattice points at once. Parallel codes for both CA and GMC algorithms are implemented. For the CA algorithm, the particle handling loop is parallelized. The GMC random walk algorithm is parallelized by the Monte Carlo trajectories distribution among the cluster cores. The efficiency of the parallel code is analyzed.

Keywords: Electron-hole transport · Drift-diffusion-Poisson equations · Global Monte Carlo · Cellular automata · MPI

1 Introduction

Semiconductor devices are widely used in modern industry. To develop novel technologies and devices, a thorough study of the phenomena in semiconductors is needed. Mathematical models and numerical simulations support physical experiments and contribute to the improvement of the device designs.

Support of the Russian Science Foundation under Grant 19-11-00019 is gratefully acknowledged.

© Springer Nature Switzerland AG 2020
L. Sokolinsky and M. Zymbler (Eds.): PCT 2020, CCIS 1263, pp. 251–265, 2020.
https://doi.org/10.1007/978-3-030-55326-5_18

One of the basic models describing a flow of charge carriers in semiconductors is the drift-diffusion-Poisson model [1]. The nonlinear drift-diffusion equations were proposed by van Roosbroeck in 1950 [2]. Many papers are devoted to the mathematical analysis of the drift-diffusion-Poisson system: the global existence and uniqueness of the solutions under different physical and geometrical conditions are proved [3,4], and asymptotic behavior of the system is studied [5]. Widely used approaches to a numerical solution of the drift-diffusion-Poisson equations are the finite-element and finite-volume discretization methods [6,7]. To solve the system of discretized equations, the Newton algorithm, the Gummel iteration method and their modifications are applied [8,9]. Here we develop a different approach to solve the drift-diffusion-Poisson equations. The new approach is based on Monte Carlo methods and random walks. The general idea is to simulate the drift-diffusion transport of electrons and holes exactly in accordance with the probabilistic representations inherently related to the drift-diffusion equations. It should be mentioned that, generally, stochastic simulation methods make it possible to get randomness, such as noise, coefficient fluctuations or space inhomogeneity, included in a model. Another hybrid, Monte Carlo finite-element method, is employed in [10,11]. The method is based on simulating the process governed by a Boltzmann equation. The relevant technique is known as the lattice-gas approach and is used in many transport and particle interaction problems. The Direct Simulation Monte Carlo (DSMC) algorithm is known as a stochastic method that solves the Boltzmann equation using a set of virtual particles. In contrast to other simulation tools, like drift-diffusion or hydrodynamic models, which involve solving coupled systems of partial differential equations, the Boltzmann equation is treated in DSMC by replacing the distribution function with a representative set of particles [12]. The lattice-gas scheme implementation is applied in some cellular automata implementations of the carrier transport simulation in semiconductors. As an example, we can mention [13–16] where similar cellular automata models of semiconductor transport are presented, and a comparison with the kinetic Boltzmann-Monte Carlo method is carried out. In [17] the hybrid lattice-gas based cellular automata and the Bolzmann-Monte Carlo approach are described as applied to a simulation of the charge transport in semiconductors.

The cellular automata approach implies variables that take an integer or Boolean values. This makes it possible to eliminate the rounding errors and may decrease the amount of required memory. Another advantage of this approach is the simplicity and locality of the modeling rules. On the other hand, the Monte Carlo methods, such as random walk on spheres [18,19] or parallelepipeds [20], speed up the calculations by replacing a lot of small simple random walking on a mesh with a single jumping inside the domain or to the domain boundary. Therefore, in this paper, we use the hybrid cellular automata and Monte Carlo approach to solve a drift-diffusion-Poisson system. The electron and hole diffusion and drift is simulated by discrete random walks implemented in the form of the cellular automaton algorithm described in [21]. The drift velocity is calculated through the derivatives of the Poisson equation solution, and calculated by Sabelfeld's global random walk algorithm [22–24]. The right-hand side of the Poisson equation depends on the electron-hole concentration, which, in its turn, depends on the drift velocity. Thus, the drift-diffusion-Poisson system is nonlinear and is solved by the iteration procedure including an alternating simulation of the drift-diffusion and the Poisson equation solving. To reduce the calculation time, a parallel implementation of the cellular automaton and global random walk algorithms is developed and its efficiency is analyzed.

The paper is organized as follows. In Sect. 2, we formulate a nonlinear boundary value problem for the system of drift-diffusion-Poisson equations. In this section, the cellular automata algorithm for modeling the drift-diffusion process is given and justified. Here, the global random walk algorithm as applied to efficiently calculate the derivatives of the Poisson equation solution is presented. In this section, an iteration procedure for solving the drift-diffusion-Poisson system of equations is described in details. Section 3 is devoted to parallel implementation of the cellular automaton and global random walk algorithms. In Section 4, we present the simulation results of modeling a drift-diffusion process governed by the system of nonlinear drift-diffusion-Poisson equations.

2 Simulation of the Electron-Hole Transport in Semiconductors

2.1 Drift-Diffusion-Poisson Equations

The transport of the electrons and holes in semiconductors is described by a nonlinear system of drift-diffusion-Poisson equations [1]. In this paper, to simplify the presentation, we consider this system of equations in a divergence-free form assuming constant diffusion coefficients and no recombinations:

$$\frac{\partial n(\mathbf{r}, t)}{\partial t} = D_n \Delta n(\mathbf{r}, t) + \mathbf{v}(\mathbf{r}, t) \cdot \nabla n(\mathbf{r}, t) + F_n(\mathbf{r}, t) , \tag{1}$$

$$\frac{\partial p(\mathbf{r}, t)}{\partial t} = D_p \Delta p(\mathbf{r}, t) - \mathbf{v}(\mathbf{r}, t) \cdot \nabla p(\mathbf{r}, t) + F_p(\mathbf{r}, t) , \tag{2}$$

$$\Delta \psi(\mathbf{r}, t) = -[n(\mathbf{r}, t) - p(\mathbf{r}, t)] . \tag{3}$$

Here, $\mathbf{r}$ is a particle coordinate, t is time, $n(\mathbf{r}, t)$ is spatial distribution of negatively charged electrons at time t, and $p(\mathbf{r}, t)$ is spatial distribution of positively charged holes, $\psi(\mathbf{r}, t)$ is a self-consistent electrostatic potential produced by the two charge carriers, electrons and holes, $\mathbf{v}(\mathbf{r}, t) = \nabla \psi(\mathbf{r}, t)$ is a drift velocity, $F_n(\mathbf{r}, t)$ and $F_p(\mathbf{r}, t)$ are sources generating electrons and holes, respectively.

This system of equations is solved in a two-dimensional domain V with Dirichlet boundary conditions: for the electron and hole densities $n|_\Gamma = 0$, $p|_\Gamma = 0$, and for the Poisson equation $\psi|_\Gamma = 0$, and initial conditions $n(\mathbf{r}, 0) = n_0(\mathbf{r})$, $p(\mathbf{r}, 0) = p_0(\mathbf{r})$.

To solve the drift-diffusion-Poisson system we use an iterative procedure. The initial electron-hole distribution is taken according to the initial conditions, for which the drift velocity is calculated by the global random walk algorithm.

Further, for each small time step, we calculate the concentrations of electrons and holes by the cellular automaton algorithm, and the drift velocity is calculated by the global random walk algorithm.

2.2 The Cellular Automata Algorithm for the Drift-Diffusion Simulation

The distribution of electrons and holes during their drift and diffusion is simulated by the random walk method as implemented in the form of cellular automaton algorithm described in [21]. This model simulates particle motion on a regular grid of cells. Each cell may contain some particles or be empty. The particles move between the neighbor cells in four directions (left, right, up, and down). In discrete time steps, the particles either can jump to one of their nearest-neighbor cells with a certain probability, or stay in the cell. The probabilities of jumping are calculated depending on the diffusion coefficients for horizontal and vertical directions D_x, D_y, and drift velocities for X and Y directions V_x, V_y:

$$
\begin{aligned}
P_x &= \frac{\tau \cdot D_x}{h^2} + \frac{\tau \cdot v_x}{2h} \;, & Q_x &= \frac{\tau \cdot D_x}{h^2} - \frac{\tau \cdot v_x}{2h} \;, \\
P_y &= \frac{\tau \cdot D_y}{h^2} + \frac{\tau \cdot v_y}{2h} \;, & Q_y &= \frac{\tau \cdot D_y}{h^2} - \frac{\tau \cdot v_y}{2h} \;,
\end{aligned}
\tag{4}
$$

where P_x and Q_x are the probabilities that a particle moves to the right and left neighbor cell, respectively, and P_y and Q_y are the probabilities that a particle moves to the top and bottom neighbor cells, h is a size of the cell, and τ is a time step. The time step should satisfy the inequality $\tau < \tau_c$, where $\tau_c = \dfrac{h^2}{2 \max(D_x, D_y)}$.

Let us assume that a point source is placed at the domain (a square) center $(x_0, y_0) = (L/2, L/2)$, where L is the length of the domain side. Based on the cellular automaton model [21], the algorithm of calculating the particle concentration in all lattice cells can be described as follows.

1. For each time step t, for all cells (x, y) of the lattice, and each particle in the cell, one of five possible events is simulated:
   - With probability P_x, the particle jumps to the right neighbor cell $(x+1, y)$, the number of particles in the cell (x, y) decreases by 1 and the number of particles in the cell $(x + 1, y)$ increases by 1.
   - With probability Q_x, the particle jumps to the left neighbor cell $(x-1, y)$, the number of particles in the cell (x, y) decreases by 1 and the number of particles in the cell $(x - 1, y)$ increases by 1.
   - With probability P_y, the particle jumps to the top neighbor cell $(x, y+1)$, the number of particles in the cell (x, y) decreases by 1 and the number of particles in the cell $(x, y + 1)$ increases by 1.
   - With probability Q_y, the particle jumps to the bottom neighbor cell $(x, y-1)$, the number of particles in the cell (x, y) decreases by 1 and the number of particles in the cell $(x, y - 1)$ increases by 1.

– With probability $1 - P_x - Q_x - P_y - Q_y$, the particle remains in the current cell (x, y), i.e., the number of particles in the cell (x, y) does not change.

2. If some particles jump to the domain boundary cell $(x^*, y^*) \in \Gamma$, then they disappear.

Thus, for each time step t, we calculate the number of particles in each cell $N(t, (x, y))$. The particle density in each cell is computed as a ratio of the number of particles in the cell at the time step t to the cell area and a total number of particles simulated in the domain: $\rho(t, (x, y)) = N(t, (x, y))/(h^2 \cdot N(0))$.

To verify the cellular automaton algorithm we compare the simulation results with the exact solution.

The particle diffusion with drift is simulated in a square domain with a side length $L = 4$ nm for the diffusion coefficients $D_x = D_y = 1$ nm^2(ns)$^{-1}$, and the drift velocity $V_x = V_y = 0.1$ nm(ns)$^{-1}$. The cell size is taken as $h = 0.1$ nm, and the time step $\tau = 0.2 \cdot \tau_c$ ns. In the paper, we use the nanometer (nm) and nanosecond (ns) scales which are characteristic units in semiconductor applications.

The source is taken as a unit instantaneous point source placed in the center of the square. The instantaneous source generates $n_0 = 10^8$ particles at time $t = 0$. The value $n_0 = 10^8$ has been chosen in order to decrease the automaton noise [25] and obtain the simulation results with sufficient accuracy.

The exact representation of the particle density in the square is given by the following formula:

$$n_{2D}(\mathbf{r}, \mathbf{r_0}, t) = n(x, x_0, t) \cdot n(y, y_0, t) , \tag{5}$$

where $\mathbf{r} = (x, y)$ is a point where the density is calculated, and $\mathbf{r_0} = (x_0, y_0)$ is the position of the point source. The one-dimensional density $n(i, i_0, t)$ at the point i with the source at i_0 is the Green function of the drift-diffusion equation which can be written explicitly in the form of a series [20]:

$$n(i, i_0, t) = \frac{2}{L} \exp\left(\frac{v}{2D}(i_0 - i) - \frac{v^2}{4D}t\right) \cdot$$
$$\cdot \sum_{m=1}^{\infty} \sin\left(\frac{\pi m i}{L}\right) \sin\left(\frac{\pi m i_0}{L}\right) \exp\left[\frac{-D\pi^2 m^2}{L^2}t\right] . \tag{6}$$

Figure 1 presents the particle density along the horizontal slice for $y = 2$ nm and $x \in [0, 4]$ at four time moments $t = 0.1, 0.01, 0.5, 1$ ns. The comparison shows that the plots of the calculated density are in a good agreement with the exact formula (5).

The cellular automaton algorithm for the drift-diffusion simulation is applied further to compute the electron-hole distribution.

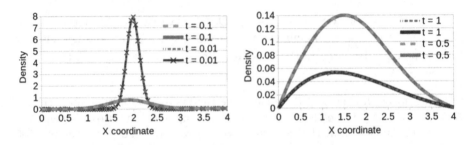

Fig. 1. Comparison of the particle density obtained by the CA model (dashed line) with the exact solution (solid line)

2.3 The Global Random Walk Algorithm for the Drift Velocity Calculation

To simulate the drift-diffusion process we need to compute the drift velocity for all cells of the 2D lattice on which the cellular automaton operates. Sabelfeld's global random walk method [23,24] is an efficient method used to calculate the solution and its derivatives in any set of points for the equations, which fundamental solution is known, e.g., for the Poisson equation: the method is mesh-free, and does not use finite-difference approximations to calculate the derivatives.

Let us consider the Poisson equation in a bounded domain V:

$$\Delta\psi(\mathbf{r}) = -f(\mathbf{r}), \qquad \mathbf{r} \in V , \qquad (7)$$

with zero Dirichlet boundary conditions $\psi(\mathbf{r}^*) = 0$, $\mathbf{r}^* \in \Gamma$.

A detailed derivation of the formulae and the algorithm are presented in [24]. According to [24], the algorithm of calculating the drift velocity for all lattice cells $\mathbf{r}_i = (x_i, y_i)$, $i = 1, \dots, m$ is as follows. Let $U_x(\mathbf{r}_i)$ and $U_y(\mathbf{r}_i)$ to be the arrays for the derivatives with respect to x and y, respectively. At the beginning, put zeros to the arrays $U_x(\mathbf{r}_i)$ and $U_y(\mathbf{r}_i)$, $i = 1, \dots, m$.

1. Sample a random cell $\boldsymbol{\xi} = (x_\xi, y_\xi)$ of the lattice from the uniform distribution $\pi(\mathbf{r})$.
2. Simulate the random walk on spheres (in our case, on circles) trajectory starting at $\boldsymbol{\xi}$ and terminating after hitting the Γ_ε-shell, and record the terminating point $\mathbf{r}^* = (x^*, y^*)$. For details, see the description of the random walk on spheres algorithm in [19].
3. Calculate the random estimators for the partial derivatives with respect to x and y by the following formulae:

$$\frac{\partial\eta}{\partial x}(\mathbf{r}_i) = \frac{f(\boldsymbol{\xi})}{\pi(\boldsymbol{\xi})}\left[\frac{x_i - x^*}{|\mathbf{r}_i - \mathbf{r}^*|^2} - \frac{x_i - x_\xi}{|\mathbf{r}_i - \boldsymbol{\xi}|^2}\right], \quad i = 1, \dots m ,$$

$$\frac{\partial\eta}{\partial y}(\mathbf{r}_i) = \frac{f(\boldsymbol{\xi})}{\pi(\boldsymbol{\xi})}\left[\frac{y_i - y^*}{|\mathbf{r}_i - \mathbf{r}^*|^2} - \frac{y_i - y_\xi}{|\mathbf{r}_i - \boldsymbol{\xi}|^2}\right], \quad i = 1, \dots m . \qquad (8)$$

Add the obtained values to the arrays U_x and U_y.

4. Go to Step 1, and start the next random trajectory. Do steps 1–4 N times.
5. The derivatives $\psi_x(\mathbf{r}_i)$ and $\psi_y(\mathbf{r}_i)$ are calculated as averages over N trajectories:

$$\psi_x(\mathbf{r}_i) \approx U_x(\mathbf{r}_i)/N, \quad \psi_y(\mathbf{r}_i) \approx U_y(\mathbf{r}_i)/N . \tag{9}$$

To validate the algorithm, as a test, we shall take the Poisson equation with the following right-hand side:

$$f(x,y) = \frac{8\pi^2}{L^2} \sin\left(\frac{2\pi x}{L}\right) \sin\left(\frac{2\pi y}{L}\right) . \tag{10}$$

The exact solution of the Poisson equation for this right-hand side has the following explicit form:

$$\psi(x,y) = \sin\left(\frac{2\pi x}{L}\right) \sin\left(\frac{2\pi y}{L}\right) . \tag{11}$$

The derivatives of the exact solution $\psi(x,y)$ are:

$$\begin{aligned}
\frac{\partial \psi(x,y)}{\partial x} &= \frac{2\pi}{L} \cos\left(\frac{2\pi x}{L}\right) \sin\left(\frac{2\pi y}{L}\right) , \\
\frac{\partial \psi(x,y)}{\partial y} &= \frac{2\pi}{L} \sin\left(\frac{2\pi x}{L}\right) \cos\left(\frac{2\pi y}{L}\right) .
\end{aligned} \tag{12}$$

The space step is taken equal to $h = 0.05$ nm, the square side length is $L = 4$ nm.

Figure 2 presents the plots for the derivatives of the solution to the Poisson equation with the right-hand side (10). The derivative with respect to x is shown for $x \in [0,4]$ at fixed $y = 1.25$ nm, while the derivative with respect to y is shown for the horizontal slice at fixed $y = 2$ nm.

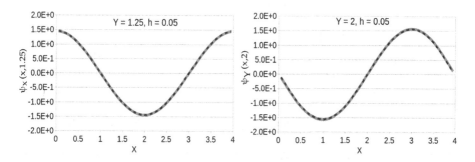

Fig. 2. Comparison of the derivatives of the solution to Poisson equation obtained by the global random walk algorithm (dashed lines) with the exact formulae (solid lines)

As seen from Fig. 2, the derivatives with respect to x and y obtained by the global random walk algorithm are in a perfect agreement with the exact representations (12).

2.4   The Iterative Algorithm for Solving the Nonlinear System of Drift-Diffusion-Poisson Equations

The drift-diffusion-Poisson system (1–3) includes the equations describing the concentrations of electrons $n(x, y)$ and holes $p(x, y)$. The electron-hole concentration distribution depends on the drift velocity, which, in turn, depends on the difference $n(x, y) - p(x, y)$. Therefore, to solve the system (1–3) we apply the iterative procedure consisting of two steps:

1. Calculation of the electron $n(x, y)$ and hole $p(x, y)$ concentrations by the cellular automaton model of drift-diffusion given above in Sect. 2.2.
2. Calculation of the drift velocity as derivatives with respect to x and y of the solution to Poisson equation by the global random walk algorithm described above in Sect. 2.3.

Let us describe the algorithm for solving the system (1–3) in more detail. We employ a regular 2D lattice presenting the domain in a discrete form.

1. Generate the initial electron-hole distribution $n_0(x, y)$, $p_0(x, y)$ on the lattice according to the given initial conditions.
   Set a number of the iteration to 1: $i = 1$.
2. Calculate the drift velocity $v_i(x, y)$ in each lattice cell by the global random walk on circles algorithm for the Poisson equation with the right-hand side $(n_{i-1}(x, y) - p_{i-1}(x, y))/(C_n(0) + C_p(0))$, $C_n(0)$ and $C_p(0)$ being the total number of electrons and holes initially generated on the lattice.
3. For the computed velocity values $v_i(x, y)$, calculate the new distribution of the electron-hole concentrations $n_i(x, y)$ and $p_i(x, y)$ using the cellular automaton model of drift-diffusion.
4. Increase the number of the iteration by 1: $i = i + 1$.
5. Calculate the current time $t = \tau \cdot i$, where τ is the time of a single iteration.
6. If the current time t is greater than the desired time T, stop the simulation process.
7. If $t < T$, continue from Step 2.

The accuracy and computational time of the iterative algorithm of solving system (1–3) depend on the accuracy and computational time of the constituent algorithms: the cellular automaton drift-diffusion model and the global random walk algorithm.

The accuracy of the multi-particle cellular automata depends on the size of a single cell and on the number of particles simulating the given density [25, 26]. The small cell size provides a good accuracy but leads to an increase in the cellular array size. A large number of particles decrease the automaton noise. This increases the accuracy but requires more calculation time.

The accuracy of the global random walk algorithm is determined by the number of trajectories. However, the greater the number of trajectories is, the longer the calculation time will be.

For example, let us consider the following test. System (1–3) with a zero boundary condition is solved in a unit square $L \times L$, $L = 1$ nm, the cell size

$h = 0.05$ nm, the diffusion coefficients for electrons and holes along x and y are equal to $D_{nx} = D_{ny} = D_n = 1$, $D_{px} = D_{py} = D_p = 1$ nm^2(ns)$^{-1}$, the time of a single iteration is $\tau = h^2/10 = 0.00025$ ns. The electrons and holes are generated by stationary point sources placed in the square center. At $t = 0$, the sources generate $C_n(0) = 10^8$ electrons and $C_p(0) = 3 \cdot 10^8$ holes. In addition, at each iteration, the sources generate $C_n^*(t) = 10^6$ electrons and $C_p^*(t) = 3 \cdot 10^6$ holes. The global random walk is modeled by $N = 10^6$ trajectories. Here, we have chosen quite small space and time steps to ensure high accuracy. Therefore, it takes 6.3 s to calculate a single iteration of the iterative algorithm for solving this problem on the processor Intel(R) Core(TM) i7-4770 CPU 3.40 GHz. To obtain a solution of the system at the time instant $T = 1$ ns, we need 4000 iterations and 7 hours of computer time. Thus, to obtain the results with good accuracy in a reasonable time, the iterative algorithm of solving the drift-diffusion-Poisson system requires parallel implementation.

3   Parallel Implementation of the Iterative Algorithm for Solving the Drift-Diffusion-Poisson System

Parallelization of the iterative algorithm for solving system (1–3) includes parallel execution of the cellular automaton drift-diffusion model and the global random walk algorithm.

The general approach to the cellular automata parallel implementation is a domain decomposition method. However, our computer experiments show that the calculation time of the drift-diffusion cellular automaton model depends more on the number of particles than on the domain size. For instance, for the test given above in Sect. 2.4, the cellular automaton computer time is $t_c = 21.28$ s. When we double the square side length $L = 2$ nm, the time is practically the same, $t_c = 21.31$ s. However, when we double the number of generated particles, the time doubles, too, $t_c = 42.37$ s. Therefore, we choose the approach where we distribute the particles among the computing nodes.

Let us describe the parallel implementation of the drift-diffusion cellular automaton model in more detail. We employ the MPI standard. Initially generated electrons and holes $C_n(0)$ and $C_p(0)$ are distributed among n_{mpi} MPI processes. Each MPI process has a copy of the domain of $L \times L$ size, where $C_n(0)/n_{mpi}$ and $C_p(0)/n_{mpi}$ particles are located. Each process computes a new electron-hole distribution of its particles and calculates the general particle distribution as a sum of the local arrays of all MPI processes using the collective communication subroutine MPI_Allreduce. The computed general array is used by the global random walk algorithm. In the case of a stationary source, the particles generated on each iteration $C_n^*(t)$ and $C_p^*(t)$ are distributed among n_{mpi} MPI processes too.

The global random walk algorithm includes two labor-consuming processes: simulation of a random exit point on the domain boundary, and calculation of the estimators for the partial derivatives of the solution to Poisson equation by formulae (8).

Estimators (8) consist of two terms. The first terms:

$$\frac{\partial \eta_1}{\partial x}(\mathbf{r}_i) = \frac{f(\boldsymbol{\xi})}{\pi(\boldsymbol{\xi})}\left[\frac{x_i - x^*}{|\mathbf{r}_i - \mathbf{r}^*|^2}\right] \ , \ \frac{\partial \eta_1}{\partial y}(\mathbf{r}_i) = \frac{f(\boldsymbol{\xi})}{\pi(\boldsymbol{\xi})}\left[\frac{y_i - y^*}{|\mathbf{r}_i - \mathbf{r}^*|^2}\right] \ , \ i = 1, \ldots m \ , \quad (13)$$

depend on the random exit points. This is simulated by the random walk on circles algorithm [19]. This algorithm is parallelized by the general approach to parallel implementation of the Monte Carlo algorithms [27,28]. We distribute N trajectories among n_{mpi} MPI processes. Each process runs N/n_{mpi} trajectories, obtains a set of random exit points for which it calculates the first terms of estimators (13).

The second terms:

$$\frac{\partial \eta_2}{\partial x}(\mathbf{r}_i) = \frac{f(\boldsymbol{\xi})}{\pi(\boldsymbol{\xi})}\left[\frac{x_i - x_{\boldsymbol{\xi}}}{|\mathbf{r}_i - \boldsymbol{\xi}|^2}\right] \ , \ \frac{\partial \eta_2}{\partial y}(\mathbf{r}_i) = \frac{f(\boldsymbol{\xi})}{\pi(\boldsymbol{\xi})}\left[\frac{y_i - y_{\boldsymbol{\xi}}}{|\mathbf{r}_i - \boldsymbol{\xi}|^2}\right] \ , \ i = 1, \ldots m \ , \quad (14)$$

depend on the starting point $\boldsymbol{\xi}$ which is uniformly sampled from the lattice cells. Thus, we can precalculate the values

$$\rho_x(i,j) = \frac{x_i - x_{\boldsymbol{\xi}_j}}{|\mathbf{r}_i - \boldsymbol{\xi}_j|^2} \ , \ \rho_y(i,j) = \frac{y_i - y_{\boldsymbol{\xi}_j}}{|\mathbf{r}_i - \boldsymbol{\xi}_j|^2}, \ i = 1, \ldots m \ , \ j = 1, \ldots m \ , \quad (15)$$

before starting the iterative procedure. Values (15) are computed for all m cells of the lattice and stored in an array of size $m \times m$. The function $f(\boldsymbol{\xi}) = (n(\boldsymbol{\xi}) - p(\boldsymbol{\xi}))$ is changed at each iteration. Therefore, the values (14) should be calculated for all m lattice cells at each iteration. To reduce the computer time, the precalculated values $\rho_x(i,j)$ and $\rho_y(i,j)$ are employed.

The partition "Broadwell" of the cluster "MVS-10P" of the Joint Supercomputer Center of RAS [29] is employed for calculations. The "Broadwell" node consists of two processors Intel Xeon CPU E5-2697A v4 2.60 GHz, each containing 16 cores with 2 threads and 40 MB SmartCache.

The performance of the parallel code is analyzed for the same parameter values as for the test given above in Sect. 2.4, except for the length of the square side taken here as $L = 4$ nm, and the number of trajectories in the global random walk algorithm, $N = 10^7$. Each cluster node runs 32 MPI processes.

The following characteristics of the iterative algorithm parallel implementation are calculated for different number n_{mpi} of the MPI processes: the computational time $T(n_{\mathrm{mpi}})$, the speedup $S(n_{\mathrm{mpi}}) = T(1)/T(n_{\mathrm{mpi}})$, and the efficiency $E(n_{\mathrm{mpi}}) = S(n_{\mathrm{mpi}})/n_{\mathrm{mpi}}$.

The plots of these characteristics are shown in Fig. 3. The speedup of the parallel code is close to a linear rate, and can be described by the relation: $S(n_{\mathrm{mpi}}) = n_{\mathrm{mpi}}^{0.86}$. The efficiency decreases to 0.87 when the number of MPI processes equals 8. The drop in efficiency requires a separate analysis. It can be assumed that the efficiency is decreasing due to the following fact. Each MPI process has its copies of cellular arrays which are stored at the memory of a single node when $n_{\mathrm{mpi}} \leq 32$. Increasing the number of MPI processes to 8 may lead to the memory bandwidth limiting. However, the efficiency of the parallel code is above 0.8 when using up to 128 MPI processes, i.e. 4 nodes. Thus, we conclude that the parallel implementation of the iterative algorithm allows us to obtain the results with the same high accuracy in significantly less computer time compared with the sequential code.

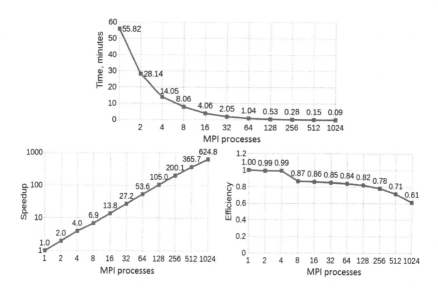

Fig. 3. The characteristic values of the iterative algorithm parallel implementation for solving the drift-diffusion-Poisson system

4   Simulation Results of the Electron-Hole Transport in Semiconductors

Let us consider the solution of system (1–3) obtained by the parallel implementation of the iterative algorithm used to solve the drift-diffusion-Poisson system for the following problem. The electron-hole transport is simulated in a square $L \times L$, $L = 2$ nm, the cell size $h = 0.1$ nm, the diffusion coefficients along x and y are equal to $D_{nx} = 4$, $D_{ny} = 3$, $D_{px} = 2$, $D_{py} = 1.5$ nm^2(ns)$^{-1}$, the time of a single iteration is $\tau = h^2/40 = 0.00025$ ns. The electrons and holes are generated by stationary point sources both placed in the square center. At $t = 0$, the sources generate $C_n(0) = 10^9$ electrons and $C_p(0) = 4 \cdot 10^9$ holes. In addition, at each iteration, the sources generate $C_n^*(t) = 10^6$ electrons and $C_p^*(t) = 4 \cdot 10^6$ holes. Zero boundary conditions are assumed, which implies that the boundary is absorbing both for electrons and holes. The global random walk algorithm employs $N = 10^6$ trajectories.

Figure 4 presents the computed electron-hole densities and drift velocities along X and Y directions for the horizontal slice at fixed $y = L/2 = 1$ nm, at different time instances. The difference between the electrons and holes concentrations leads to a non-zero drift velocity. Under the influence of diffusion and drift, the particles reach the domain boundary and are adsorbed in it, and as a result, the electron and hole densities decrease. However, at each iteration, the stationary sources generate a new portion of particles. Thus, the system reaches a steady state when the electron and hole densities and the drift velocity are not changed with time anymore.

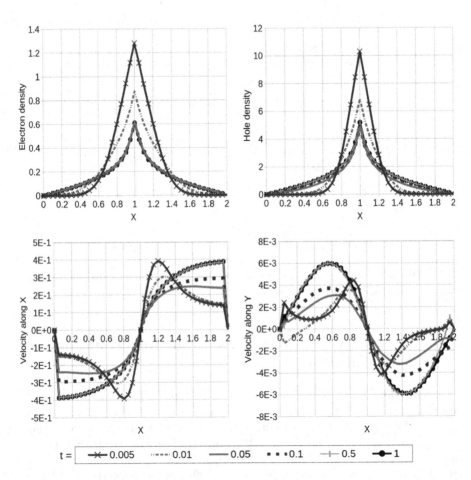

Fig. 4. The electron and hole densities (upper panels), and velocities (lower panels) obtained by the parallel code of the iterative algorithm for solving the drift-diffusion-Poisson system

In addition, the electron-hole space distributions and velocity fields for the steady state are shown in Fig. 5.

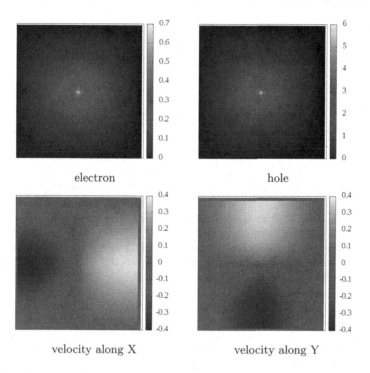

electron   hole

velocity along X   velocity along Y

Fig. 5. The electron and hole distribution (upper panels), and the velocity fields (lower panels) obtained at time instant $t = 0.4$ ns

5   Conclusion

In this paper, parallel implementation of the iterative algorithm for solving the drift-diffusion-Poisson system is presented. The distribution of electrons and holes undergoing the drift and diffusion motion is simulated by a discrete random walk implemented in the form of the cellular automaton model. The parallel implementation of this model employs distribution of the particles among the MPI processes. The drift velocity for all cells of the 2D lattice is calculated by the global random walk on spheres algorithm. This algorithm is parallelized by the distribution of the Monte Carlo trajectories among the MPI processes. The parallel code has been executed on the partition "Broadwell" of the cluster "MVS-10P" of the RAS Joint Supercomputer Center. We have obtained a speedup of the parallel code close to the linear and the efficiency above 0.8 when using up to 4 cluster nodes.

Using the parallel implementation of the iterative algorithm for solving the nonlinear system of drift-diffusion-Poisson equations we solved a test problem with zero boundary conditions and stationary electron and hole sources. The parallel code allows us to obtain the electron-hole densities and drift velocity fields for different time instants and reach the steady-state distribution.

References

1. Markowich, P.A., Ringhofer, C.A., Schmeiser, C.: Semiconductor Equations. Springer, Vienna (1990). https://doi.org/10.1007/978-3-7091-6961-2
2. Van Roosbroeck, W.: Theory of flow of electron and holes in germanium and other semiconductors. Bell. Syst. Tech. J. **29**(4), 560–607 (1950)
3. Wu, H., Jie, J.: Global solution to the drift-diffusion-Poisson system for semiconductors with nonlinear recombination-generation rate. Asympt. Anal. **85**, 75–105 (2011). https://doi.org/10.3233/ASY-131176
4. Wu, H., Markowich, P.A., Zheng, S.: Global existence and asymptotic behavior for a semiconductor drift-diffusion-Poisson Model. Math. Models Methods Appl. Sci. **18**(3), 443–487 (2007). https://doi.org/10.1142/S0218202508002735
5. Biler, P., Dolbeault, J., Markowich, P.A.: Large time asymptotics of nonlinear drift-diffusion systems with Poisson coupling. Transp. Theory Statist. Phys. **30**(4–6), 521–536 (2001). https://doi.org/10.1081/TT-100105936
6. Baumgartner, S., Heitzinger, C.: A one-level FETI method for the drift-diffusion-Poisson system with discontinuities at an interface. J. Comput. Phys. **243**, 74–86 (2013). https://doi.org/10.1016/j.jcp.2013.02.043
7. Chainais-Hillairetm, C., Liu, J.-G., Peng, Y.-J.: Finite volume scheme for multidimensional drift-diffusion equations and convergence analysis. Math. Model. Numer. Anal. **37**(2), 319–338 (2003). https://doi.org/10.1051/m2an:2003028
8. Gummel, H.K.: A self-consistent iterative scheme for one-dimensional steady state transistor calculations. IEEE Trans. Electr. Devices **11**(11), 455–465 (1964). https://doi.org/10.1109/T-ED.1964.15364
9. Fiori, G., Iannaccone, G.: Three-dimensional simulation of one-dimensional transport in silicon nanowire transistors. IEEE Trans. Nanotechnol. **6**(5), 524–529 (2007). https://doi.org/10.1109/tnano.2007.896844
10. Taghizadeh, L., Khodadadian, A., Heitzinger, C.: The optimal multilevel Monte-Carlo approximation of the stochastic drift-diffusion-Poisson system. Comput. Meth. Appl. Mech. Eng. **318**, 739–761 (2017). https://doi.org/10.1016/j.cma.2017.02.014
11. Carrillo, J.A., Gamba, I.M., Muscato, O., Shu, C.-W.: Comparison of monte carlo and deterministic simulations of a silicon diode. In: Abdallah, N.B., et al. (eds.) Transport in Transition Regimes. The IMA Volumes in Mathematics and its Applications, vol. 135, pp. 75–84. Springer, New York (2004). https://doi.org/10.1007/978-1-4613-0017-5_4
12. Muscato, O., Di Stefano, V., Wagner, W.: Numerical study of the systematic error in Monte Carlo schemes for semiconductors. ESAIM Math. Model. Numer. Anal. **44**, 1049–1068 (2010). https://doi.org/10.1051/m2an/2010051
13. Zandler, G., Di Carlo, A., Kometer, K., Lugli, P., Vogl, P., Gornik, E.: A comparison of Monte Carlo and cellular automata approaches for semiconductor device simulation. IEEE Electr. Device Lett. **4**(2), 77–79 (1993). https://doi.org/10.1109/55.215114
14. Liebig, D.: Cellular automata simulation of GaAs-IMPATT-diodes. In: Ryssel, H., Pichler, P. (eds.) Simulation of Semiconductor Devices and Processes. Springer, Vienna (1995). https://doi.org/10.1007/978-3-7091-6619-2_17
15. Kometer, K., Zandler, G., Vogl, P.: Lattice-gas cellular-automaton method for semiclassical transport in semiconductors. Phys. Rev. B. **46**(3), 1382–1394 (1992). https://doi.org/10.1103/physrevb.46.1382

16. Vogl, P., Zandler, G., Rein, A., Saraniti, M.: Cellular automaton approach for semiconductor transport. In: Scholl, E. (ed.) Theory of Transport Properties of Semiconductor Nanostructures. Electronic Materials Series, vol. 4, pp. 103–126. Springer, Boston (1998). https://doi.org/10.1007/978-1-4615-5807-1_4

17. Saraniti, M., Goodnick, S.M.: Hybrid fullband cellular automaton/Monte Carlo approach for fast simulation of charge transport in semiconductorsd. IEEE Trans. Electr. Devices **47**(10), 1909–1916 (2000). https://doi.org/10.1109/16.870571

18. Sabelfeld, K.K.: Splitting and survival probabilities in stochastic random walk methods and applications. Monte Carlo Methods Appl. **22**(1), 55–72 (2016). https://doi.org/10.1515/mcma-2016-0103

19. Sabelfeld, K.K.: Random walk on spheres method for solving drift-diffusion problems. Monte Carlo Methods Appl. **22**(4), 265–275 (2016). https://doi.org/10.1515/mcma-2016-0118

20. Sabelfeld, K.K.: Random walk on rectangles and parallelepipeds algorithm for solving transient anisotropic drift-diffusion-reaction problems. Monte Carlo Methods Appl. **25**(2), 131–146 (2019). https://doi.org/10.1515/mcma-2019-2039

21. Karapiperis, T., Blankleider, B.: Cellular automation model of reaction-transport processes. Physica D **78**(1–2), 30–64 (1994). https://doi.org/10.1016/0167-2789(94)00093-X

22. Sabelfeld, K.K.: Monte Carlo Methods in Boundary Value Problems. Springer, Heidelberg (1991)

23. Sabelfeld, K.K.: A global random walk on spheres algorithm for transient heat equation and some extensions. Monte Carlo Methods Appl. **25**(1), 85–96 (2019). https://doi.org/10.1515/mcma-2019-2032

24. Sabelfeld, Karl K., Kireeva, A.: A new Global Random Walk algorithm for calculation of the solution and its derivatives of elliptic equations with constant coefficients in an arbitrary set of points. Appl. Math. Lett. **107**, 106466 (1–9) (2020). https://doi.org/10.1016/j.aml.2020.106466

25. Medvedev, Yu.: Automata noise in diffusion cellular-automata models. Bull. Nov. Comp. Center, Comp. Sci. **30**, 43–52 (2010)

26. Kireeva, A., Sabelfeld, K.K., Kireev, S.: Synchronous multi-particle cellular automaton model of diffusion with self-annihilation. In: Malyshkin, V. (ed.) PaCT 2019. LNCS, vol. 11657, pp. 345–359. Springer, Cham (2019). https://doi.org/10.1007/978-3-030-25636-4_27

27. Rosenthal, J.S.: Parallel computing and Monte Carlo algorithms. Far East J. Theor. Stat. **4**, 207–236 (2000)

28. Esselink, K., Loyens, L.D.J.C., Smit, B.: Parallel Monte Carlo Simulations. Phys. Rev. E. **51**(2), 1560–1568 (1995). https://doi.org/10.1103/physreve.51.1560

29. MVS-10P cluster, JSCC RAS. http://www.jscc.ru

Application of High Performance Computations for Modeling Thermal Fields Near the Wellheads

Elena N. Akimova[1,2] , Mikhail Yu. Filimonov[1,2(✉)] ,
Vladimir E. Misilov[1,2] , and Nataliia A. Vaganova[1,2]

[1] Krasovskii Institute of Mathematics and Mechanics, Ural Branch of RAS,
16 S. Kovalevskaya Street, Ekaterinburg, Russia
`aen15@yandex.ru`,{`fmy`,`vna`}`@imm.uran.ru`
[2] Ural Federal University, 19 Mira Street, Ekaterinburg, Russia
`v.e.misilov@urfu.ru`

Abstract. The paper is devoted to construction and study of parallel algorithms for solving the problem of modeling non-stationary thermal fields from two producing wells in Arctic oil and gas fields. A computational algorithm and parallel programs for multicore processors using OpenMP technology are developed. The results of a number of numerical experiments and evaluation of the parallel algorithm effectiveness are presented.

Keywords: Heat and mass transfer · Simulation · Parallel computing · OpenMP

1 Introduction

The most of Russian oil and gas fields are located in the zone of permafrost with its own specifics. There are certain requirements to designing the well pads of such fields. For example, between two production wells there should be a distance twice greater than the thawing zone around one well, which will be reached in 30 years of the well operation. The simulation of such nonstationary fields in a three-dimensional computational domain is described by the heat equation with the initial and boundary conditions considering possible phase transitions. At the same time, the boundary condition on the day surface is non-linear because that solar radiation is taken into account in the mathematical model. After discretization of the equation on an orthogonal grid using finite differences, we applied the method of splitting in spatial variables to solve the problem. The problem is approximated by an implicit central-difference three-point scheme and reduced to a system of linear algebraic equations with a three-diagonal matrix at each time and spatial step. Solving a problem with real data requires

The work was supported by Russian Foundation for Basic Research (project no. 19-07-00435).

L. Sokolinsky and M. Zymbler (Eds.): PCT 2020, CCIS 1263, pp. 266–278, 2020.
https://doi.org/10.1007/978-3-030-55326-5_19

computations on spatial grids of large size and large time intervals. Preliminary calculations revealed that for some types of soil it is desirable to consider a model consisting of two wells, since the propagation of zero isotherms between a couple of active wells exceeds the propagation of the zero isotherm from a single well with no other heat sources. It should be noted that the numerical solution of such problems can take a long time—up to several days—depending on the size of the computational domain. It is necessary to choose sufficiently large dimensions of the computational domain in order to avoid the influence of boundary conditions at the boundary of the computational domain.

One way to reduce the computing time and increase the efficiency of the problem solving is to use parallelization of the solver algorithm and computational systems with multicore processors. Many papers and books of Russian and foreign authors are devoted to the problems of algorithm parallelizations [1–3]. To implement the algorithms on multicore processors, OpenMP parallel programming technology is used for developing multi-threaded applications on multiprocessor systems with shared memory [4].

To study the results of the parallelization and compare the performance of the serial and parallel algorithms, the speedup and efficiency coefficients are introduced: $S_m = T_1/T_m$, $E_m = S_m/m$, where T_1 and T_m are execution times of the serial algorithm on a single processor and of the parallel algorithm on the multicore or multi-processor system, respectively, where the number of processors is $m, m > 1$. T_m includes the calculation time and overhead.

This paper aims to the construct an effective parallel algorithm to simulate the propagation of unsteady thermal fields near the heads of two production wells located in an Arctic oil and gas field. A numerical algorithm and a parallel software package for the multicore processors using OpenMP technology are developed. We present the results of the computational experiments and estimates of the parallel algorithm effectiveness.

2 Problem Statement and Mathematical Model

Permafrost occupies about 25% of the total land area of the world [5]. In Russia, the main oil and gas reserves are located in the cryolithozone, which occupies about 60% of the entire territory. The development of this region is accompanied by a great influence on the permafrost. Technical systems used in oil and gas fields, and various engineering structures are sources of heat that affect the permafrost and decrease its strength properties [6,7]. Because of such processes, buildings may collapse, and transport communications, production wells and pipelines may be damaged. It can lead to major environmental disasters [8,9]. A conservation principle of frozen soil foundations is usually used to construct buildings in permafrost zones [10]. More than 75% of all Russian buildings and structures in the permafrost zone are constructed and operated based on this principle. The construction pads can also be prepared by preliminary thermal stabilization of the soil. This requires using cooling devices to freeze the soil.

Thus, we have a system of heat (or cold) sources, which needs to minimize the effect of thermal fields on the frozen ground over a long period. To solve this problem, we need an adequate mathematical model that takes into account the most important technological (specifics of technical systems), physical (soil lithology) and climatic (change in air temperature, solar radiation, etc.) factors [11,12]. These factors are taken into account, for example, in [13].

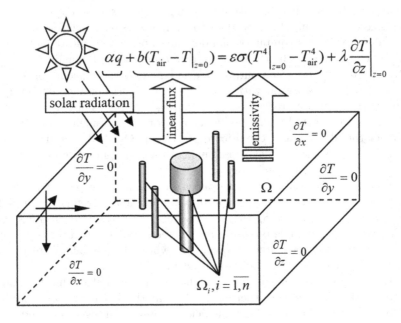

$$\underbrace{\alpha q}_{} + \underbrace{b(T_{\text{air}} - T|_{z=0})}_{} = \underbrace{\varepsilon\sigma(T^4|_{z=0} - T_{\text{air}}^4)}_{} + \lambda \frac{\partial T}{\partial z}\Big|_{z=0}$$

solar radiation   linear flux   emissivity

$\frac{\partial T}{\partial x} = 0$

$\frac{\partial T}{\partial y} = 0$

Ω

$\frac{\partial T}{\partial y} = 0$

$\frac{\partial T}{\partial z} = 0$

$\frac{\partial T}{\partial x} = 0$

$\Omega_i, i = \overline{1,n}$

Fig. 1. The main heat flows and boundary conditions

Let us describe a mathematical model based on which parallel algorithms are developed to model the long-term influence of a technical system consisting of two producing wells with different isolation options provided that they both operate in the frozen soil. We will further use the following notations: (**2W**), (**2Wi**), and (**2WI**) are the systems of two producing wells with no isolated shells, with a simple insulation, and with a complex two-level insulating shell, respectively. Also, the seasonal cooling devices (SCDs) may be inserted into the system. They are used to thermally stabilize the soil top layer. Then the complicated systems will be denoted by (**2WS**), (**2WiS**), and (**2WIS**), respectively.

Let $T = T(t, x, y, z)$ be the soil temperature at point (x, y, z) at instant t. A three-dimensional simulation of unsteady thermal fields, such as oil and gas fields (the well pads, SCDs) located in the permafrost area, is required to take into account different technological and climatic factors. The computational domain is a three-dimensional box, where x and y axes are parallel to the ground surface and the z axis is directed downwards (see Fig. 1). We assume that the size of the domain Ω is defined by positive numbers L_x, L_y, L_z: $-L_x \leqslant x \leqslant Lx$, $-L_y \leqslant y \leqslant L_y$, $-L_z \leqslant z \leqslant 0$. The main heat flow associated with the climatic factors on the surface $z = 0$ is shown in Fig. 1. $T_{\text{air}} = T_{\text{air}}(t)$ denotes the temperature in the surface layer of air, which varies from time to time in accordance with the annual temperature cycle; $\sigma = 5.67 \cdot 10^{-8}\,[\mathrm{W}/(\mathrm{m}^2 \cdot \mathrm{K}^4)]$ is Stefan–Boltzmann constant; $b = b(t,x,y)$ is the heat transfer coefficient; $\varepsilon = \varepsilon(t,x,y)$ is the coefficient of emissivity. The coefficients of heat transfer and emissivity depend on the type and condition of the soil surface. Total solar radiation $q(t)$ is the sum of direct solar radiation and diffuse radiation. The soil absorbs just a part of the total radiation. This part equals to $\alpha q(t)$, where $\alpha = \alpha(t,x,y)$ is the energy part that is formed to heat the soil, which in general depends on atmospheric conditions, the incidence angle of solar radiation, i.e. latitude and time. Ω can include two wells, same SCDs and layers of riprap on the surface of the ground. The Ω is supposed to contain n objects being the heat sources (producing wells) and cold sources (SDCs). We denote the surface of these objects by $\Omega_i = \Omega_i(x,y,z)$, $i = 1, ..., n$.

To account the heat from each of the Ω_i the equation of the contact (diffusion) heat conductivity with inhomogeneous coefficients is used as a basic mathematical model including a localized heat of phase transition—an approach to solve the problem of Stefan type, without the explicit separation of the phase transition [14,15]. The heat of the phase transformation is introduced with using Dirac δ-function as a concentrated heat of phase transition in the specific heat ratio. The obtained discontinuous function is "shared" with respect to temperature, and does not depend on the number of measurements, phases, and fronts. Thus, the modeling of thawing in the soil is reduced to the solution in Ω of the equation

$$\rho \left(c_v(T) + k\delta(T - T^*) \right) \frac{\partial T}{\partial t} = \mathrm{div}(\lambda(T)\mathrm{grad}\,T), \tag{1}$$

where $\rho = \rho(x,y,z)$ is density [kg/m^3], $T^* = T^*(x,y,z)$ is the temperature of phase transition,

$$c_v(T) = \begin{cases} c_1(x,y,z), \text{for } T < T^*, \\ c_2(x,y,z), \text{for } T > T^*, \end{cases} \quad \text{is specific heat [J/(kg} \cdot \text{K)]},$$

$$\lambda(T) = \begin{cases} \lambda_1(x,y,z), \text{for } T < T^*, \\ \lambda_2(x,y,z), \text{for } T > T^*, \end{cases} \quad \text{is thermal conductivity [Wt/(m} \cdot \text{K)]},$$

$k = k(x,y,z)$ is the specific heat of phase transition, δ is Dirac delta function. The coefficients included in Eq. (1) may vary at different points in the computational domain Ω because of the soil heterogeneity and a presence of engineering structures. The ground surface $z = 0$ is the main formation zone of the natural thermal fields. On this surface the equation of balance of flows is used as a boundary condition considering the main climate factors, such as air temperature and solar radiation. On the surfaces Ω_i, bounding the objects in Ω, a set of the temperatures $T_i(t)$, $i = 1, ..., n$, is given. The bottom surface ($z = -L_z$) and lateral faces ($x = \pm L_x, y = \pm L_y$) of the parallelepiped Ω (Fig. 1) is assumed that the heat flux is equal to zero. Thus it is necessary to solve Eq. (1) in area Ω with the initial condition

$$T(0, x, y, z) = T_0(x, y, z) \tag{2}$$

and boundary conditions

$$\alpha q + b(T_{\text{air}} - T|_{z=0}) = \varepsilon\sigma(T^4_{z=0} - T^4_{\text{air}}) + \lambda\frac{\partial T}{\partial z}\bigg|_{z=0}, \tag{3}$$

$$T|_{\Omega_i} = T_i(t), i = 1, ..., n, \tag{4}$$

$$\frac{\partial T}{\partial x}\bigg|_{x=\pm L_x} = 0, \ \frac{\partial T}{\partial y}\bigg|_{y=\pm L_y} = 0, \ \frac{\partial T}{\partial z}\bigg|_{z=-L_z} = 0. \tag{5}$$

Condition (2) determines the initial distribution of soil temperature at the instant from which we plan to start the numerical calculation. Condition (3) is obtained from the balance of the heat fluxes on the ground surface $z = 0$. Conditions (4) appear in the case of multiple underground objects with their temperature being different from the surrounding soil. For SCDs the temperature $T_i(t)$ is determined by the temperature of air T_{air}. Conditions (5) are necessary to carry out the numerical calculations in the given area.

To solve Eq. (1) in a three-dimensional domain, we use the finite-difference method with splitting by spatial variables. Computations are carried out on an orthogonal grid. In the layers near the boundaries a priory condensed meshes are constructed. Piecewise nonuniform grids are used to satisfy the engineering construction locations.

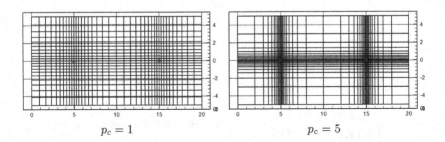

Fig. 2. Two types of the computational grid condensing to the points in a horizontal layer

Moreover, dummy nodes of the inner boundary are introduced to accurately determine the position and intensity of a heat flux generated by inner boundaries Ω_i. To construct a condensed mesh a parameter of condensation (p_c) is used. Let $[0, L]$ have N nodes with a condensation to L. The nodes distribution $[x_0 = 0, x_1, ..., x_N - 1 = L]$ is described by the following relations: $x_i = L - L(p_c - i \cdot h/(1 + i \cdot h)p_c), h = p_c/(N-1), i = 0, ..., N$. Figure 2 shows two types of the nodes condensing for $p_c = 1$ and $p_c = 5$. To check the convergence of the method and to choose the computational grid, a series of methodological computations is carried out using a sequence of condensing grids. The number of nodes of the computational grid is selected in such a way that while it increases, the accuracy of calculations is about 2%.

In Fig. 3 the computational grid in a horizontal layer is presented. The right figure shows the inner boundaries of the SCD and well pipes as blue and red points, respectively. The colored circles denote the positions of insulation shells, which nave to be taken into account during the grid construction.

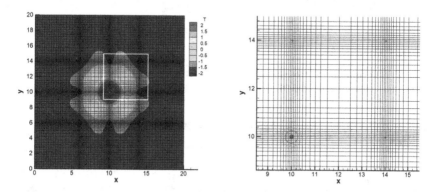

Fig. 3. The computational grid in a horizontal layer. The right figure shows the inner boundaries of the SCD and wells pipes (Color figure online)

3   Numerical Results

To solve the problem (1)–(5), a finite-difference method is used. At present there are some difference methods for solving Stefan type problems, such as the method of front localization by the difference grid node, the method of front straightening, the method of smoothing coefficients and schemes of through computations [15]. The method of the front localization in the mesh node is used only for one-dimensional single-front problems, and the method of front straightening is used for the multi-front problems. A basic feature of these methods is that the difference schemes are constructed with an explicit separation of phase transformation front.

We note that the methods with the explicit separation of the unknown boundary of the phase transformation are not suitable for the case of cyclic temperature changes on the boundary, because the number of non-monotonically moving fronts may be more than one, and some of them may merge with each other or disappear.

In [15], an effective scheme with a smoothing of discontinuous coefficients in the equation of thermal conductivity by temperature in the neighborhood of the phase transformation was developed. This scheme is characterized by that the boundary of phase transition is not explicitly separated, and the homogeneous

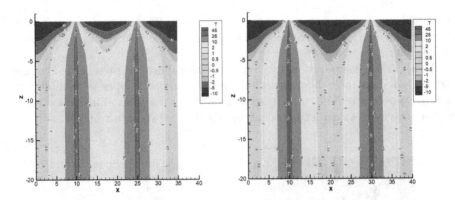

Fig. 4. The thermal fields around (**2W**) system after 9 years of operation. The distance between the wells is 15 m (left figure) and 20 m (right figure), respectively

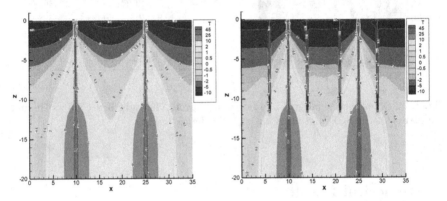

Fig. 5. Thermal fields around of (**2WI**) and (**2WIS**) systems after 9 years of operation

difference schemes may be used. The phase transformation heat is introduced by using the Dirac δ-function as a concentrated heat of the phase transition in the specific heat ratio. Thus, the obtained discontinuous function is then "shared" with respect to temperature, and does not depend on the number of measurements and phases.

An implicit additive locally-1D finite difference method with splitting by spatial variables in three-dimensional domain is used with $O(\tau + |h|^2)$ in a uniform grid and $O(\tau + |h|)$ in non-uniform one [15]. The system of linear difference equations has a three-diagonal form and may be solved by a sweep method. On the upper boundary $z = 0$ the algebraic equation of the fourth degree is solved by Newton's method. The solvability of the implicit difference equations has been previously proved by the authors.

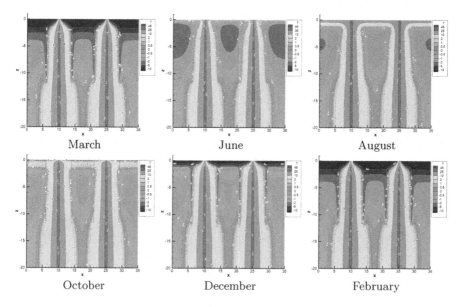

March   June   August

October   December   February

Fig. 6. Thermal fields around system (**2WS**) during the 5th year of operation

The choice of the parameters entering into the nonlinear boundary condition (3) allows us to adapt the developed algorithm to some given geographic coordinates and indirectly consider the influence of additional climatic factors such as snow cover, rainfall, a number of sunny days, etc. It should be noted that the size of the computational domain $\Omega(L_x, L_y, L_z)$ is chosen so that the effect of the lateral boundaries in the numerical calculations has no significant effect on the temperature distribution around the objects, bounded by surfaces Ω_i.

Let us consider numerical results for the following two-well systems: (**2W**), (**2WI**), (**2WS**), (**2WIS**). We will consistently increase the complexity of the technical system having two production wells (Fig. 4). Figure 5 shows the two-well system equipped with a two-level insulating shell (the left figure) and 8 SCDs in addition (the right figure). The systems are (**2WI**) and (**2WIS**), respectively. Figure 6 presents an effect of SCDs on non-insulated wells. In Fig. 7 thermal fields are presented during system (**2WIS**) operation in different months. The calculations allow us to estimate the effectiveness of the wells insulation, the top layer of ripraps and the required number of SCDs for the soil thermal stabilization.

We compared the results of the numerical calculations with the experimental data to determine the zero isotherm from the operating well after 3 years, and there is a good agreement between these values (an error takes about 5%) [13].

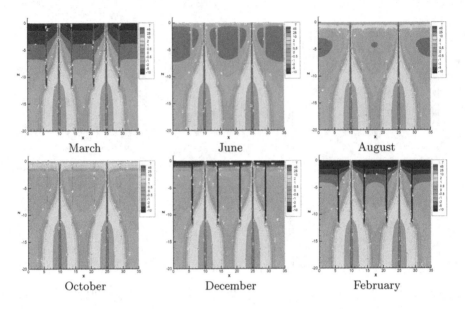

March   June   August

October   December   February

Fig. 7. Thermal fields around system (**2WIS**) during the 5th year of operation

4   Constructing the Parallel Algorithm and Studying Its Efficiency

Numerical simulations focused on a long-term forecasting of changes in the permafrost boundaries due to thermal effects of technical systems require a lot of computing time, which significantly increases when new objects are added to such a system and the number of nodes in such a computational grid is increased. Therefore, it is important to use parallel technologies to solve these problems [16–18].

To analyze the serial program, Intel VTune Amplifier was used. The results presented in Fig. 8 show that these three procedures take over 90% of the computing time:

- forming and solving the temperature field equations (*comp_on_mesh*) 46%
- updating the soil thermophysical parameters (*update_params_soil*) 9%
- updating the wells thermophysical parameters (*update_params_hole*) 38%.

The method for parallelizing the temperature computation using the OpenMP technology is described in works [19,20]. SLAEs for the one-dimensional scheme do not depend on each other, so we can distribute them between OpenMP threads.

The thread load is slightly unbalanced because some of the formed SLAEs correspond to the "scanlines" passing through wells or SCDs causing additional computations. We used dynamic scheduling to compensate this, which reduced the computation time by 10%.

Function	CPU Time: Total ▼ ⊡
comp_on_mesh	46.5%
update_params_hole	37.9%
update_params_soil	9.4%

Fig. 8. Intel VTune Amplifier results for the serial program

The procedures for updating domain parameters just cycle between cells in 3D grid, so we just use '**omp parallel for**' to parallelize it.

Figure 9 shows the analysis results of the parallel program running on 6 cores. The profiler shows good multicore CPU utilization.

Figure 10 shows the computing time for different variants of the test problem. The spatial grid size is $189 \times 91 \times 51$ nodes. The time interval is 1 year with a one day step. It can be seen that the computing time of the variants grows nonlinearly with an increasing number of objects included in the model and the complexity of design of the objects, though the size of the computational grid is fixed.

Table 1 contains the computing times, speedup S, and efficiency E of the parallel program obtained on the six-core AMD Ryzen 5 1600X CPU for different variants of the test problem (see Sect. 2). The obtained values are consistent with the theoretical speedup calculated according to Amdahl law $\bar{S}_m = 1/(\alpha + (1 - \alpha)/m)) = 3$, with the proportion of serial code $\alpha = 0.2$.

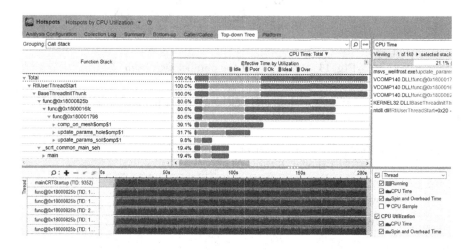

Fig. 9. The analysis results of the parallel program running on 6 cores

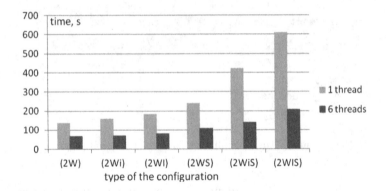

Fig. 10. Computing time for different variants of the test problem

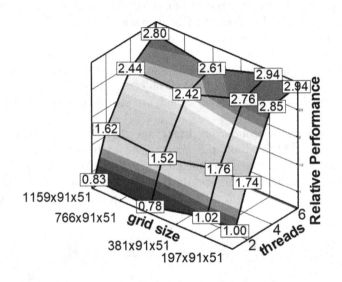

Fig. 11. Scalability of the parallel algorithm

Table 1. Computing time, speedup, and efficiency of the parallel algorithm for 197 × 91 × 51 grid

Variants	Computing time, sec		Speedup S	Efficiency E
	1 thread	6 threads		
(2W)	138	68	2.03	0.34
(2Wi)	158	69	2.29	0.38
(2WI)	184	82	2.25	0.37
(2WS)	240	107	2.24	0.37
(2WiS)	422	141	2.99	0.50
(2WIS)	608	207	2.94	0.49

Table 2. The relative speedup for the weak scaling experiment

Number of threads	Grid size			
	$197 \times 91 \times 51$	$381 \times 91 \times 51$	$766 \times 91 \times 51$	$1159 \times 91 \times 51$
1	1	1.02	0.78	0.83
2	1.74	1.76	1.52	1.62
4	2.85	2.76	2.42	2.44
6	2.94	2.94	2.61	2.8

Table 2 and Fig. 11 show the results of the experiments investigating the scalability of the parallel algorithm. The simulation of system (**2WIS**) was carried out with various numbers of OpenMP threads and various sizes of the computational grid. The parallel algorithm shows good strong and weak scaling. Profiling of the parallel programs shows that the overhead of the parallelization is near 10%. The high overhead and reduction of performance for larger grids are caused by a non-linear increase of the problem's computational complexity due to a non-uniformity of the grid and inhomogeneity of the simulated system.

Remark. To obtain the results presented in Sect. 3, we used the $381 \times 381 \times 201$ grid. It took over 20 h using 6 cores to simulate the system with two wells and SCDs for 10 years.

5   Conclusion

A parallel software package for multicore processors was developed using the OpenMP technology for modeling thermal processes in frozen soils from two production wells and cooling devices. The programs are based on the mathematical model, which makes it is possible to refine the model parameters containing climatic data and geophysical research results obtained for a specific geographic point. The numerical results demonstrate the ability to solve the applied problems associated with the design and operation of petroleum fields in the zone of permafrost. If the complexity of the simulated technical systems increases, the computing time in the studied problems rises up to 6 times. The parallel technologies allowed us to reduce the computing time up to 3 times. The investigation of the parallel algorithm shows a good efficiency and scalability. The purpose of the future research will be to balance the workload for threads, to compensate the problem heterogeneity.

References

1. Voevodin, V.V., Antonov, S.A., Dongarra, J.: AlgoWiki: an open encyclopedia of parallel algorithmic features. Supercomput. Front. Innov. **2**(1), 4–18 (2015)
2. Ortega, J.M.: Introduction to Parallel and Vector Solution of Linear Systems. Springer, Boston (2013). https://doi.org/10.1007/978-1-4899-2112-3

3. Rodrigue, G. (ed.): Parallel Computations, vol. 1. Elsevier, Amsterdam (2014)
4. Chandra, R., Dagum, L., Kohr, D., Menon, R., Maydan, D., McDonald, J.: Parallel Programming in OpenMP. Morgan Kaufmann, New York (2001)
5. Zhang, T., Barry, R.G., Knowles, K., Heginbottom, J.A., Brown, J.: Statistics and characteristics of permafrost and ground ice distribution in the Northern Hemisphere. Polar Geogr. **23**(2), 132–154 (1999)
6. Filimonov, M., Vaganova, N.: Permafrost thawing from different technical systems in Arctic regions. IOP Conf. Ser.: Earth Environ. Sci. **72**, 012006 (2017)
7. Romanovsky, V.E., Smith, S.L., Christiansen, H.H.: Permafrost thermal state in the polar northern 430 hemisphere during the international polar Year 2007–2009: a synthesis permafrost and periglacial 431. Processes **21**, 106–116 (2010)
8. Nelson, F.E., Anisimov, O.A., Shiklomanov, N.I.: Subsidence risk from thawing permafrost. Nature **410**, 889–890 (2001)
9. Nelson, F.E., Anisimov, O.A., Shiklomanov, N.I.: Climate change and hazard zonation in the Circum-Arctic permafrost regions. Nat. Hazards **2**, 203–225 (2002)
10. Stepanov, S.P., Sirditov, I.K., Vabishchevich, P.N., Vasilyeva, M.V., Vasilyev, V.I., Tceeva, A.N.: Numerical simulation of heat transfer of the pile foundations with permafrost. In: Dimov, I., Faragó, I., Vulkov, L. (eds.) NAA 2016. LNCS, vol. 10187, pp. 625–632. Springer, Cham (2017). https://doi.org/10.1007/978-3-319-57099-0_71
11. Kong, X., Doré, G., Calmels, F.: Thermal modeling of heat balance through embankments in permafrost regions. Cold Reg. Sci. Technol. **158**, 117–127 (2019)
12. Gornov, V.F., Stepanov, S.P., Vasilyeva, M.V., Vasilyev, V.I.: Mathematical modeling of heat transfer problems in the permafrost. In: AIP Conference Proceedings, vol. 1629, pp. 424–431 (2014)
13. Vaganova, N.A., Filimonov, M.Y.: Computer simulation of nonstationary thermal fields in design and operation of northern oil and gas fields. In: AIP Conference Proceedings, vol. 1690, p. 020016 (2015)
14. Samarskii, A.A., Moiseyenko, B.D.: An economic continuous calculation scheme for the Stefan multidimensional problem. USSR Comput. Math. Math. Phys. **5**(1), 43–58 (1965)
15. Samarsky, A.A., Vabishchevich, P.N.: Computational Heat Transfer. The Finite Difference Methodology, vol. 2. Wiley, Chichester (1995)
16. Pavlova, N.V., Vabishchevich, P.N., Vasilyeva, M.V.: Mathematical modeling of thermal stabilization of vertical wells on high performance computing systems. In: Lirkov, I., Margenov, S., Waśniewski, J. (eds.) LSSC 2013. LNCS, vol. 8353, pp. 636–643. Springer, Heidelberg (2014). https://doi.org/10.1007/978-3-662-43880-0_73
17. Pepper, D.W., Lombardo, J.M.: High-performance computing for fluid flow and heat transfer. In: Minkowycz, W.J., Sparrow, E.M. (eds.) Advances in Numerical Heat Transfer, vol. 2. Routledge, Abingdon (2018)
18. Orgogozo, L., et al.: Water and energy transfer modeling in a permafrost-dominated, forested catchment of central Siberia: the key role of rooting depth. Permafrost Periglac. Process. **30**(2), 75–89 (2019)
19. Akimova, E.N., Filimonov, M.Y., Misilov, V.E., Vaganova, N.A.: Simulation of thermal processes in permafrost: parallel implementation on multicore CPU. In: CEUR Workshop Proceedings, vol. 2274, pp. 1–9 (2018)
20. Akimova, E.N., Filimonov, M.Yu., Misilov, V.E., Vaganova, N.A.: Supercomputer modelling of thermal stabilization processes of permafrost soils. In: 18th International Conference on Geoinformatics: Theoretical and Applied Aspects (2019)

Algorithm for Numerical Simulation of the Coastal Bottom Relief Dynamics Using High-Performance Computing

Alexander I. Sukhinov[1], Alexander E. Chistyakov[1], Elena A. Protsenko[2],
Valentina V. Sidoryakina[2(✉)], and Sofya V. Protsenko[1]

[1] Don State Technical University, Rostov-on-Don, Russia
`sukhinov@gmail.com,cheese_05@mail.ru,rab55555@rambler.ru`
[2] Taganrog Institute named after A.P. Chekhov (branch of Rostov State University of Economics), Taganrog, Russia
`eapros@rambler.ru,cvv9@mail.ru`

Abstract. The present research considers a non-stationary 2D model of the transport of bottom sediments in the coastal zone of shallow water bodies with respect to the following physical processes and parameters: bottom material porosity, critical tangential stress of the commencement of the bottom materials' transport, dynamically changing bottom surface geometry caused by the movement of the aquatic environment, and turbulent exchange. A discrete sediment transport model is proposed and investigated, generated as a result of approximation of the corresponding linearized continuous model. Since sediment transport forecasting should be done in real or accelerated time scales, on grids of 106–109 nodes, parallel algorithms for hydrodynamic problems on systems with mass parallelism need to be developed. For the proposed model of hydrodynamic processes, parallel algorithms are developed to be implemented as a complex of programs. For the model problems of bottom sediment transport and bottom topography transformation, numerical experiments were performed to align the results with real physical experiments.

Keywords: Distributed computing · High-performance computing ·
Parallel programming · Mathematical model · Sediment dynamics ·
Bottom topography

1 Introduction

The current environmental condition of the coastal systems is largely determined by the diversity of incoming solid particles of both mineral and organic origin. The suspended matter particles are transformed and deposited and, as a consequence, the bottom sediment is formed under the influence of complex of intra-water processes. The coastal systems' sediments are a complex heterogeneous

The reported study was funded by RFBR, project number 19-01-00701.

L. Sokolinsky and M. Zymbler (Eds.): PCT 2020, CCIS 1263, pp. 279–290, 2020.
https://doi.org/10.1007/978-3-030-55326-5_20

physical and chemical system with the ongoing processes that are especially relevant to study now [1,2]. It is relevant to assess the pollutant flows and determine the amount of sediments associated with the development of industrial and recreational activities in the coastal areas [3]. As a rule, this area is researched with the help of mathematical models aligned with the real processes and making it possible to predict the suspended matter distribution in the aquatic environment [4].

The complexity of the sediment transport processes in the aquatic environment of the coastal systems and the need to model them on detailed grids of 10^6–10^9 nodes requires the use of parallel computing systems (more than 100 teraflops) to do the operational forecasting of these processes. In this work, the authors proposed and investigated a discrete spatially two-dimensional model of sediment transport, produced by approximating the corresponding linearized continuous model and supplemented with the Navier–Stokes and continuity equations, and the aquatic environment condition equation. The model takes into account the following physical parameters and processes: soil porosity; the critical shear stress of the commencement of the sediment movement; turbulent exchange; dynamically changing bottom geometry and elevation function; wind currents; bottom friction [5–8]. To achieve a joint numerical solution of the sediment transport and wave hydrodynamics problems using a supercomputing system with distributed memory with a relatively small number of cores (up to 2048), using parallel algorithms is considered. The authors also investigated the issues of parallelizing the numerical solution processes on massively parallel systems that provide high efficiency algorithms for the systems containing many tens of thousands of cores. To solve the system of grid equations, an adaptive modified alternately triangular iterative method was used. The results of numerical experiments are presented.

2   Continuous 2D Model of Sediment Transport

Water flows carry a large amount of solid sediment particles, moving particles of clay, mud, gravel, pebbles, sand, loess, carbonate compounds, mineral oil emulsions, petroleum products and other components. The surfaces of the sediment particles are capable of absorbing various pollutants, including heavy metals and pesticides which make a negative impact on the ecological situation of the water body [9,10]. Sediment can be carried by the flow in a suspended state (suspended sediment), and can be moved in the bottom layer of the stream by rolling, sliding, saltation (entrained sediment). Particles of the flow-carried sediment can transform from the suspended state to the entrained state and vice versa, entrained particles can stop moving, and motionless particles can begin moving. The nature of the movement of suspended and entrained sediments are determined by the flow velocity, depth, and other hydraulic elements of the water flow.

3 Mathematical Description of the Initial Boundary Problem of Sediment Transport

Let us consider the sediment transport equation [11,12]:

$$(1 - \varepsilon)\frac{\partial H}{\partial t} = div\left(k\frac{\tau_{bc}}{\sin\varphi_0}gradH\right) - div\left(k\boldsymbol{\tau}_b\right), \tag{1}$$

where $H = H(x, y, t)$ is the depth of the pond; ε is the porosity of bottom materials; $\boldsymbol{\tau}_b$ is the tangential stress vector at the bottom of the reservoir; τ_{bc} is the critical tangential stress value; $\tau_{bc} = a\sin\varphi_0$, φ_0 is the angle of the soil repose in the pond; $k = k(H, x, y, t)$ is a nonlinear coefficient defined by the relation:

$$k = \frac{A\tilde{\omega}d}{((\rho_1 - \rho_0)gd)^{\beta}}\left|\boldsymbol{\tau}_b - \frac{\tau_{bc}}{\sin\varphi_0}gradH\right|^{\beta-1}, \tag{2}$$

where ρ_1, ρ_0 are the particle densities of the bottom material and the aqueous medium, respectively; g an acceleration of gravity; $\tilde{\omega}$ is wave frequency; A and β are the dimensionless constants; d are the characteristic sizes of the soil particles.

We supplement Eq. (1) with the initial condition assuming that the function of the initial conditions belongs to the corresponding smoothness class:

$$H(x, y, 0) = H_0(x, y), H_0(x, y) \in C^2(D) \cap C(\overline{D}),$$
$$grad_{(x,y)}H_0 \in C(\overline{D}), (x, y) \in \overline{D}. \tag{3}$$

Let us formulate the region $\bar{D}$ boundary conditions as:

$$|\overrightarrow{\tau_b}|\big|_{y=0} = 0, \tag{4}$$

$$H(L_x, y, t) = H_2(y, t), \ 0 \le y \le L_y. \tag{5}$$

$$H(0, y, t) = H_1(y, t), \ 0 \le y \le L_y, \tag{6}$$

$$H(x, 0, t) = H_3(x), \ 0 \le x \le L_x. \tag{7}$$

$$H(x, L_y, t) = 0, \ 0 \le x \le L_x. \tag{8}$$

In addition to the boundary conditions (5)–(8), assume that their smoothness conditions are satisfied, and continuous derivatives of the H function on the boundary of the region D exist:

$$grad_{(x,y)}H \in C(\overline{CY}_T) \cap C^1(CY_T). \tag{9}$$

The condition of non-degeneracy operator has the form:

$$k \ge k_0 = const > 0, \ \forall(x, y) \in \overline{D}, \ 0 < t \le T. \tag{10}$$

The vector of tangential stress at the bottom is expressed using unit vectors of the coordinate system in a natural way:

$$\boldsymbol{\tau}_b = \boldsymbol{i}\tau_{bx} + \boldsymbol{j}\tau_{by}, \tau_{bx} = \tau_{bx}(x, y, t), \tau_{by} = \tau_{by}(x, y, t). \tag{11}$$

4   Construction of the Linearized Initial-Boundary Value Problem of Sediment Transport

We construct linearized chain of sediment transport problems [13] on a uniform time grid $\omega_\tau = \{t_n = n\tau, \; n = 0, 1, ..., N, \; N\tau = T\}$, approximating with an error $O(\tau)$ in Hilbert space $L_1(D \times [0, T])$ initial initial-boundary value problem (1)–(8).

Introduce the following notation:

$$k^{(n-1)} \equiv \frac{A\tilde{\omega}d}{\left((\rho_1 - \rho_0)\, gd\right)^\beta} \left| \overrightarrow{\tau}_b - \frac{\tau_{bc}}{\sin\varphi_0} gradH^{(n-1)}(x, y, t_{n-1}) \right|^{\beta-1}, \qquad (12)$$

$$n = 1, 2, ..., N.$$

Then Eq. (1) after linearization, will take the form:

$$(1 - \varepsilon)\frac{\partial H^{(n)}}{\partial t} = div\left(k^{(n-1)}\frac{\tau_{bc}}{\sin\varphi_0} gradH^{(n)}\right) - div\left(k^{(n-1)}\overrightarrow{\tau}_b\right), \qquad (13)$$

$$t_{n-1} < t \leq t_n, \; n = 1, ..., N$$

and add the initial conditions:

$$H^{(1)}(x, y, t_0) = H_0(x, y), \; H^{(n)}(x, y, t_{n-1}) = H^{(n-1)}(x, y, t_{n-1}), \qquad (14)$$

$$(x, y) \in \overline{D}, \; n = 2, ..., N.$$

$div\left(k^{(n-1)}\overrightarrow{\tau}_b\right)$ is a known function of the right-hand side with such linearization; boundary conditions (4)–(8) are assumed to be satisfied for all time intervals $t_{n-1} < t \leq t_n, n = 1, 2, ..., N$.

The coefficients $k^{(n-1)}, n = 1, 2, ..., N$ depend on spatial variables x, y and time variable $t_{n-1}, n = 1, 2, ..., N$, determined by the choice of grid step τ, i.e. $k^{(n-1)} = k^{(n-1)}(x, y, t_{n-1}), \; n = 1, 2, ..., N$.

In [14], the conditions for the existence and uniqueness of the sediment transport problem under conditions of smoothness of the solution function were studied

$$H(x, y, t) \in C^2(CY_T) \cap C\left(\overline{CY}_T\right), \; gradH \in C\left(\overline{CY}_T\right)$$

and the necessary smoothness of the region boundary, as well as the a priori estimate of the solution in the norm of the space L_1 depending on the integral estimates of the right-hand side, boundary conditions and the norm of the initial condition [18, 19] were considered. The results of the convergence of the linearized problem solution to the solution of the initial nonlinear initial-boundary problem of sediment transport in the norm of the space were presented L_1 with the time grid step of the linearization tending to zero [15].

5   Spatially Heterogeneous 3D Model of Hydrodynamics

The input data for the sediment transport model is the velocity vector of the aquatic environment. To calculate tangential stresses in the sediment transport model, the velocity vector data for the water medium at the reservoir bottom is needed. The initial hydrodynamic model equations are [16,17]:

– equation of motion (Navier-Stokes):

$$u'_t + uu'_x + vu'_y + wu'_z = -\frac{1}{\rho}P'_x + (\mu u'_x)'_x + (\mu u'_y)'_y + (\nu u'_z)'_z,$$

$$v'_t + uv'_x + vv'_y + wv'_z = -\frac{1}{\rho}P'_y + (\mu v'_x)'_x + (\mu v'_y)'_y + (\nu v'_z)'_z, \qquad (15)$$

$$w'_t + uw'_x + vw'_y + ww'_z = -\frac{1}{\rho}P'_z + (\mu w'_x)'_x + (\mu w'_y)'_y + (\nu w'_z)'_z + g;$$

– continuity equation for variable density cases:

$$\rho'_t + (\rho u)'_x + (\rho v)'_y + (\rho w)'_z = 0, \qquad (16)$$

where $V = \{u, v, w\}$ is the velocity vector of the water current in a shallow water body; ρ is the aquatic environment density; P is the hydrodynamic pressure; g is the gravitational acceleration; ; μ, ν are coefficients of turbulent exchange in the horizontal and vertical directions; n is the normal vector to the surface describing the boundary of the computational domain.

Add boundary conditions to system (1)–(2):

– entrance (left border): $\mathbf{V} = \mathbf{V}_0$, $P'_n = 0$,
– bottom border: $\rho\mu(\mathbf{V}_\tau)'_n = -\boldsymbol{\tau}$, $\mathbf{V}_n = 0$, $P'_n = 0$,
– lateral border: $(\mathbf{V}_\tau)'_n = 0$, $\mathbf{V}_n = 0$, $P'_n = 0$,
– upper border: $\rho\mu(\mathbf{V}_\tau)'_n = -\boldsymbol{\tau}$, $w = -\omega - P'_t/\rho g$, $P'_n = 0$,
– surface of the structure: $\rho\mu(\mathbf{V}_\tau)'_n = -\boldsymbol{\tau}$, $w = 0$, $P'_n = 0$,

where ω is the liquid evaporation intensity, $\mathbf{V}_n$, $\mathbf{V}_\tau$ are the normal and tangential components of the velocity vector, $\boldsymbol{\tau} = \{\tau_x, \tau_y, \tau_z\}$ is the tangential stress vector.

Let $\boldsymbol{\tau} = \rho_a Cd_s |w| w$, where w is the wind velocity relative to water, ρ_a is the atmosphere density, $Cd_s = 0.0026$.

Let us set the tangential stress vector for the bottom taking into account the movement of water as follows: $\boldsymbol{\tau} = \rho Cd_b |\mathbf{V}| \mathbf{V}$, $Cd_b = gk^2/h^{1/3}$, where $k = 0,04$ is the group roughness coefficient in the Manning formula, considered in the range of $0,025$–$0,2$. Note that the magnitude of the coefficients Cd_s, Cd_b is influenced by many parameters, including wind speed, stratification, age of sea waves, wind direction, sea depth, roughness of the bottom surface, the shape of the bottom, siltation and erosion processes, the presence of obstacles, etc. factors. In the coastal part of the reservoir, as a rule, an increase in the ratio Cd_s compared with values in the deep sea, due to a decrease of the phase velocity of the waves, increasing their steepness, fast changing wave field, a growing number of collapses, bottom topography, nature of the shoreline. It is believed that with a decrease in depth, the value of the coefficient Cd_b also increases. Taking into account the influence of the above factors on the nature of the coefficients Cd_s, Cd_b is quite difficult and, therefore, they are the result of processing numerous experimental data obtained experimentally.

6   Discrete Model

Let construct a finite-difference scheme approximating problem (13), (14), (4)–(8). Cover the area D uniform rectangular calculation grid $\omega = \omega_x \times \omega_y$, assuming that the time grid ω_τ previously defined.

$$\omega_x = \{x_i = ih_x,\ 0 \le i \le N_x - 1,\ l_x = h_x\,(N_x - 1)\},$$
$$\omega_y = \{y_j = jh_y,\ 0 \le j \le N_y - 1,\ l_y = h_y\,(N_y - 1)\},$$

where n, i, j are indices of grid nodes constructed on a temporary Ot and spatial Ox, Oy directions, respectively, τ, h_x, h_y are grid steps in temporal and spatial directions, respectively, N_t, N_x, N_y are the number of nodes in the temporal and spatial directions, respectively.

The balance method was used to obtain a difference scheme. We integrate both sides of Eq. (13) over the region D_{txy}:

$$D_{txy} \in \left\{ t \in [t_n, t_{n+1}],\ x \in \left[x_{i-1/2}, x_{i+1/2}\right],\ y \in \left[y_{j-1/2}, y_{j+1/2}\right] \right\},$$

as a result, we obtain the following equality:

$$
\begin{aligned}
\iiint_{D_{txy}} (1 - \varepsilon)\, H_t^{(n)'}\, dt\,dx\,dy &+ \iiint_{D_{txy}} \left(k^{(n-1)} \tau_{b,x}\right)_x' dt\,dx\,dy \\
+ \iiint_{D_{txy}} \left(k^{(n-1)} \tau_{b,y}\right)_y' dt\,dx\,dy & \\
= \iiint_{D_{txy}} \left(k^{(n-1)} \frac{\tau_{bc}}{\sin \varphi_0} H_x^{(n)'}\right)_x' dt\,dx\,dy & \\
+ \iiint_{D_{txy}} \left(k^{(n-1)} \frac{\tau_{bc}}{\sin \varphi_0} H_y^{(n)'}\right)_y' dt\,dx\,dy. &
\end{aligned}
\tag{17}
$$

In equality (17), we calculate the approximate integrals from the rectangle formulas, divide the resulting equation by the product of multipliers τ, h_x, h_y, and, replacing the approximate equality with the exact one, we obtain a difference scheme approximating the linearized continuous problem:

$$
\begin{aligned}
(1 - \varepsilon)\, &\frac{H_{i,j}^{(n+1)} - H_{i,j}^{(n)}}{\tau} \\
+ \frac{k_{i+1/2,j}^{(n)}(\tau_{b,x})_{i+1/2,j}^{(n)} - k_{i-1/2,j}^{(n)}(\tau_{b,x})_{i-1/2,j}^{(n)}}{h_x} &+ \frac{k_{i,j+1/2}^{(n)}(\tau_{b,y})_{i,j+1/2}^{(n)} - k_{i,j-1/2}^{(n)}(\tau_{b,y})_{i,j-1/2}^{(n)}}{h_y} \\
= \frac{\tau_{bc}}{\sin \varphi_0} \left(k_{i+1/2,j}^{(n)} \frac{H_{i+1,j}^{(n+\sigma)} - H_{i,j}^{(n+\sigma)}}{h_x^2} \right. &\left. - k_{i-1/2,j}^{n} \frac{H_{i,j}^{(n+\sigma)} - H_{i-1,j}^{(n+\sigma)}}{h_x^2} \right) \\
+ \frac{\tau_{bc}}{\sin \varphi_0} \left(k_{i,j+1/2}^{(n)} \frac{H_{i,j+1}^{(n+\sigma)} - H_{i,j}^{(n+\sigma)}}{h_y^2} \right. &\left. - k_{i,j-1/2}^{n} \frac{H_{i,j}^{(n+\sigma)} - H_{i,j-1}^{(n+\sigma)}}{h_y^2} \right),
\end{aligned}
\tag{18}
$$

where

$$
(\tau_{b,x})^{(n)}_{i+1/2,j} = \frac{(\tau_{b,x})^{(n)}_{i+1,j} + (\tau_{b,x})^{(n)}_{i,j}}{2}, \; (\tau_{b,y})^{(n)}_{i,j+1/2} = \frac{(\tau_{b,y})^{(n)}_{i,j+1} + (\tau_{b,y})^{(n)}_{i,j}}{2},
$$

$$
k^{(n)}_{i+1/2,j} = \frac{A \varpi d \left| (\vec{\tau}_b)^{(n)}_{i+1/2,j} - \frac{\tau_{bc}}{\sin \varphi_0} (gradH)^{(n)}_{i+1/2,j} \right|}{((\rho_1 - \rho_0) \, gd)^{\beta}} h \left(\left| (\vec{\tau}_b)^{(n)}_{i+1/2,j} - \frac{\tau_{bc}}{\sin \varphi_0} (gradH)^{(n)}_{i+1/2,j} \right| - \tau_{bc} \right).
$$

The value $gradH|_{(x_{i+1/2}, y_j)}$ is written as

$$
(gradH)_{i+1/2,j} = \frac{H_{i+1,j} - H_{i,j}}{h_x} \vec{i} + \frac{H_{i+1/2,j+1} - H_{i+1/2,j-1}}{2h_y} \vec{j}.
$$

Similarly, you can get the following approximation:

$$
(gradH)_{i,j+1/2} = \frac{H_{i+1,j+1/2} - H_{i-1,j+1/2}}{2h_x} \vec{i} + \frac{H_{i,j+1} - H_{i,j}}{h_y} \vec{j}.
$$

Difference schemes are investigated for stability using the grid maximum principle. The constructed difference schemes are stable under the following restriction on the time step:

$$
\tau < \frac{\sin \varphi_0 \, (1 - \varepsilon)}{\tau_{bc} \, (1 - \sigma) \max\limits_{0 \le m \le N-1} \{k(t_m)\} \left(\frac{2}{h_x^2} + \frac{2}{h_y^2} \right)}.
$$

The upper bound for the grid function is the solution of the difference problem in the grid norm $C_{h,\tau}$ formulated as:

$$
\| H^n \|_{C_{h,\tau}} \le \| H^0 \|_{C_h} + \max \left(\| H_1 \|_{C_{h,\tau}}, \| H_2 \|_{C_{h,\tau}}, \| H_3 \|_{C_{h,\tau}} \right)
$$

$$
+ \frac{\tau}{1 - \varepsilon} \sum_{m=0}^{n} \left\| (k(t_m) \tau_{b,x})^m_0 + (k(t_m) \tau_{b,y})^m_0 \right\|_{C_{h,\tau}},
$$

which guarantees the stability of the constructed difference scheme with respect to the function of the right-hand side, the boundary and initial conditions.

The approximation error of the discrete sediment transport model was found that an order of magnitude $O\left(\tau + h_x^2 + h_y^2\right)$.

7 Parallel Algorithm Description

A software package in C++ is designed to build turbulent flows of an incompressible velocity field of the aquatic environment on high-resolution grids for predicting sediment transport and possible scenarios of changing the geometry of the bottom region of shallow water bodies. Parallel algorithms implemented in the software package for solving systems of grid equations arising during the discretization of model problems were developed using MPI technology.

To solve this problem, the adaptive modified alternating-triangular method of minimum corrections was used. In parallel implementation, decomposition methods of grid domains were applied to computationally time-consuming diffusion-convection problems with respect to the architecture and parameters of a multiprocessor computing system. The calculated two-dimensional region was decomposed with two spatial variables x, y. The peak performance of the multiprocessor computing system is 18.8 teraflops. As computing nodes, 128 HP ProLiant BL685c homogeneous 16-core Blade servers of the same type were used, each being equipped with four 4-core AMD Opteron 8356 2.3 GHz processors and 32 GB RAM.

Figure 1 shows the dependence of acceleration on the number of processors needed to solve the model problem on various grids. The numbering of the graphs corresponds to the following dimensions of the calculation grids: 1–100 × 100, 2–200 × 200, 3–500 × 500, 4–1000 × 1000, 5–2000 × 2000, 6–5000 × 5000.

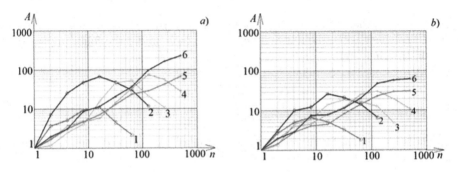

Fig. 1. Graph of the acceleration of the parallel algorithm (a – based on an explicit scheme, b – based on an implicit scheme

Figure 1a shows that when using a parallel algorithm based on an explicit scheme, the maximum acceleration was achieved on 200 × 200 grids. In this case, a superlinear acceleration arises due to the fact that with an increase in the number of computers, the total amount of their RAM and cache grows. Therefore, most of the task data is located in RAM and, moreover, is "placed" in the cache. In Fig. 1b shows a graph of the acceleration of the parallel algorithm based on the implicit scheme using the parallel version of the alternating triangular method of minimal corrections. The constructed graphs show that for each of the computational grids, the acceleration assumes the greatest value at a certain value of calculators, and with a further increase in the number of computing cores, the acceleration only decreases. This is due to a significant increase in time spent on data exchanges between computers. Parallel algorithms based on an explicit scheme are better parallelized than the methods of a parallel algorithm based on an implicit scheme using a parallel version of the alternating-triangular method of minimal corrections. Despite this, a parallel version of the modified alternating triangular method of minimal corrections is preferable for ill-conditioned problems, since it requires a significantly smaller number of iterations when solving the problem of transport of bottom materials.

Figure 2.A presented the graph of acceleration versus the number of processors needed to solve the problems of hydrodynamics is presented. To calculate the velocity vector of the aqueous medium, computational grids of size $50 \times 250 \times 40$.

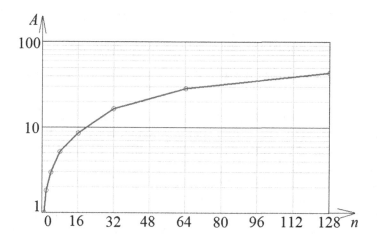

Fig. 2. Acceleration schedule of the hydrodynamic problems' parallel solution algorithm

The maximum acceleration of 43.668 was achieved on 128 cores, each running an MPI process. On 256 computers, a drop in acceleration was observed.

8 Numerical Experiments for Modeling Sediment Transport and Bottom Topography Dynamics

After the software package had been developed, a series of numerical experiments was performed to simulate the dynamically changing bottom topography of various configurations. In the model problems, the subsequent movement of the aquatic environment was calculated with respect to the irregularities of the bottom surface (hilly terrain, boulders, terraces, underwater valleys, underwater breakwaters, dams, etc.).

The paper presents the results of modeling the bottom change dynamics for the case of presence of pointed structures, such as discontinuous dams, acting as obstacles on the bottom surface. Retaining the sediment, dams do not only stop the movement of the material carried by the waves along the shore, but also contribute to its deposition.

The simulation section under consideration has dimensions of 55 m by 55 m horizontally and 2 m vertically (in depth), with the peak point rising 1 m above

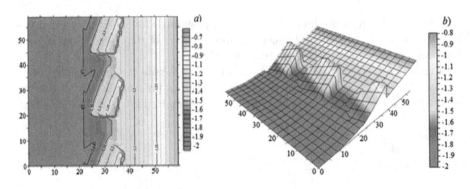

Fig. 3. The geometry of the computational domain at the initial moment

sea level. Suppose that the liquid is at rest at the initial time. The computational grid size is 110 by 110, the step in spatial variables is 0.05 m, the time step is 0.01 s, the wind is directed from left to right at the speed of 5 m/s.

Figure 3 represents the initial position of the contour lines of the depth function (top) and bottom topography (bottom) featuring three pointed structures with an uneven surface. Structures of this kind can be, for example, moles, boulders, dumps, breakwaters. Fluctuations in the isolines of the depth function are observed in the central part of the computational domain. Being below the level of 0.2 m from the free surface, these structures are located at a distance of about 10 m from each other.

Sediment transport process modeling showed that smoothing of surface irregularities, sediment formation, and the decreasing repose angle of the coastal zone bottom that occur over time cause a gradual shallowing of the considered reservoir zone. For this reason, after 5 min, in the center of the computational domain the depth function contours took on a more tortuous shape, the pointed structures deformed and took the form of gentle slides, and the depth of the coastal zone reduced (Fig. 4).

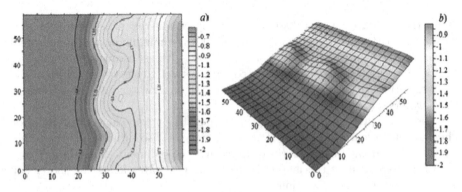

Fig. 4. The geometry of the computational domain after 5 min from the start of the simulation.

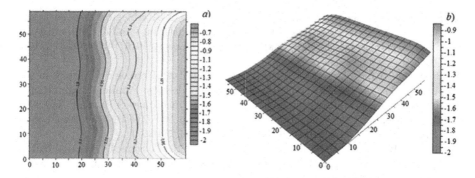

Fig. 5. The computational domain geometry 15 min after the modeling start (a - isolines of the depth function, b - bottom relief)

According to Fig. 5, the result of the described processes gets more visible at the process simulation time of 15 min. The depth function contours acquire a soft wave-like shape in the center of the computational domain, including the region of peak values. Along the coast, there is a zone of sediment formation and a decrease in the level of depth. The height of the slides decreases, and the slides themselves acquire a more smoothed appearance.

9 Conclusion

The experiment results provide the materials to analyze the dynamics of changes in the bottom surface shape, the function the aquatic environment elevation, the formation of ridges, sediments and the wave form of the functions of the bottom and the water surface. The proposed mathematical model and the developed software package can be used to predict the dynamics of the bottom surface behavior, the shape of marine braids and ridges, their growth and transformation. Therefore, the proposed mathematical models for sediment transport have been verified by the numerical experiment results.

References

1. Leontyev, I.O.: Coastal dynamics: waves, moving streams. Deposits drifts, GEOS, San Moscow (2001). (in Russian)
2. Liu, X., Qi, S., Huang, Y., Chen, Y., Du, P.: Predictive modeling in sediment transportation across multiple spatial scales in the Jialing River Basin of China. Int. J. Sedim. Res. **30**(3), 250–255 (2015)
3. Aksoy, H., Kavvas, M.L.: A review of hillslope and watershed scale erosion andsediment transport models. Catena **64**(2–3), 247–271 (2005). https://doi.org/10.1016/j.catena.2005.08.008
4. Ouda, M., Toorman, E.A.: Development of a new multiphase sediment transport model for free surface flows. Int. J. Multiph. Flow **117**, 81–102 (2019)

5. Sukhinov, A.I., Sukhinov, A.A.: Reconstruction of 2001 ecological disaster in the Azov Sea on the basis of precise hydrophysics models. In: Parallel Computational Fluid Dynamics, pp. 231–238 (2005). https://doi.org/10.1016/B978-044452024-1/50030-0

6. Sukhinov, A.I., Sukhinov, A.A.: Reconstruction of 2001 ecological disaster in the Azov Sea on the basis of precise hydrophysics models. In: Parallel Computational Fluid Dynamics, Multidisciplinary Applications, Proceedings of Parallel CFD 2004 Conference, Las Palmas de Gran Canaria, Spain, pp. 231–238. Elsevier, Amsterdam (2005). https://doi.org/10.1016/B978-044452024-1/50030-0

7. Alekseenko, E., Roux, B., Sukhinov, A., Kotarba, R., Fougere, D.: Coastal hydrodynamics in a windy lagoon. Nonlinear Processes Geophys. 20(2), 189–198 (2013). https://doi.org/10.1016/j.compfluid.2013.02.003

8. Alekseenko, E., Roux, B., Sukhinov, A., Kotarba, R., Fougere, D.: Nonlinear hydrodynamics in a mediterranean lagoon. Comput. Math. Math. Phys. 57(6), 978–994 (2017). https://doi.org/10.5194/npg-20-189-2013

9. Francke, T., López-Tarazón, J.A., Vericat, D., Bronstert, A., Batalla, R.J.: Flood-based analysis of high-magnitude sediment transport using a non-parametric method. Earth Surf. Proc. Land. 33(13), 2064–2077 (2008). https://doi.org/10.1002/esp.1654

10. Karaushev, A.N.: Theory and methods for river load calculation. Gidrometeoizdat, Leningrad (1977). (in Russian)

11. Alekseevskii, N.I.: Hydrophysics. Akademiya, Moscow (2006). (in Russian)

12. Goloviznin, V.M., Chetverushkin, B.N.: New generation algorithms for computational fluid dynamics. Comput. Math. Math. Phys. 58(8), 1217–1225 (2018). https://doi.org/10.1134/S0965542518080079

13. Sukhinov, A.I., Chistyakov, A.E., Protsenko, E.A.: Mathematical modeling of sediment transport in the coastal zone of shallow reservoirs. Math. Models Comput. Simul. 6(4), 351–363 (2014). https://doi.org/10.1134/S2070048214040097

14. Sidoryakina, V.V., Sukhinov, A.I.: Well-posedness analysis and numerical implementation of a linearized two-dimensional bottom sediment transport problem. Comput. Math. Math. Phys. 57(6), 978–994 (2017). https://doi.org/10.7868/S0044466917060138

15. Sukhinov, A.I., Sidoryakina, V.V.: Convergence of linearized sequence tasks to the nonlinear sediment transport task solution. Matematicheskoe modelirovanie 29(11), 19–39 (2017)

16. Belotserkovskii, O.M., Gushchin, V.A., Shchennikov, V.V.: Decomposition method applied to the solution of problems of viscous incompressible fluid dynamics. Compu. Math. Math. Phys. 15, 197–207 (1975)

17. Favorskaya, A.V., Petrov, I.B.: Numerical modeling of dynamic wave effects in rock masses. Doklady Math. 95(3), 287–290 (2017). https://doi.org/10.1134/S1064562417030139

18. Sukhinov, A.I., Chistyakov, A.E., Shishenya, A.V.: Error estimate for diffusion equations solved by schemes with weights. Math. Models Comput. Simul. 6(3), 324–331 (2014). https://doi.org/10.1134/S2070048214030120

19. Nikitina, A.V., Sukhinov, A.I., Ugolnitsky, G.A., Usov, A.B., Chistyakov, A.E., Puchkin, M.V., Semenov, I.S.: Optimal control of sustainable development in the biological rehabilitation of the Azov Sea. Math. Models Comput. Simul. 9(1), 101–107 (2017). https://doi.org/10.1134/S2070048217010112

Computer Design of Hydrocarbon Compounds with High Enthalpy of Formation

V. M. Volokhov[1], T. S. Zyubina[1], A. V. Volokhov[1], E. S. Amosova[1], D. A. Varlamov[1(✉)], D. B. Lempert[1], and L. S. Yanovskiy[1,2]

[1] Institute of Problems of Chemical Physics of the Russian Academy of Sciences, Chernogolovka, Moscow Region, Russia
{vvm,zyubina,aes,dima,lempert}@icp.ac.ru, vav@icp.ru
[2] The P. I. Baranov Central Institute of Aviation Motor Development, Moscow, Russia
Yanovskiy@ciam.ru

Abstract. New approaches are being actively developed to create promising new-generation rocket fuels that meet the increased requirements for the energy content of their components. One of such approaches is the computer design of new substances that, though not synthesized yet, for various reasons present a great potential for creating new fuel components. In this paper, the thermochemical properties of a number of compounds similar in structure to 1,4-Diethynylbenzene (DEB) $C_{10}H_6$: $C_{18}H_8$, $C_{26}H_{10}$, $C_{19}H_{10}$, $C_{16}H_8$ are studied. Quantum-chemical calculations of thermochemical properties are carried out with the GAUSSIAN 09 software package for getting high-accuracy enthalpy values for the studied substances. A part of GAUSSIAN 09, the composite method G4 was used as the main method for calculating the enthalpy of molecular formation. Due to the high computational complexity, the calculation takes from several hours to months on multi-node configurations.

Keywords: Quantum chemical calculations · High enthalpy compounds · Enthalpy of formation

1 Introduction

The main parameter defining the energy content of the substance is enthalpy of formation ΔH^o_f in the state of matter it is intended for use. The operating characteristics (specific impulse of fuels, flight range, and others) of energy-intensive compounds and compositions based on them in a fairly wide range of values depend approximately linearly on the change in ΔH^o_f. To get the most reliable evaluation of the prospects of a particular compound in energy compositions for

This work was supported by RFBR according to the research project No. 20-07-00319, and by Government of the Russian Federation according to the contract AAAA-A19-119120690042-9.

various purposes, it is necessary to have the maximally ΔH_f^o values of the components, especially those acting as the main substance in the composition. One of the few experimental methods for determining enthalpy of formation is burning substance in a calorimeter: researchers determine the specific heat of combustion in oxygen, and then calculate the enthalpy of formation. On modern equipment, the calorimetric measurement of the actual thermal effect is made with high accuracy, subject to 100% combustion of the substance to water, nitrogen and CO_2. As nitrogen oxides are also often formed in the combustion products together with nitrogen, their amount needs to be measured and the formation heat needs to be corrected. The most precise experiments usually give average enthalpy of formation values with an accuracy of 40–50 kJ/kg. For example, calculating the detonation velocity of HMX (octogen), such accuracy in determining the enthalpy of formation value gives an error of ∼0.2%, which is negligible. But using HMX as a fuel component with a 50% content, the indicated deviation in the enthalpy of formation results in a specific impulse deviation of 0.4 s, which is significant. The purity of the substance which is usually not high enough also presents a challenge in the experimental determination of the enthalpy of formation. Moreover, as a rule, the trace elements present in the sample and their quantity are never known in advance. The presence of 0.1% water during the calorimetric measurement of the HMX enthalpy of formation gives an error of 9.5 kJ/kg (where the enthalpy of formation of HMX is 295.15 kJ/kg). If the impurity is a substance with a higher specific heat of combustion than HMX, e.g. benzene, then every 0.1% of benzene impurity will increase the enthalpy of formation by 32.3 kJ/kg.

The above examples show the importance of using direct methods for calculating the energy and thermochemical parameters of high-energy substances, such as quantum-chemical calculations based on *ab initio* approaches.

Moreover, quantum chemical methods allow designing and calculating the expected thermochemical parameters of hypothetical, not yet synthesized compounds with high accuracy. Therefore, the range of substances planned for synthesis can be significantly narrowed, speeding up the new fuel creation process. Paper [1] compares the G4 method accuracy with the G2 and G3 methods and presents the experimental data for the set of 454 substances. It shows that the average deviation of the G4 method-based calculation results from the experimental data is 0.83 kcal/mol.

In [2], authors used the combined W4-17 method to calculate the atomization reactions of 200 molecules and radicals consisting of atoms of the first (Li-F) and second (Na-Cl) row with no more than eight non-hydrogen atoms. Analysis of the investigated compounds showed that a standard deviation of the calculation by the combined G4, Wn, Wn-F12, WnX, ccCA-PS3 methods from the experiment ranges from 2.3 to 4.0 kJ/mol, while for G3, Gn(MP2) and CBS methods it exceeds 4.2 kJ/mol. The combined G4 method gives the best results (standard deviations are 4.0 kJ/mol), but for high fluorinated and chlorinated systems (such as SF_6, PF_5, PF_3, BF_3, CCl_4, C_2Cl_2, C_2Cl_4, and C_2Cl_6), these deviations reach 4–20 kJ/mol. G4(MP2) is the best among the combined methods of the Gn(MP2) type (standard deviations are 5.4 kJ/mol). The G4(MP2)6X, ROG4(MP2)-6X, G3(MP2) and G3(MP2)B3 methods are not so good. The standard deviations of their results range from 6.9 to 9.3 kJ/mol. The combined methods such as CBS (CBS-QB3, ROCBS-QB3 and CBS-4M) manifest standard deviations in the range of 8.5–17.7 kJ/mol. Therefore, in this paper these methods were not used to calculate the atomization energy. The ccCA-PS3 method provides results close to those acquired with the G4 method (standard deviation 3.8 kJ/mol). With the W1 method, the standard deviation is reduced to 3.1 kJ/mol. W1-F12 and W1X-n methods give results comparable with W1 in much shorter computational time, namely, they present a standard deviation ranging from 2.8 (W1X-1) to 3.3 (W1X-2) kJ/mol. More computation-intensive W2-type methods give somewhat better results with standard deviations of 2.3 (W2-F12), 2.4 (W2) and 2.6 (W2X) kJ/mol. However, methods like Wn are only available so far for small molecules. Thus, the combined G4 and G4(MP2) methods were selected as operating ones for the present research.

Based on the GAUSSIAN software package using the G4 method as the main method for calculating the molecule enthalpy of formation, quantum-chemical calculations of the thermochemical properties of substances of a number of high enthalpy ethinyl hydrocarbons were carried out to acquire high-accuracy the enthalpy values for the studied substances and to make preliminary estimates of the prospects of the components, as well as to develop the most effective methodology for further research.

Table 1. Compounds of the DEB group (composition, structural formula, name)

DEB $C_{10}H_6$		1,4-diethinylbenzene
$C_{18}H_8$		1,4,5,8-tetraethynylnaphthalene
$C_{26}H_{10}$		1,4,5,8,9,10-hexaethynylanthracene
$C_{19}H_{10}$		2,5,8-triethinyl-1H-phenalene
$C_{16}H_8$		1,3,5,7-tetraethynylcycloocta-1,3,5,7-tetraen

In this work we study thermochemical properties of a number of compounds similar to 1,4-Dienthinylbenzene (DEB) $C_{10}H_6$, which, as previously shown in [3], increases the flight range of the flying vehicle when used as a fuel component for a ramjet engine. In this work we also study homologues of DEB: $C_{18}H_8$, $C_{26}H_{10}$, $C_{19}H_{10}$, $C_{16}H_8$ (Table 1), with high enthalpy ethinyl groups content comparable with DEB. It should be noted that all other substances in the considered series, except DEB, have not been synthesized yet.

2   Choice of the Calculation Method of the Enthalpy of Formation in the Gas Phase

The enthalpy of formation in the solid phase is of great interest for evaluating the energy intensity of substances. The most effective way to calculate this value is to find the enthalpy of formation of a molecule in the gas phase and then make corrections for the sublimation energy. There are several methods of varying complexity and accuracy for calculating the molecule enthalpy of formation ΔH_f^o. These include empirical and semi-empirical methods based on the assumption that the enthalpy of formation of a complex molecule from the groups of atoms with known formation enthalpy is additive [4]. Methods based on DFT calculations are also used [5,6]. Calculations based on the combined G3 method [7–9] (G3(MP2, CC)//B3LYP/6-311G(d,p)) presented in [10–21] suggest the accuracy of comparative energies within 10 kJ/mol in the study of the decomposition of structures of the C_mH_n type.

In this work we calculated the enthalpy of formation of a number of gaseous molecules of the DEB group ($C_{18}H_8$, $C_{26}H_{10}$, $C_{19}H_{10}$, $C_{16}H_8$) by atomization of the studied molecules using a high-level quantum chemical theory, namely, G4 and G4(MP2) methods [1,22] within the Gaussian 09 software package. Over the past several decades, a number of combined methods have been developed and improved especially for thermochemical calculations. One of the best-known methods are the Gaussian-N (Gn) family [22]. They envisage approximation of the high-level calculation with a series of lower-level calculations to achieve higher accuracy in predicting the thermochemical characteristics of the system. In our work the G4 (Gaussian-4) method was used, introduced in 2007 by L. Curtiss and colleagues [1]. In this paper an array of 454 substances was used to demonstrate that compared with G3 [7,23], the G4 method significantly improves the standard deviation from 1.13 kcal/mol for G3 to 0.83 kcal/mol for G4. In many respects, this determined the choice of this method for our quantum chemical calculations.

The G4 method for approximating the energies of more accurate calculations combines the CCSD (T) method with a sufficiently high level of electron correlation and a medium-sized basis set (6-31G (d)) with the energies from the calculations of a lower level of the theory (MP4 and MP2) with large basis sets. In addition, several empirical corrections, independent of the investigated molecule, are included in the estimation of the remaining errors.

There are several modifications of the G4 method used in the present work the G4(MP2) method [22]. In this case, the MP4 calculation is replaced by the MP2 calculation. The G4(MP2) method requires less computer capacity; therefore, it was used in the present work when G4 calculation was impossible to do due to the limited computing capacity. This method is slightly less accurate compared to the G4 method. To test it, the method developers used a set of 270 experimental formation enthalpies. The total mean absolute deviation in this test set was 3.3 kJ/mol for G4 and 4.1 kJ/mol for the G4 (MP2) method. X. He and colleagues [24] in 2012 presented the calculation results for ΔH^{o}_{f298}(g) for 63 nitrogen-containing compounds (amines, amides, nitroesters, nitro compounds, etc.). The standard deviation of the G4 method was 3.1 kJ/mol. According to [22], total mean absolute deviation for the G4 method is 3.1 kJ/mol and 3.9 kJ/mol for the G4(MP2) method. Currently, the G4 method can be used to acquire the most reliable values of the enthalpy of formation by atomization reactions for large (more than 10 non-hydrogen atoms) organic molecules.

After choosing the methods and calculation bases, we carried out a quantum chemical simulation of the investigated high-energy $C_m H_n$ compounds. The created models showed good alignment with the experimental data and made it possible to calculate a number of energy parameters for substances that had not been simulated before.

3   Calculation Result

3.1   Enthalpy of Formation

Table 2. Enthalpy of formation (in kJ/kg and kJ/mol))

Structural formula	Calculation method	Enthalpy, kJ/kg	Enthalpy, kJ/mol
$C_{19}H_{10}$	B3LYP/6-311+G(2d,p)	4511.1	1074.9
	G4(MP2)	3750.0	893.6
	G4	3792.3	903.7
$C_{10}H_6$ (DEB)	B3LYP/6-311+G(2d,p)	4962.3	626.0
experimental data:		(+681.9)	(+86.0)
4280.4 kJ/kg	G4(MP2)	4360.2 (+79.8)	550.1 (+10.1)
540 kJ/mol [26]	G4	4403.1 (+122.7)	555.5 (+15.5)
$C_{18}H_8$	B3LYP/6-311+G(2d,p)	5783.3	1297.0
	G4(MP2)	5023.7	1127.0
	G4	5067.7	1136.5
$C_{26}H_{10}$	B3LYP/6-311+G(2d,p)	6194.0	1996.8
	G4(MP2)	5338.5	1721.0
	G4	5380.7	1734.6
$C_{16}H_8$	B3LYP/6-311+G(2d,p)	6922.0	1386.1
	G4(MP2)	6167.4	1235.0
	G4	6207.0	1242.9

– We used atom enthalpies of formation from [27]: $\Delta H^o_{f298} = 716.68 \pm 0.45$ (C) and $\Delta H^o_{f298} = 217.998 \pm 0.006$ (H)
– For DEB, the deviation of the calculated value from the experimental value is given in parentheses.
– Molar mass: $M(C_{19}H_{10}) = 0.238288$ kg/mol, $M(C_{10}H_66) = 0.126158$ kg/mol, $M(C_{18}H_8) = 0.224262$ kg/mol, $M(C_{26}H_{10}) = 0.322365$ kg/mol, $M(C_{16}H_8) = 0.200240$ kg/mol.

Table 2 shows the calculated enthalpies of formation of the studied compounds (kJ/kg and kJ/mol). For DEB, the deviation of the calculated value of the enthalpy of formation from the experimental data was accepted to be +79.88 kJ/kg at the G4(MP2) level and +122.71 kJ/kg at the G4 level, which is 2% and 3% respectively. Figure 1 presents calculated enthalpies of formation at different levels (B3LYP/6-311+G(2d,p), G4(MP2), G4) and the experimental data for DEB.

The graph in Fig. 1 and Table 2 show that the results obtained at the G4 (MP2) and G4 calculation levels are very close. Although the B3LYP/6311+G(2d,p) calculation level, which was taken for comparison, gives significantly overestimated values (by 16%), being a tendency toward an increase in the enthalpy of formation in the series of the considered molecules.

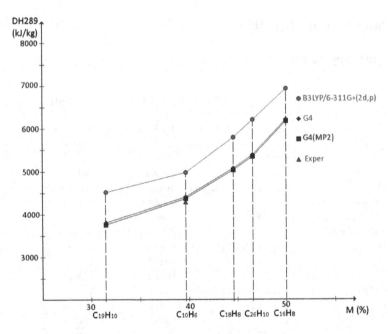

Fig. 1. Enthalpies of formation (in kJ/kg) of molecules in the gaseous state ($C_{19}H_{10}$, $C_{10}H_6$, $C_{18}H_8$, $C_{26}H_{10}$, $C_{16}H_8$) in dependence from the mass fraction of – CCH ethinyl groups (M, %): B3LYP/6-311+G(2d,p) - rounds, G4(MP2) - squares, G4 - diamonds; experimental data for DEB - triangle.

3.2   Structure and IR Spectra of the Calculated Compounds

Figure 2 presents the IR absorption spectra and the molar absorption coefficients of the calculated compounds (at the B3LYP/6-311+G(2d,p) level). Figures 3, 4, 5, 6 and 7 show the displacements of atoms for the most intensive vibrations. In all spectra, the most intense vibrations with a frequency of 3459–3461 cm^{-1} correspond to displacements of the end hydrogen atoms along the −CCH group (Figs. 3, 4, 5, 6 and 7). The vibrations of these atoms perpendicularly and in the plane of the ring bound to the −CCH group are less intensive by 2.4–1.4 times for $C_{10}H_6$, $C_{18}H_8$, $C_{26}H_{10}$, $C_{16}H_8$ (Figs. 4, 5, 6 and 7) and 3.5 times for $C_{19}H_{10}$

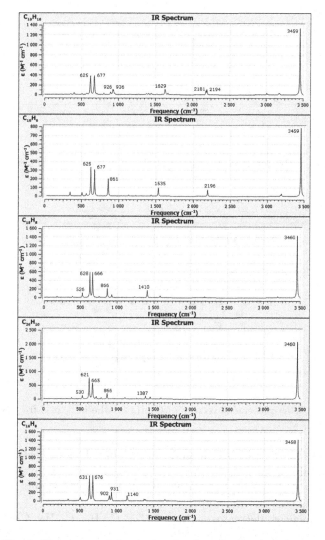

Fig. 2. IR spectra of absorption for the calculated compounds

(Fig. 3). The former vibrations correspond to frequencies of 622–631 cm^{-1} the latter ones to frequencies of 665–677 cm^{-1}.

Vibrations of hydrogen atoms associated with the rings are not very intense and correspond to frequencies of 861–866 cm^{-1} (Figs. 4, 5 and 6), 902 cm^{-1} (Fig. 7) and 927 cm^{-1} (Fig. 3) for vibrations perpendicular to the ring plane. Vibrations in the area of 2181–2196 cm^{-1} (Fig. 3, 4) correspond to asynchronous displacements of the end hydrogen atom and its adjacent carbon atom along the −CCH group. Asynchronous displacements of hydrogen and carbon atoms around the central atom of the CCH group are characterized with frequencies of 526–531 cm^{-1} with low intensities.

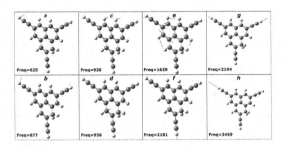

Fig. 3. a–h. Displacement vectors of $C_{19}H_{10}$ for the indicated frequencies

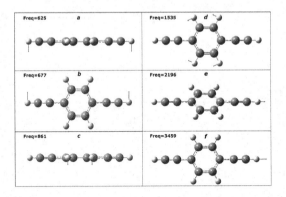

Fig. 4. a–f. Displacement vectors of $C_{10}H_6$ for the indicated frequencies

Vibrations of central hydrogen atoms in the ring plane (asynchronously with the carbon atoms of the associated ring), corresponding to frequencies of 1140–1630 cm^{-1} (Fig. 3, 4, 5, 6 and 7), automatically cause extension of the distances from the ring to the −CCH group.

These frequencies decrease in the $C_{19}H_{10}$, $C_{10}H_6$, $C_{18}H_8$, $C_{26}H_{10}$, $C_{16}H_8$ series symbatically with an increase in the mass fraction of −CCH fragments. See more detailed information on the most intense vibrations in Table 3.

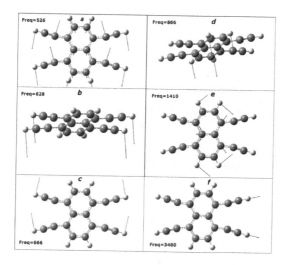

Fig. 5. a–f. Displacement vectors of $C_{18}H_8$ for the indicated frequencies

Table 3. Vibration frequencies and intensities

Freq. cm^{-1}	Intensity (km/mol)	Freq. cm^{-1}	Intensity (km/mol)	Freq. cm^{-1}	Intensity (km/mol)	Freq. cm^{-1}	Intensity (km/mol)	Freq. cm^{-1}	Intensity (km/mol)
$C_{19}H_{10}$		$C_{10}H_6$		$C_{18}H_8$		$C_{26}H_{10}$		$C_{16}H_8$	
401	11.3	345	13.5	526	31.3	381	17.3	496	10.9
619	44.9	509	12.6	628	168.9	530	41.4	508	12.2
625	58.5	625	96.2	666	167.3	621	176.2	630	37.1
625	38.0	677	88.0	866	61.3	622	12.2	631	92.7
673	51.3	861	58.1	927	19.1	624	45.3	638	10.1
677	65.5	1535	27.8	1410	49.6	652	11.2	676	66.6
677	16.2	2196	21.8	1585	13.2	665	109.9	677	36.9
893	13.9	3459	228.3	3460	30.1	670	103.3	902	17.4
926	28.3			3460	385.3	718	27.0	930	15.4
936	26.2					785	15.4	931	22.9
1629	28.6					866	57.2	1140	17.4
1636	19.1					1109	18.4	3458	101.8
1669	10.3					1387	32.2	3458	154.5
2181	23.5					1452	18.7		
2194	23.9					1596	12.7		
3007	11.0					3460	30.2		
3459	164.2					3460	291.9		
3459	201.9					3460	38.5		
3460	28.4					3461	241.4		
						3462	17.7		

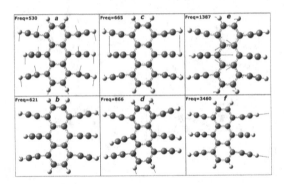

Fig. 6. a–f. Displacement vectors of $C_{26}H_{10}$ for the indicated frequencies

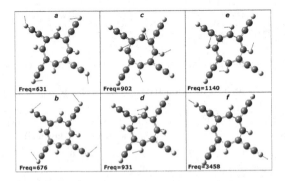

Fig. 7. a–f. Displacement vectors of $C_{16}H_8$ for the indicated frequencies

4   Details of Computational Calculations

During the quantum chemical simulation, a number of computational configurations (Table 4) based on various Intel Xeon processors, provided pools of computational cores, RAM and disk memory, GPU availability, and versions of the Gaussian application package (https://gaussian.com) were used.

The computational complexity of the tasks is rather high; depending on the basis used and the calculated temperatures, the average calculation time for the indicated structures varies from 20 h to 30 days. The computation time can be reduced by using more modern versions of processors. However, computation time for more complex systems (up to 50 atoms) on 24 physical core workstation in a pseudo-single-task mode reached 1–1.5 month (for example, $C_{26}H_{10}$ solution).

The processors' support of the **avx2** and **sse42** instructions is critical to the speed of calculations, especially the former one that can benefit by 8–10 times on some tasks using processors with a close clock rate [28]. Unfortunately, at the time of this writing we were not able to accurately assess the possibilities of using GPU accelerators. It is highly desirable to perform calculations on SSD disks or high-speed SAS disks with a large amount of allocated disk memory, since

Table 4. Computational combinations used

Computational resource	Processors/cores	Usage
Lomonosov-2 'compute'	Intel Xeon® E5-2697 v3@2.60 GHz, 14 cores, RAM 64 Gb; Tesla K40s	Up to 104 cores, 64 Gb per node
Computational cluster of the IPCP RAS	Intel Xeon®5450-5670@ 3 GHz, 4–6 cores, 8 and 12 Gb RAM	1 to 50 CPU (up 200 cores), 1–12 Gb per node
Workstations IPCP RAS (two WS)	Intel Xeon X5675@3.46 GHz, 26 cores, 48 Gb RAM, Nvidia Tesla C2075	Up to 12 cores, +1 GPU
Workstation Godwin, IEM RAS	Intel Xeon E5-2690v3, 2 × 12 cores, RAM 256 Gb, SSD+4 Tb HDD	1–24 cores, 4–248 Gb RAM per task

the package creates giant intermediate files up to 2 TB during the calculations. It can take up to 35–50 min to record them on an SSD disk and, of course, significantly longer on SATA arrays. The different publications report that the speed of calculations is greatly influenced by the availability of the latest versions of the Gaussian package (as compared to g9 installed on the IPCP cluster), which makes maximum use of the hardware capabilities provided by the new series of processors, accelerating calculations for most of the used bases by up to 7–8 times. The authors did not intend to elaborate upon the parallelization degree of the performed calculations (despite the fact that the Gaussian package usually uses its own Linda parallelization software). However, steady acceleration on pools up to 12 cores was observed, but further this effect reduced. It also depends on the amount of the allocated memory task (but it should be not less than 4 GB per physical core).

5 Conclusions

We used quantum chemical calculations at the B3LYP/6-311+G(2d,p), G4(MP2) and G4 levels within the GAUSSIAN 09 software package on high-performance computing resources to evaluate the thermochemical properties of a number of gaseous substances with the structural formula C_mH_n. Comparability of the calculated data with the available experimental data for DEB is good (within 3%), which allows using the technique for further quantum chemical calculations. For some substances, thermochemical data were collected in this work for the first time. These data identify the most promising groups of substances for use as high-energy components of promising fuels.

It was shown that the enthalpy of formation of high enthalpy polyethylene derivatives in the $C_{10}H_6$ - $C_{18}H_8$ - $C_{26}H_{10}$ series rises with as the mass fraction of $-CCH$ fragments increases, while the number of cross-links of benzene rings in the $C_{18}H_8$ - $C_{16}H_8$ and $C_{10}H_6$ - $C_{19}H_{10}$ series tends to reduce.

Using the G4(MP2) calculation level for the selected class of molecules was demonstrated to produce very close results (within 1%) to the G4 level, reducing calculation time by 3–8 times.

The work demonstrated the possibility of computer design of hypothetically promising molecules that have not been synthesized yet, opening new opportunities for effective selection of promising molecules.

Acknowledgments. The work was performed using the equipment of the Center for Collective Use of Super High-Performance Computing Resources of the Lomonosov Moscow State University [25, 29, 30] (project Enthalpy-2065). We also thank the Institute of Experimental Mineralogy of the Russian Academy of Sciences for providing pools of computational resources for some of the calculations. This work was carried out in accordance with the state assignment, state registration No. 00892019-0017 [theme AAAA-A19-119120690042-9], and also with the support of the RFBR grant No. 20-07-00319.

References

1. Curtiss, L.A., Redfern, P.C., Raghavachari, K.: Gaussian-4 theory. J. Chem. Phys. **126**, 084108 (2007). https://doi.org/10.1063/1.2436888
2. Karton, A., Sylvetsky, N., Martin, J.M.L.: W4-17: A diverse and high-confidence dataset of atomization energies for benchmarking high-level electronic structure methods. J. Comput. Chem. **38**, 2063–2075 (2017). https://doi.org/10.1002/jcc.24854
3. Lempert D., Kazakov A., Yanoskiy L.: Potential capability of some ethynyl derivatives as fuel dispersants for solid fuel ducted rockets. In: Proceedings of 22nd International Seminar "New Trends in Research of Energetic Materials, Pardubice, Czech Republic, pp. 527–532 (2019)
4. Muthurajian, H., Sivabalan, R., Talavar, M.B., Anniyappan, M., Venugopalan, S.: Prediction of heat of formation and related parameters of high energy materials. J. Hazardous Mater. **133**, 30 (2006). https://doi.org/10.1016/j.jhazmat.2005.10.009
5. Ryce, B.M., Hai, S.V., Hare, J.: Predicting heats of formation of energetic materials using quantum mechanical calculations. Combust. Flame **118**, 445 (1999). https://doi.org/10.1016/S0010-2180(99)00008-5
6. Byrd, E.F.C., Ryce, B.M.: Improved prediction of heats of formation of energetic materials using quantum mechanical calculations. J. Phys. Chem. A **110**, 1005 (2006). https://doi.org/10.1021/jp0536192
7. Curtiss, L.A., Raghavachari, K., Redfern, P.C., Rassolov, V., Pople, J.A.: Gaussian-3 (G3) theory for molecules containing first and second-row atoms. J. Chem. Phys. **109**, 7764–7776 (1998). https://doi.org/10.1063/1.477422
8. Curtiss, L.A., Raghavachari, K., Redfern, P.C., Baboul, A.G., Pople, J.A.: Gaussian-3 theory using coupled cluster energies. Chem. Phys. Lett. **314**, 101–107 (1999). https://doi.org/10.1016/S0009-2614(99)01126-4
9. Baboul, A.G., Curtiss, L.A., Redfern, P.C., Raghavachari, K.: Gaussian-3 theory using density functional geometries and zero-point energies. J. Chem. Phys. **110**, 7650–7657 (1999). https://doi.org/10.1063/1.478676

10. Belisario-Lara, D., Mebel, A.M., Kaiser, R.I.: Computational study on the uni-molecular decomposition of JP-8 Jet Fuel Surrogates III: Butylbenzene Isomers (n-, s-, and t-$C_{14}H_{10}$). J. Phys. Chem. A **122**(16), 3980–4001 (2018). https://doi.org/10.1021/acs.jpca.8b01836

11. Morozov, A.N., Mebel, A.M., Kaiser, R.I.: A theoretical study of Pyrolysis of exo-Tetrahydrodicyclopentadiene and its primary and secondary unimolecular decomposition products. J. Phys. Chem. A **122**(22), 4920–4934 (2018). https://doi.org/10.1021/acs.jpca.8b02934

12. Zhao, L., et al.: A vacuum ultraviolet photoionization study on high-temperature decomposition of JP-10 (exo-tetrahydrodicyclopentadiene). Phys. Chem. Chem. Phys. **19**, 15780–15807 (2017). https://doi.org/10.1039/C7CP01571B

13. Zhang, F.T., Kaiser, R.I., Kislov, V.V., Mebel, A.M., Golan, A., Ahmed, M.: A VUV Photoionization study of the formation of the Indene Molecule and its Isomers. J. Phys. Chem. Lett. **2**(14), 1731–1735 (2011). https://doi.org/10.1021/jz200715u

14. Zhang, F.T., Kaiser, R.I., Golan, A., Ahmed, M., Hansen, N.: A VUV Photoionization study of the combustion-relevant reaction of the Phenyl Radical (C_6H_5) with Propylene (C_3H_6) in a high temperature chemical reactor. J. Phys. Chem. A **116**, 3541–3546 (2012). https://doi.org/10.1021/jp300875s

15. Kaiser, R.I., Belau, L., Leone, S.R., Ahmed, A., Wang, Y., Braams, B.J., Bowman, J.M.: A combined experimental and computational study on the ionization energies of the cyclic and linear C_3H Isomers. ChemPhysChem **8**, 1236–1239 (2007). https://doi.org/10.1002/cphc.200700109

16. Kaiser, R.I., Mebel, A., Kostko, O., Ahmed, M.: On the ionization energies of C_4H_3 isomers. Chem. Phys. Lett. **485**, 281–285 (2010). https://doi.org/10.1016/j.cplett.2009.12.027

17. Kaiser, R.I., et al.: Untangling the chemical evolution of Titan's atmosphere and surface-from homogeneous to heterogeneous chemistry. Faraday Discuss. **147**, 429–478 (2010). https://doi.org/10.1039/c003599h

18. Kaiser, R.I., et al.: An experimental and theoretical study on the ionization energies of polyynes (H-$(C{\equiv}C)_n$-H; $n = 1$–9). Astrophys. J. **719**, 1884–1889 (2010). https://doi.org/10.1088/0004-637X/719/2/1884

19. Kostko, O., Zhou, J., Sun, B.J., Lie, J.S., Chang, A.H.H., Kaiser, R.I., Ahmed, M.: Determination of ionization energies of C_nN ($n = 4$–12): vacuum ultraviolet photoionization experiments and theoretical calculations. Astrophys. J. **717**, 674–682 (2010). https://doi.org/10.1088/0004-637X/717/2/674

20. Kaiser, R.I., Krishtal, S.P., Mebel, A.M., Kostko, O., Ahmed, M.: An experimental and theoretical study of the ionization energies of SiC_2H_x ($x = 0, 1, 2$) isomers. Astrophys. J. **761**, 178–184 (2012). https://doi.org/10.1088/0004-637X/761/2/178

21. Golan, A., Ahmed, M., Mebel, A.M., Kaiser, R.I.: A VUV photoionization study of the multichannel reaction of phenyl radicals with 1,3-butadiene under combustion relevant conditions. Phys. Chem. Chem. Phys. **15**, 341–347 (2013). https://doi.org/10.1039/C2CP42848B

22. Curtiss, L.A., Redfern, P.C., Raghavachari, K.: Gn theory. Comput. Mol. Sci. **1**, 810–825 (2011). https://doi.org/10.1002/wcms.59

23. Curtiss, L.A., Redfern, P.C., Raghavachari, K.: Assessment of Gaussian-3 and density-functional theories on the G3/05 test set of experimental energies. J. Chem. Phys. **123**, 124107 (2005). https://doi.org/10.1063/1.2039080

24. He, X., Zhang, J., Gao, H.: Theoretical thermochemistry: enthalpies of formation of a set of nitrogen-containing compounds. Int. J. Quant. Chem. **112**, 1688–1700 (2012). https://doi.org/10.1002/qua.23163

25. Voevodin, V.I., et al.: Supercomputer Lomonosov-2: large scale, deep monitoring and fine analytics for the user community. Supercomput. Front. Innov. **6**(2), 4–11 (2019). https://doi.org/10.14529/jsfi190201
26. Lias, S.G., Bartmess, J.E., Liebman, J.F., Holmes, J.L., Levin, R.D., Mallard, W.G.: Gas-phase ion and neutral thermochemistry. J. Phys. Chem. Ref. Data **1**, 421 (1988)
27. NIST-JANAF Thermochemical tables. https://janaf.nist.gov/
28. Grigorenko, B., Mironov, V., Polyakov, I., Nemukhin, A.: Benchmarking quantum chemistry methods in calculations of electronic excitations. Supercomput. Front. Innov. **5**(4), 62–66 (2019). https://doi.org/10.14529/jsfi180405
29. Voevodin, V.V., et al.: Practice of "Lomonosov" supercomputer. Otkrytye sistemy [Open Syst.] **7**, 36–39 (2012). (in Russian)
30. Nikitenko, D., Voevodin, V., Zhumatiy, S.: Deep analysis of job state statistics on "Lomonosov-2" supercomputer. Supercomput. Front. Innov **5**(2), 4–10 (2019). https://doi.org/10.14529/jsfi18020110.14529/jsfi18020110.14529/jsfi180201

Optimizing Numerical Design of a Multi-mode Aero-Cooled Climatic Wind Tunnel Nozzle on a PNRPU High-performance Computational Complex

Stanislav L. Kalyulin$^{(\boxtimes)}$ ⓘ, Vladimir Ya. Modorskii ⓘ, Anton O. Mikryukov,
Ivan E. Cherepanov ⓘ, Danila S. Maksimov, and Vasily Yu. Petrov

Perm National Research Polytechnic University, Perm, Russia
{ksl,modorsky}@pstu.ru

Abstract. Creation of multi-mode aero-cooled climatic wind tunnels (ACCWT) for testing the icing of aircraft structures is an urgent task of modern aircraft construction. The ACCWT nozzle is one of its most complex elements in terms of design. The article presents the results of parallel multi-criteria optimization of the nozzle profile in two ACCWT operating modes of the gas dynamics of the watered flow and icing on the walls, which is performed by parallel calculations. The results were obtained on the PNRPU supercomputer (peak performance of 24 TFLOPS). The proposed algorithm scalabilities are estimated, the acceleration is compared on various grid models.

Keywords: Parallel algorithm · Numerical simulation · Gas hydrodynamics · Icing · Multicriteria optimization · Aero-cooling climatic wind tunnel

1 Introduction

According to the publication in the ICAO Journal [1], as per ECCAIRS, since 1970, at least 323 events have occurred that are related to the in-flight icing of aircraft engines and other components.

Since icing conditions are quite common, to ensure a safe flight, the aircraft components are protected by anti-icing systems. The majority of modern aircraft anti-icing systems are based on the utilizing the air or power from an aircraft engine. Thus, increasing the protection system efficiency and, as a consequence, flight safety under icing conditions, inevitably leads to a decrease in the operational, economic and environmental characteristics of the aircraft engine, which is a negative factor.

At the same time, a situation has developed in the aviation industry when, on the one hand, the requirements for flight safety in icing conditions are being

© Springer Nature Switzerland AG 2020
L. Sokolinsky and M. Zymbler (Eds.): PCT 2020, CCIS 1263, pp. 305–320, 2020.
https://doi.org/10.1007/978-3-030-55326-5_22

tightened expanding the list of negative impact conditions, and, on the other hand, the requirements for the economy and environmental friendliness of aircraft engines are increasing. These requirements are, in a sense, mutually exclusive, and to meet the current level of requirements for both flight safety in icing conditions and the minimum impact on the aircraft engine operation, the creation of innovative energy-efficient icing protection systems is required.

The development and creation of a set of measures to simulate the ice growth and to prevent the aircraft structure icing is a resource-intensive interdisciplinary scientific and technical task. So, to obtain a highly accurate final solution, it is necessary to obtain high-quality solutions to aerodynamic and hydraulic problems of subsonic and supersonic air flows, conjugate heat transfer, phase transition, and the problems of a deformable structure interaction with a multiphase airflow.

The leading world aviation science centers (ONERA, CIRA, Cranfield University, NASA Glenn Research Center, NRC, etc.) actively conduct experimental studies of such technologies [2–7], however, there are not enough works related to developing the methods for mathematical modeling of icing and mechanical ice removal. Experimental studies of this problem in the existing large-sized climatic wind tunnels are complicated by a high test cost. At the same time, the methods for creating small-sized and energy-efficient climatic wind tunnels do not exist, although such tunnels are necessary for a large amount of experimental work aimed to verify numerical icing models.

Thus, the development of mathematical models and scalable algorithms allowing for a parallel icing modeling for the optimization of climate wind tunnel designs would significantly reduce an anti-icing system cost and development time.

The aircraft icing has long been the subject of research. The works of Prikhodko Yu. M., Klimenkov G. P., Puzyrev L. N., Kharitonov A. M., Babulin A. A., Bolshunov K. Yu., Alekseenko S. V., Gaifullin A. M., Zubtsov A. V., Amelyushkin I. A., Wright W. B, Gent R. W., Guffond D., Hedde T., Henry R., Tran P., Brahimi M. T., Paraschivoiu I. P., Tezok F., Mingione G., Brandi V., Dillingh J. E., Hoeijmakers H. W. M., Beaugendre H., Morency F., Habashi W. G., Pueyo A., Chocron D., Kafyeke F., Hannat R., Reggio M., Ilinca A., Broeren A. P., Bragg M. B., Addy H. E., Lee S., Moens F., Wang Y. B., Xu Y. M., Huang Q., Lei Y., Wang Z., as well as many other researchers, are known [2–24].

Since 1980, mathematical models that describe the ice growth under various initial gas-dynamic conditions have been constantly improved in the world [15]. There are several foreign two-dimensional software implementations of these mathematical icing models: LEWICE 2D (USA) [2], ONERA (France) [3], TRAJICE2D (United Kingdom) [4], CANICE (Canada) [5], CIRA (Italy) [6], and 2DFOIL-ICE (Netherlands) [8].

Researchers from France (ONERA) [7,23], modified their algorithm several times in 2D and 3D, but did not receive the solutions that were qualitatively consistent with the ice shape in the physical experiment, although the algorithm simulated the droplet capture correctly. In the LEWICE software [2], in comparison with other algorithms, the surface tension is additionally taken into account, which makes it possible to estimate the amount of water through the Weber number.

Some of the software is refined to three-dimensional models, but they are all based on the shield method for calculating the potential flow and cannot determine the flow separation points, therefore, they give less accurate results compared to the FENSAP-ICE 3D algorithm, which is based on the Navier-Stokes equations [9]. However, they allow to obtain a solution much faster. The bisector method proposed in [10] is used to model the ice boundary motion.

Icing modeling in FENSAP-ICE 3D did not gain much popularity due to the need to conclude additional licensing agreements and insufficient verification studies, although an example of solving the icing problem in FENSAP-ICE 3D together with ANSYS CFX was presented by researchers from Louisiana in [16]. But, at the same time, the FENSAP-ICE 3D algorithm is mentioned in the ICAO Journal [1], as it is specially promising among the numerical modeling of solid body icing.

In this regard, in the research practice, ANSYS CFX or ANSYS FLUENT gas-dynamic systems of engineering analysis are usually used, where data on temperature, humidity, speed and other gas-dynamic parameters can be used to predict the icing since the direct simulation of water-ice phase transition is not implemented in this software.

In addition, there are foreign studies on solving the gas-dynamic problem in ANSYS CFX or ANSYS FLUENT with the subsequent use of accurate gas-dynamic parameters to solve the icing problem by semi-empirical or partially analytical methods [24].

The authors of this article have published a number of articles devoted to parallel numerical modeling of icing and gas-dynamic processes [25–34]. There is also a work on optimizing gas-dynamic parameters of a chilled drop and airflow, co-authored with some scientists from Lobachevsky State University of Nizhny Novgorod. This work presents the discovered combination of the droplet velocities and temperatures, as well as the airflow velocities, at which the gas-dynamic flow temperature in the working part tends to its maximum [35, 36].

This article presents the results of optimizing the nozzle profile with an explicit calculation of the gas-dynamics of the flow and icing of the nozzle walls in a three-dimensional non-stationary formulation at each optimization iteration, using parallel computation within the PNRPU high-performance computing complex (PNRPU HPC).

2 Conceptual and Mathematical Statements for Describing Gas-Dynamics of a Flow with Drops in the Nozzle and the Icing on the Nozzle Walls

The concept for describing gas-dynamics of a flow with drops in the nozzle and the icing on the nozzle walls is formulated as the following set of hypotheses:

- the behavior of a multiphase flow is considered viscous compressible gas – incompressible fluid;
- the motion dynamics and the deposition of liquid droplets on the nozzle walls are taken into account;
- the processes of ice formation on the nozzle walls are considered to be a set of sequential gas-dynamic, liquid, and ice-forming processes;
- the nozzle walls are assumed to be impermeable, adiabatic and rough (it is assumed that all roughness elements protrusions lie inside the viscous layer) with particles sticking;
- the gas-dynamic processes are considered in a three-dimensional quasistationary formulation;
- the liquid processes are considered in a three-dimensional quasistationary formulation;
- the processes of ice formation are considered in a three-dimensional nonstationary formulation;
- deposited liquid droplets on the nozzle walls are considered a thin liquid film;
- the liquid film can move along the flow due to shear stresses created by the gas-dynamic flow movement;
- processes of accretion and sublimation of ice, evaporation of a liquid film are taken into account.

The mathematical formulation fully describing the icing physical processes is implemented in 3 stages:

- gas-dynamic stage;
- liquid stage;
- ice formation stage.

The totality of these stages determines the mathematical model of the icing processes with the corresponding closure by the boundary and initial conditions.

Under the accepted hypotheses, a mathematical model is used to describe the gas-dynamic processes. The model is based on a system of Reynolds-averaged Navier-Stokes equations, closed by state equations of an ideal compressible gas and turbulence, initial and boundary conditions. The Navier-Stokes system of equations is a combination of the following equations:

- The equation of continuity, or mass balance in the Euler formulation

$$\frac{\partial \rho_{air}}{\partial t} + \nabla \cdot (\rho_{air} \boldsymbol{V}_{air}) = 0, \tag{1}$$

where ρ_{air} is the air density, $\boldsymbol{V}_{air}$ is the velocity vector of the gas-dynamic flow and t is the time.
- Eulerian impulse balance equations

$$\frac{\partial (\rho_{air} \boldsymbol{V}_{air})}{\partial t} + \nabla \cdot (\rho_{air} \boldsymbol{V}_{air} \otimes \boldsymbol{V}_{air}) = -\nabla P + \nabla \cdot \boldsymbol{\tau}, \tag{2}$$

where P is the static pressure and τ is the viscous stress tensor defined by the relation

$$\tau = \mu_{air}\left(\nabla V_{air} + V_{air}\nabla - \frac{2}{3}\delta_K\nabla\cdot V_{air}\right), \tag{3}$$

where μ_{air} is the dynamic viscosity of air, and δ_K is the Kronecker delta.

- Law of energy conservation

$$\frac{\partial(\rho_{air}H^*)}{\partial t} - \frac{\partial P}{\partial t} + \nabla\cdot(\rho_{air}V_{air}H^*) = \nabla\cdot(\lambda\nabla T_{air} + V_{air}\cdot\tau), \tag{4}$$

where H^* is the total air braking enthalpy, T_{air} is the static air temperature and λ is the specific thermal conductivity.

The system of equations is closed by determining state relations for the enthalpy and air pressure

$$P = \rho_{air}R_{air}T_{air}, \tag{5}$$

$$dH = c_p dT_{air}, \tag{6}$$

where c_p is the isobaric heat capacity of air and R_{air} is the air gas constant.

Turbulence models based on the equation for the dissipation rate predict the flow separation onset much lower downstream than is observed in the experiments, and the estimated recirculation zone size and its movement intensity are underestimated. At the same time, another class of models, which are based on the equation for $\omega = \varepsilon/k$ (*turbulence frequency* – the reciprocal of the lifetime of large vortices) instead of the transport equation, leads to better agreement with the experiment for the small zone of separation. The $k - \omega$ model was developed in [37] and formed a basis of the combined shear stress transport model (SST), created later [38] and combining a linear combination of the $k - \varepsilon$ model away from surfaces and the $k - \omega$ model. To calculate the in-nozzle gas dynamics, the SST model of turbulence was used.

The mathematical formulation for the liquid phase, which describes droplet capture by the nozzle walls and determine droplet concentration, is represented by a set of equations:

- The equation of continuity, or mass balance in the Euler formulation

$$\frac{\partial\alpha}{\partial t} + \nabla\cdot(\alpha V_{drop}) = 0, \tag{7}$$

where α is the average volume fraction of the liquid phase and V_{drop} is the drop velocity.

- Eulerian impulse balance equations

$$\begin{aligned}\frac{\partial(\alpha V_{drop})}{\partial t} + \nabla\cdot[\alpha V_{drop}\otimes V_{drop}] \\ = \alpha\frac{C_D Re_{drop}}{24K}(V_{air} - V_{drop}) + \frac{\alpha}{Fr^2}\left(1 - \frac{\rho_{air}}{\rho_{drop}}\right),\end{aligned} \tag{8}$$

where C_D is the drag coefficient, Re_{drop} is the Reynolds number of drops, K is the inertial drop parameter, Fr is the local Froude number and ρ_{drop} is the drop density.

To be able to proceed to the ice formation stage, it is necessary to establish a dependence for the liquid film motion.

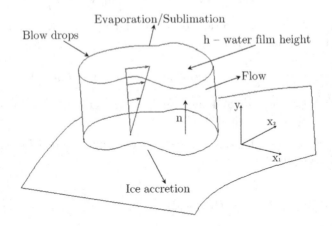

Fig. 1. A scheme of physical processes during ice formation on nozzle walls

The speed of a liquid film V_f near the surface of a streamlined body is a function of spatial coordinates in the plane of the surface under consideration $x = x(x_1, x_2)$ and the normal to the streamlined surface y (Fig. 1).

Let us introduce a linear profile for the liquid film velocity, which is an oriented normal to the nozzle wall, near which the zero velocity is realized. Then the velocity of the liquid film can be determined as follows:

$$V_f(x, y) = \frac{y}{\mu_f} \tau_{air}(x), \tag{9}$$

where μ_f is the liquid film viscosity and $\tau_{air}(x)$ is the shear stress vector by gas-dynamic flow, which is the main driving force for the liquid film.

Averaging the speed of the liquid film over its thickness, we obtain the expression necessary to solve the next stage:

$$\overline{V}_f(x, y) = \frac{1}{h_f} \int_0^h V_f(x, y) dy = \frac{h_f}{2\mu_f} \tau_{air}(x) \Big|_{wall}, \tag{10}$$

where h_f is the liquid film thickness on a streamlined body surface.

The mathematical formulation for the ice formation stage, which describes the ice growth on structures, is represented by a set of balance equations that are resolved on the surfaces of the conforming structures:

– Eulerian mass balance equation

$$\rho_{drop} \left[\frac{\partial h_f}{\partial t} + \nabla \cdot (\overline{V}_f h_f) \right] = V_{air,\infty} \varphi \beta - \dot{m}_{vapor} - \dot{m}_{ice}, \tag{11}$$

where φ is the liquid water content (absolute humidity), β is the efficiency of local deposition of the liquid phase on the structure surface, $\dot{m}_{vapor}$ is the local rate of change in the mass of the liquid phase evaporation under convection flow conditions and $\dot{m}_{ice}$ is the local rate of change in the ice mass.
 - Law of energy conservation

$$
\begin{aligned}
\rho_{drop} &\left[\frac{\partial h_f c_f T_f}{\partial t} + \nabla \cdot (\overline{V}_f h_f c_f \widetilde{T}_f) \right] = \left[c_f (\widetilde{T}_{air,\infty} - \widetilde{T}_f) \right. \\
&+ \frac{\|V_{drop}\|^2}{2} \right] V_{air,\infty} \varphi \beta - L_{vapor} \dot{m}_{vapor} \\
&+ (L_{melt} - c_{ice}\widetilde{T}_{ice})\dot{m}_{ice} - \lambda_m (\widetilde{T}_f - \widetilde{T}_{ice}),
\end{aligned}
\tag{12}
$$

where c_f is the specific heat of the liquid phase, c_{ice} is the specific heat of ice, $\widetilde{T}_f$ is the liquid temperature, $\widetilde{T}_{ice}$ is the equilibrium temperature at the phase boundary (air/liquid/ice/nozzle), L_{vapor} is the specific heat of liquid phase vaporization, L_{melt} is the specific heat of melting ice and λ_m is the coefficient of molecular thermal conductivity.

To close the system of equations, it is necessary to define them with a system of compatibility relations:

$$
\begin{cases}
h_f \geq 0, \\
\dot{m}_{ice} \geq 0, \\
h_f \widetilde{T}_f \geq 0, \\
\dot{m}_{ice}\widetilde{T}_{ice} \leq 0.
\end{cases}
\tag{13}
$$

System of inequalities (13) allows us to exclude the possibility of liquid phase formation on the structure surface when the equilibrium temperature is below the dew point (less than $273.15\,\mathrm{K}$) and the situation of the ice growth if the temperature of the liquid film is higher than $273.15\,\mathrm{K}$ during further modeling.

3   The Optimization Algorithm Description. Problem Formulation

To carry out the optimization design of the ACCWT nozzle profile, the domestic IOSO PM software [39] was used, which selects a set of Pareto-optimal solutions.

The choice of the Pareto set is as follows. All alternatives are compared in pairs under all criteria. When comparing the alternatives, if it turns out that one of them is not better than the other by any criterion, then it is excluded from the consideration. The excluded alternative does not need to be compared with the other alternatives since it is unpromising. The central place of the method is the presentation of each iteration of the search for the Pareto set in two stages: the construction of functions that approximate the objective functions in a certain region and the search for the extrema of these approximation functions.

The optimization algorithm can be described as follows:

1. Select in the current experimental design all Pareto-optimal points distinguishable by the criterion (subset A, the number of such points is $n \geq 1$).

2. Randomly choose from a subset A one Pareto-optimal x_i^P point.
3. Construct a subset of the experimental design $X^M \in X^W$ (X^M consists of M points closest to the x_i^P point in a linear metric). Determine the current search region $B \in X$, which is the closure of the subset X^M.
4. In the current search area, construct B functions that approximate y_j particular optimization criteria, $j = \overline{1, m}$ and limited parameters.
5. The solution in the current B search region m optimization tasks for approximation functions (the result of this step are m points x_j and $j = \overline{1, m}$, in which extreme values of particular optimization criteria for the current search region are expected).
6. At the x_j and $j = \overline{1, m}$ points, find true values of the particular optimization criteria by direct appeals to the problem of determining the watered gas-dynamics in the nozzle and icing on the walls. Add x_j and $j = \overline{1, m}$ points, to the experimental design.
7. Check the stop criterion.
8. Generate n points $x_j \rightarrow N(x_i^P, \sigma)$ in the neighborhood of the current Pareto optimal point according to the normal law. Add points x_j and $j = \overline{1, n}$ to the experimental design.
9. Repeat step 1.

For optimization, a parametric design model of the nozzle was constructed (Fig. 2).

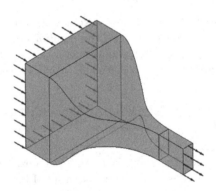

Fig. 2. Parametric design nozzle model

As optimized parameters, it is proposed to consider the nozzle length (L), 4 coordinates in the vertical plane on the curve (Y_2, Y_3, Y_4, Y_5) and 4 coordinates in the horizontal plane on the curve (Z_2, Z_3, Z_4, Z_5) (Fig. 3).

Three optimization criteria have been adopted, which are necessary to be minimized in the problem solving process:

1. Loss of total pressure in the nozzle

$$dP = P_{inlet}^* - P_{outlet}^*, \tag{14}$$

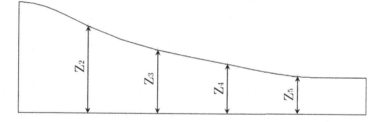

Fig. 3. Sketch nozzle profile in the horizontal plane

where P^*_{inlet} is the mass average total pressure at the nozzle inlet and P^*_{outlet} is the mass average total pressure at the nozzle exit.

2. The degree of uneven flow at the nozzle exit

$$Unev = \frac{V_{max} - \langle V \rangle}{\langle V \rangle} \cdot 100\%, \tag{15}$$

where V_{max} is the maximum flow rate at the nozzle exit and $\langle V \rangle$ is the mass average flow rate at the nozzle exit.

3. Mass of ice m_{ice}.

At each iteration of the optimization algorithm, 2 gas-dynamic calculations of the watered flow in the nozzle and 2 icing calculations on the nozzle walls were carried out for 2 operation modes of the air-cooled climatic wind tunnel.

The objective functions were defined as the arithmetic mean values of the relevant objective functions for 2 gas-dynamic modes for each optimization iteration. The interrupt criterion for the gas-dynamic calculation is a decrease in the standard deviation of the total pressure loss by 1 Pa over the last iterations depending on the time step.

Figure 4 shows a block diagram of the nozzle profile optimization with an explicit calculation of the gas-dynamics of the watered flow and icing on the walls.

4   Scalability. Grid Convergence Estimation

In solving the multicriteria optimization problem, the following steps are implemented in parallel:

1. Determining the gas-dynamics of the in-nozzle watered flow and icing on the nozzle walls, 3-stage implementation (gas-dynamic, liquid and ice formation) – parallelism at the solution stage (up to 248 Intel Xeon E5-2680 cores).
2. Parallel gas-dynamics and icing calculations simultaneously for 2 modes within the framework of one optimization iteration.
3. Parallel calculations of several structures (up to 8 threads). One calculation is understood as the gas-dynamics and icing solutions for 2 operating modes of an air-cooled climate wind tunnel.

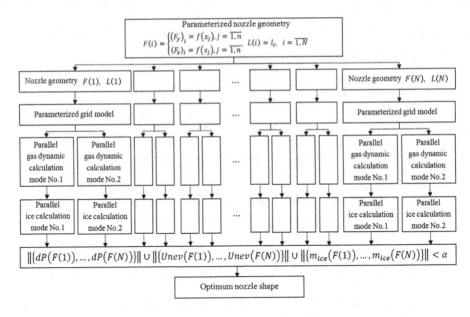

Fig. 4. The optimization algorithm

In total, more than 500 calculations were carried out with a converged computational grid of 2 to 4 million finite elements, depending on the nozzle geometry. All calculations were carried out on the PNRPU high-performance computing complex (PNRPU HPC) [40]. Below are the main technical characteristics of PNRPU HPC:

- 95 computational nodes;
- 128 four-core processors "Barcelona-3" (total number of cores is 512);
- 62 eight-core processors "Intel Xeon E5-2680" (total number of cores is 480);
- 24,096 TFlops peak performance;
- 78% performance in Linpack test packet;
- 27 TB information storage system capacity;
- −5888 GB RAM installed (32 GB/node with processors "Barcelona-3", 128 GB/node with processors "Intel Xeon E5-2680");
- 12 computing modules GPU NVIDIA Tesla M2090 (512 cores, 6 GB).

The grid model was parameterized using the TCL scripting language. The Ogrid topology was applied with the implementation of a multi-block method for constructing a structured computational grid consisting entirely of hexahedrons – a topologically similar model was constructed that was projected onto the parameterized nozzle geometry. The size of the first parietal element was 105 m, which provides the dimensionless parameter $y+$ up to 5 in accordance with the applied SST model of turbulence. The growth rate was 1.2.

The grid convergence was estimated for a tetragonal unstructured computational grid and a hexagonal structured grid with the implementation of a multi-block construction method. Figure 5 shows the dependence of the difference norm in the total pressure losses (deviation of the subsequent resulting point from the previous one) on the number of finite elements for two grid models.

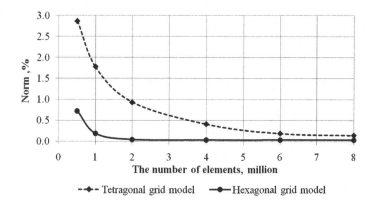

Fig. 5. Dependence of the total pressure loss difference norm on the number of the finite elements

It can be noted that the hexagonal grid model in comparison with the tetragonal model more accurately describes the flow behavior with fewer finite elements. Figure 6 shows a converging hexagonal structured grid nozzle model containing 2 million finite elements.

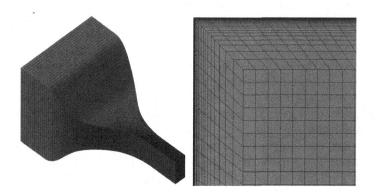

Fig. 6. The grid nozzle model

The convergence of the gas-dynamic problem on a hexagonal structured grid is already observed at 120–150 iterations, while on a tetragonal unstructured grid it is only 300–400. To increase the calculation speed and improve the solution accuracy, the scalability of the model was estimated with a hexagonal computational grid.

The acceleration of the solution to the problem was estimated depending on the number of cores, a graph of the required estimated time versus the number of used Intel Xeon E5-2680 processor cores at the PNRPU HPC was presented (Fig. 7).

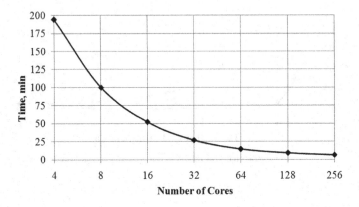

Fig. 7. Graph of the estimated time depending on the number of Intel Xeon E5-2680 processor cores

5   Optimization Results

Based on the optimization results, it is possible to form a front of Pareto-optimal solutions according to the two objective functions of the total pressure loss and the flow unevenness degree, while deriving the ice mass values for each of the Pareto-optimal solutions (Fig. 8).

By analyzing the data obtained, it is possible to unambiguously determine the optimal nozzle profile according to the criteria for total pressure loss, uneven flow and ice mass on the walls for two gas-dynamic operating modes of ACCWT. Figure 9 shows the velocity distribution fields for horizontal and vertical sections of the optimal nozzle profile.

For the obtained optimal design of the nozzle, the total pressure loss in the nozzle in mode 1 is 557 Pa, in mode 2–501 Pa. The uneven flow in the outlet cross-section is 1.52% for mode 1 and 1.60% for mode 2.

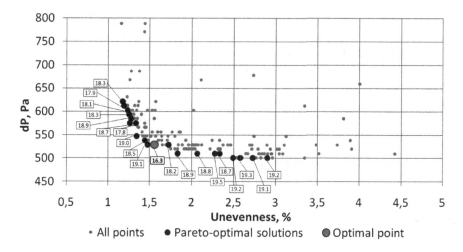

Fig. 8. FFront of Pareto-optimal solutions for two objective functions with the indication of ice mass in grams

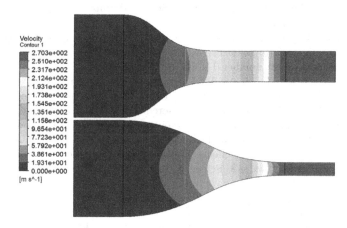

Fig. 9. Distribution of flow velocity in horizontal and vertical sections of the optimal nozzle profile

6 Conclusion

The authors implemented a single parallel algorithm for calculating the following tasks: gas-dynamics taking into account the liquid water content, icing and multicriteria optimization in several streams. Parallel software implementation at the stage of gas-hydrodynamic solving and icing, as well as at the stage of parallelizing the optimization algorithm will allow scaling the proposed approach to various computer architectures. Within the framework of the complex task considered, a thirty-fold increase in the calculation speed was achieved using 256 Intel Xeon E5-2680 cores. The presented method enables designing optimal nozzle profiles for any flow section of the wind tunnel working part in a wide range of operating modes. As a result of solving the multicriteria optimization problem, an optimal nozzle design was obtained for 2 operating modes of an aero-cooled climatic wind tunnel, which has minimal total pressure loss and minimal flow unevenness with minimal icing on the nozzle walls. A universal parallel scalable approach to design is proposed and implemented, in which the coordinate-wise optimization of an air-cooled climatic wind tunnel nozzle is carried out. The resulting optimal profiles are polynomials of n degrees.

Acknowledgements. The study was supported by the Russian Foundation for Basic Research (RFBR), project No. 17-47-590017.

References

1. Runway Safety. International Civil Aviation IKAO, vol. 66, no. 2 (2011). (in Russian)
2. Wright, W.B.: Users manual for the improved NASA lewis ice accretion code LEWICE 1.6. National Aeronautical and Space Administration (NASA), Contractor Report Number 185129, p. 95 (1995)
3. Gent, R.W.: TRAJICE2, a combined water droplet and ice accretion prediction program for aerofoil. Royal Aerospace Establishment (RAE), Farnborough, Hampshire, Technical report number TR90054, p. 83 (1990)
4. Guffond, D., Hedde, T., Henry, R.: Overview of icing research at ONERA, Advisory Group for Aerospace Research and Development. In: Fluid Dynamics Panel (AGARD/FDP) Joint International Conference on Aircraft Flight Safety - Actual Problems of Aircraft Development, Zhukovsky, Russia, p. 7 (1993)
5. Tran, P., Brahimi, M.T., Paraschivoiu, I.P., Tezok, F.: Ice accretion on aircraft wings with thermodynamic effects. In: American Institute of Aeronautics and Astronautics, 32nd Aerospace Sciences Meeting & Exhibit, Reno, Nevada, AIAA Paper, art. no. 0605 (1994)
6. Mingione, G., Brandi, V.: Ice accretion prediction on multielements airfoils. J. Aircr. **35**(2), 240–246 (1998)
7. Hedde, T., Guffond, D.: ONERA 3-dimensional icing model. AIAA J. **6**, 1038–1045 (1995)
8. Dillingh, J.E., Hoeijmakers, H.W.M.: Accumulation of ice accretion on air-foils during flight. In: Federal Aviation Administration In-flight Icing and Aircraft Ground De-icing Conference, Chicago, Illinois, p. 13 (2003)
9. Beaugendre, H., Morency, F., Habashi, W.G.: ICE 3D, FENSAP-ICE'S 3D in-flight ice accretion module. In: American Institute of Aeronautics and Astronautics, 40th Aero-space Sciences Meeting & Exhibit, Reno, Nevada, AIAA Paper, art. no. 0385, p. 20 (2002)
10. Pueyo, A., Chocron, D., Kafyeke, F.: Improvements to the ice accretion code CANICE. In: Proceedings of the 8th Canadian Aeronautics and Space Institute (CASI), Aerodynamic Symposium, Toronto, Canada, p. 9 (2001)
11. Afanasiev, V.A., Barsukov, V.S., Gofin, M.Y., Zakharov, Y.V., Strelchenko, A.N., Shalunov, N.P.: Experimental testing of spacecraft. MAI, Moscow (1994). (in Russian)

12. Goryachev, A.V., Mezhzil, E.K., Petrov, S.B., Syrov, V.A., Harlamov, A.V., Chivanov, S.V.: A way of ground testing objects of aircraft, subject to icing, and a device for its implementation. Patent RF, no. 2345345 (2007)
13. Klemenkov, G.P., Prihodko, Y.M., Puzyrev, L.N., Haritonov, A.M.: Modelling of aircraft icing processes in aeroclimatic tubes. Thermophys. Aeromech. **15**(4), 563–572 (2008). (in Russian)
14. Alekseenko, S.V., Prihodko, A.A.: Numerical simulation of icing cylinder and profile. Models review and calculation results. TsAGI Sci. J. **44**(6), 25–57 (2013). (in Russian)
15. Prihodko, A.A., Alekseenko, S.V.: Numerical simulation of icing aerodynamic surfaces in the presence of large overcooled water drops. JETP Lett. **40**(19), 75–82 (2014). (in Russian)
16. Hannat, R., Morency, F.: Numerical validation of conjugate heat transfer method for anti-/de-icing piccolo system. J. Aircr. **51**(1), 104–116 (2014). https://doi.org/10.2514/1.c032078
17. Villalpando, F., Reggio, M., Ilinca, A.: Prediction of ice accretion and anti-icing heating power on wind turbine blades using standard commercial software. Energy **114**, 1041–1052 (2016). https://doi.org/10.1016/j.energy.2016.08.047
18. Lynch, F.T., Khodadoust, A.: Effects of ice accretions on aircraft aerodynamics. Prog. Aeosp. Sci. **37**(8), 669–767 (2001). https://doi.org/10.1016/s0376-0421(01)00018-5
19. Bragg, M.B., Broeren, A.P., Blumenthal, L.A.: Iced-airfoil aerodynamics. Prog. Aeosp. Sci. **41**(5), 323–362 (2005). https://doi.org/10.4271/2003-01-2098
20. Wang, Y.B., Xu, Y.M., Huang, Q.: Progress on ultrasonic guided waves de-icing techniques in improving aviation energy efficiency. Renew. Sustain. Energy Rev. **79**, 638–645 (2017)
21. Wang, Y., Xu, Y., Lei, Y.: An effect assessment and prediction method of ultrasonic de-icing for composite wind turbine blades. Renew. Sustain. Energy Rev. **118**, 1015–1023 (2018)
22. Wang, Z.: Recent progress on ultrasonic de-icing technique used for wind power generation, high-voltage transmission line and aircraft. Energy Build. **140**, 42–49 (2017)
23. Guffond, D., Hedde, T.: Prediction of ice accretion - comparison between the 2D and 3D codes. Recherche Aerospatiale **2**, 103–115 (1994)
24. Reggio, M., Ilinca, A.: Prediction of ice accretion and anti-icing heating power on wind turbine blades using standard commercial software. Energy **114**, 1041–1052 (2016). https://doi.org/10.1016/j.energy.2016.08.047
25. Kalyulin, S.L., Modorskii, V.Ya., Cherepanov, I.E.: Numerical modeling of the influence of the gas-hydrodynamic flow parameters on streamlined surface icing. In: Fomin, V. (ed.) ICMAR 2018, AIP Conference Proceedings, vol. 2027, art. no. 030180 (2018). https://doi.org/10.1063/1.5065274
26. Kalyulin, S.L., Modorskii, V.Ya., Maksimov, D.S.: Physical modeling of the influence of the gas-hydrodynamic flow parameters on the streamlined surface icing with vibrations. In: Fomin, V. (ed.) ICMAR 2018, AIP Conference Proceedings, vol. 2027, art. no. 040090 (2018). https://doi.org/10.1063/1.5065364
27. Shchapov, V.A., Pavlinov, A.M., Popova, E.N., Sukhanovskii, A.N., Kalyulin, S.L., Modorskii, V.Ya.: Supercomputer real-time experimental data processing: technology and applications. In: RuSCDays 2018, Communications in Computer and Information Science, vol. 965, pp. 641–652 (2018). https://doi.org/10.1007/978-3-030-05807-4_55

28. Kalyulin, S.L., Modorskii, V.Ya., Petrov, V.Y., Masich, G.F.: Computational and experimental modeling of icing processes by means of PNRPU high-performance computational complex. In: Journal of Physics: Conference Series, vol. 965, art. no. 012081 (2018). https://doi.org/10.1088/1742-6596/1096/1/012081

29. Kalyulin, S.L., Modorskii, V.Y., Paduchev, A.P.: Numerical design of the rectifying lattices in a small-sized wind tunnel. In: Fomin, V. (ed.) ICMAR 2016, AIP Conference Proceedings, vol. 1770, art. no. 030110 (2016). https://doi.org/10.1063/1.4964052

30. Kalyulin, S.L., Modorskii, V.Ya., Shmakov, A.F.: Numerical coupled 2FSI analysis of gas-dynamic and deformation processes in the discharger of the model compressor of a gas transmittal unit. In: Fomin, V. (ed.) ICMAR 2018, AIP Conference Proceedings, vol. 2027, art. no. 030168 (2018). https://doi.org/10.1063/1.5065262

31. Shmakov, A.F., Modorskii, V.Ya.: Energy conservation in cooling systems at metallurgical plants. Metallurgist, 882–886 (2016). https://doi.org/10.1007/s11015-016-0188-8

32. Kozlova, A.V., Modorskii, V.Y., Ponik, A.N.: Modeling of cooling processes in the variable section channel of a gas conduit. Rus. Aeronaut. **53**(4), 401–407 (2010). https://doi.org/10.3103/s1068799810040057

33. Modorskii, V.Y., Sipatov, A.M., Babushkina, A.V., Kolodyazhny, D.Y., Nagorny, V.S.: Modeling technique for the process of liquid film disintegration. In: Fomin, V. (ed.) ICMAR 2016, AIP Conference Proceedings, vol. 1770, art. no. 030109 (2016). https://doi.org/10.1063/1.4964051

34. Gaynutdinova, D.F., Modorsky, V.Y., Masich, G.F.: Infrastructure of data distributed processing in high-speed process research based on hydroelasticity tasks. Procedia Comput. Sci. **66**, 556–563 (2015). https://doi.org/10.1016/j.procs.2015.11.063. Sloot, P. (ed.) YSC

35. Kalyulin, S.L., Shavrina, E.V., Modorskii, V.Y., Barkalov, K.A., Gergel, V.P.: Optimization of drop characteristics in a carrier cooled gas stream using ANSYS and globalizer software systems on the PNRPU high-performance cluster. In: Sokolinsky, L., Zymbler, M. (eds.) PCT 2017. CCIS, vol. 753, pp. 331–345. Springer, Cham (2017). https://doi.org/10.1007/978-3-319-67035-5_24

36. Modorskii, V.Y., Gaynutdinova, D.F., Gergel, V.P., Barkalov, K.A.: Optimization in design of scientific products for purposes of cavitation problems. In: Simos, T.E. (ed.) ICNAAM 2015, AIP Conference Proceedings, vol. 1738, art. no. 400013 (2016). https://doi.org/10.1063/1.4952201

37. Morkovin, M.: Effects of compressibility on turbulent flows. In: Favre, A. (ed.) The Mechanics of Turbulence, pp. 367–380. Gordon and Breach, New York (1964)

38. ATAAC - Advanced turbulence simulation for aerodynamic application challenges. http://cfd.mace.manchester.ac.uk/twiki/bin/view/ATAAC/WebHome

39. Egorov, I.N., Kretinin, G.V., Leshchenko, I.A., Kuptzov, S.V.: Robust design optimization strategy of IOSO technology. In: Proceedings of Fifth World Congress on Computational Mechanics, Vienna, Austria, pp. 1–8 (2002)

40. Modorskii, V.Y., Shevelev, N.A.: Research of aerohydrodynamic and aeroelastic processes on PNRPU HPC system. In: Fomin, V. (ed.) ICMAR 2016, AIP Conference Proceedings, vol. 1770, art. no. 020001 (2016). https://doi.org/10.1063/1.4963924

Applying Parallel Calculations to Model the Centrifugal Compressor Stage of a Gas Transmittal Unit in 2FSI Statement

Ivan E. Cherepanov$^{(\boxtimes)}$ ⓘ, Vladimir Ya. Modorskii ⓘ, Stanislav L. Kalyulin ⓘ, Anton O. Mikryukov, Danila S. Maksimov, and Anna V. Babushkina

Perm National Research Polytechnic University, Perm, Russia
`cherepanovie@sbiw.ru`, {`modorsky,ksl`}`@pstu.ru`

Abstract. The problem of vibrations arising during the operation of gas transmittal unit compressors is an urgent task in designing a rotor constructor. Existing approaches do not allow predicting the vibration probability in every operating mode. The article presents a new 2FSI approach allowing a designer to take into account the previously omitted factor of gas and structure interaction. The current level of high-performance computing systems allows using the 2FSI approach for complex systems such as a gas transmittal unit centrifugal compressor. The article presents the numerical calculation results of the gas transmittal unit compressor stage. The results were, obtained using the PNRPU high-performance computing complex (peak performance 24 TFLOPS). In addition, the paper provides the estimated computation time when using different core numbers.

Keywords: Centrifugal compressor · GTU · Vibrations · Numerical simulation · 2FSI · Spectral analysis

1 Introduction

Modern trends in technology development are aimed at increasing the power of aggregates and the rotation speed of their individual elements, at improving mass-geometric characteristics, which leads to a decrease in the structure rigidity. In this case, vibration may occur, which leads to a decreased reliability of the gas transmittal unit (GTU). Therefore, the problem of vibration today is becoming increasingly relevant for the entire industry of turbomachinery, in particular for the GTU, which emergency stop leads to enormous losses.

There are simplified approaches aimed at reducing vibrations by balancing rotor imbalances arising from inaccurate machining and assembly errors. They are reduced to acceptable values both by application of calculation methods [1], and experimental approaches, including the improvement of balancing techniques [2] and subsequent vibration diagnostics [3].

© Springer Nature Switzerland AG 2020
L. Sokolinsky and M. Zymbler (Eds.): PCT 2020, CCIS 1263, pp. 321–335, 2020.
https://doi.org/10.1007/978-3-030-55326-5_23

Despite the minimization of residual imbalances, there are situations of a GTU emergency shutdown in terms of vibration. In such cases, the influence of the gas-dynamic component becomes the key factor determining the vibrational state of the structure. The closest analogy for this phenomenon may be blade flutter.

Currently, the prediction of possible blade flutter occurrence is based on a probabilistic-statistical approach, which is a generalization of experimental data by the methods of mathematical statistics and the construction of stability areas depending on various factors [4]. This approach is limited by the need for multiple tests for various geometries and gas-dynamic flow parameters.

Another approach is based on numerical modeling and obtaining the forms and frequencies of natural vibrations of individual structural elements [5–7]. From the obtained values, a Campbell frequency diagram is constructed, which reflects the change in the natural frequency of the oscillations from the rotor speed. According to the diagram, the relative position of the curves is estimated and a conclusion is made about oscillation occurrence.

In article [8], the authors consider an approach in which additional elastic elements are introduced into the rotor system. They are described by the stiffness and damping coefficients of the labyrinth seals. The simulation results made it possible to obtain a subharmonic component in the vibrational spectrum.

The described approaches explicitly do not take into account the influence of the gas-dynamic flow on the structure. This assumption is valid in the case when the loads are not high, the structure has significant rigidity and disturbances caused by structural vibrations in the stream do not have a significant inverse effect.

Modern supercomputer technology allows for the numerical simulation of unsteady processes in gas and construction at the same time. However, the joint calculation of gas-dynamics and the stress-strain state of the structure is currently a complex computational task. To simplify it, several approaches have been developed that differ in the degree to which gas-dynamics is taken into account when calculating vibrations.

One approach determines the frequencies of the structure natural vibrations. The obtained values are used to specify structure wall displacements according to the harmonic law for calculating gas dynamics. According to the calculation results, the work of gas-dynamic forces is performed by gas pressure forces during the period of natural oscillations. A positive operation value corresponds to an unstable mode of operation.

With another approach, an unsteady gas-dynamic calculation is initially performed without taking into account the movement of the structure walls. The resulting gas pressure fields are transferred in the form of forces to the contact surfaces of the structure to calculate the stress-strain state. The results allow us to determine the nature of the change in time movements.

A more accurate and complete approach that allows taking into account the effects of interaction is the simultaneous calculation of gas-dynamics and stress-strain state [9]. This approach is called Two-Way Fluid-Structure Interaction

(2FSI). A direct analysis of rotor vibration displacements, taking into account all structural features and flow gas-dynamics, is an incredibly complex computational task requiring the use of specially designed high-performance computer systems. Due to the high demands on computing resources, the application of this method has so far been limited and reduced to solving model problems in two-dimensional or axisymmetric formulations [10–17]. The PNRPU high-performance computing complex (PNRPU HPC) allowed us to use the 2FSI approach to perform calculations for the GTU centrifugal compressor stage, including an impeller, a shaft, an unloading device (UD), and a labyrinth seal (LS) on the shroud (Fig. 1). The results of the 2FSI computational experiments are presented in this article.

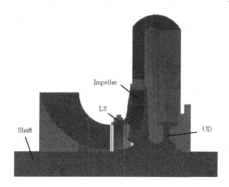

Fig. 1. Calculation model section of the centrifugal compressor stage

2   Mathematical Model

2.1   Problem Statement

The 2FSI task involves two subdomains – gas-dynamic and solid-state. Each subregion has its own conceptual model.

The gas-dynamic subdomain is a flow path of a compressor filled with gas. For the gas-dynamics problem, it is assumed that the gas in the flow path is single-phase and is described by the equation of an ideal gas state. The calculations take into account gas viscosity effects. The walls of the gas-dynamic region are smooth, impermeable, adiabatic and deformable.

The solid-state subdomain is a rotor, which consists of a shaft in bearings with an impeller of the third stage mounted on it and an unloading device. The rotor rotates. The gas-dynamic force acting on the gas side leads to deformation of the rotor structure. In this case, the rotor is operated under external influences (kinematic and force) in the elastic region, which does not imply the plastic flow of the material during the GTU compressor operation. Therefore, in the physical sense, the rotor material is linear; it is assumed to be a continuous and homogeneous medium [18]. The journal bearings are simulated by elastic elements with constant radial stiffness. Axial structure displacement is limited at one of the shaft ends.

2.2 Mathematical Model of the Gas-Dynamic Subdomain

In accordance with the accepted conceptual model, a mathematical model is developed, which is based on non-stationary equations of gas-dynamics and mechanics of a deformable solid. The gas-dynamics equations include the law of conservation of mass, momentum, energy, are closed by the equation of state of an ideal gas and a model of turbulence [19], as well as initial and boundary conditions. The mathematical model of the gas-dynamic problem includes the following relations:

Mass conservation equation (1):

$$\frac{\partial \rho_g}{\partial t} + \nabla \cdot (\rho_g \boldsymbol{V}) = G, \tag{1}$$

where ρ_g is the gas density, $\boldsymbol{V}$ is the gas flow velocity vector, G is the gas mass flow and t is the time.

Momentum conservation equation (2):

$$\frac{\partial (\rho_g \boldsymbol{V})}{\partial t} + \nabla \cdot (\rho_g \boldsymbol{V} \otimes \boldsymbol{V}) = -\nabla P + \nabla \cdot \boldsymbol{\tau}, \tag{2}$$

where P is the static pressure. The ratio for determining viscous stresses is 3):

$$\boldsymbol{\tau} = \mu_g \left(\nabla \boldsymbol{V} + \boldsymbol{V} \nabla - \frac{2}{3} \delta \nabla \cdot \boldsymbol{V} \right), \tag{3}$$

where μ_g is the gas dynamic viscosity and δ is the Kronecker delta.

Energy conservation equation (4):

$$\frac{\partial (\rho_g H^*)}{\partial t} - \frac{\partial P}{\partial t} + \nabla \cdot (\rho_g \boldsymbol{V} H^*) = \nabla \cdot (\lambda \nabla T + \boldsymbol{V} \cdot \boldsymbol{\tau}), \tag{4}$$

where $H^* = H + \frac{|V|^2}{2}$ is the total enthalpy, T is the gas static temperature and λ is the thermal conductivity.

The described equations are supplemented by defining state relations for the density and gas enthalpy. For ideal gas, the following equations are valid (5–6):

$$\rho_g = \frac{PM}{R_0 T}, \tag{5}$$

$$dH = c_p dT, \tag{6}$$

where M is the ideal gas molar mass, R_0 is the universal gas constant and c_p is the gas isobaric heat capacity.

The finite volume solver included in the ANSYS CFX software has been used to carry out the simulation. The SST $k - \omega$ double equation turbulence model is adopted to make equations closed. The equations are solved in a relative coordinate system, rotating with the angular velocity of the compressor rotor. This approach does not require rebuild the grid model, which leads to less computational intensiveness. The need to coordinate gas-dynamic parameters at the boundaries of the rotating and stationary regions constitutes the approach disadvantage. For non-stationary calculations, the most accurate results can be obtained using the Transient Rotor-Stator (TRS) interpolation model [20].

2.3   Mathematical Model of the Solid-State Subdomain

The rotor motion of the GTU compressor is described by differential equations in displacements U in the framework of the linear theory of elasticity. Limited to this theory to describe the rotor motion, we accept the smallness ε of possible deformations that are determined through the gradients of displacements (7):

$$\varepsilon = \frac{1}{2}(\nabla U + U \nabla), \tag{7}$$

where U is the field of displacements, represented as functions of changing Euler (spatial) coordinates x_i (8):

$$U = U(x_1, x_2, x_3, t). \tag{8}$$

Since in the process of elastic deformation associated with inertial forces acting on structural elements at the rotation moment and an external action due to the influence of gas mass transfer in GTU, forces arise on the surface of a solid body, called stresses σ. The relation between the resulting stresses σ and strains ε is described by the Hooke's isotropic law in finite form [18] (9):

$$\sigma = \lambda I_1(\varepsilon)\varepsilon + 2\mu\varepsilon, \tag{9}$$

where λ and μ are the Lame parameters, $I_1(...)$ is the trace of the tensor determined by the sum of the main (diagonal) values, I is the unit tensor.

Lame parameters are determined by the following relations (10–11):

$$\lambda = \frac{E\nu}{(1+\nu)(1-2\nu)}, \tag{10}$$

$$\mu = \frac{E}{2(1+\nu)}, \tag{11}$$

where E is Young's modulus and ν is Poisson's ratio. Since relation (9) is realized within the framework of the linear theory of elasticity, the Lame parameters are constant values.

The differential equations of continuous medium motion, which is the GTU rotor compressor, follow from the equations of static equilibrium when volume inertia forces are taken into account (12):

$$\nabla \cdot \sigma + \rho_s f_m + f_e = \rho_s \frac{\partial^2 U}{\partial t^2}, \tag{12}$$

where ρ_s is the construction material density, $\boldsymbol{f}_m$ is the density of mass (internal) forces, and $\boldsymbol{f}_e$ is the density of external forces.

In this case, the density of the structure material is an invariable value since the rotor material is assumed to be homogeneous, i.e. its physical and mechanical properties are the same at all points of the body. The density of mass forces, in the general case, is a function of spatial coordinates and is formulated from physical considerations about the processes occurring inside the body (13):

$$\boldsymbol{f}_m = \boldsymbol{f}_m(x_1, x_2, x_3, t). \tag{13}$$

However, within the framework of this article, the presence of internal mass sources, i.e. $\boldsymbol{f}_m = 0$. The external forces are determined from the boundary conditions due to external interaction.

Thus, substituting expressions (7, 9–11) for relation (12), we obtain the differential equations of GTU rotor compressor motion in displacements (14):

$$(\lambda + \mu)\nabla\nabla \cdot \boldsymbol{U} + \mu\Delta\boldsymbol{U} + \boldsymbol{f}_e = \rho_s\frac{\partial^2 \boldsymbol{U}}{\partial t^2}, \tag{14}$$

where Δ is the Laplace operator.

The resulting equation system is solved using a Finite element method.

2.4   Boundary and Initial Conditions

Each of the subdomains (gas-dynamic and solid-state) was assigned rotation. As the boundary conditions for the gas-dynamic region, the total inlet pressure and the outlet flow rate are determined. The walls are impermeable, adiabatic, without slipping. The radial force is proportional to the radial displacement and acts on the solid model in the journal surfaces. At the shaft end, axial displacement is established (Fig. 2).

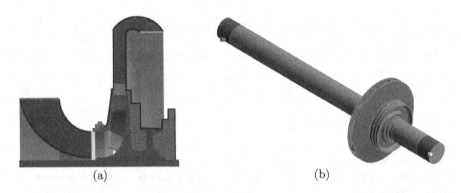

(a)   (b)

Fig. 2. Boundary conditions: (a) gas-dynamic region; (b) solid-state region

Two-way interaction between the elements of the step is determined by 16 surfaces (Fig. 3). Table 1 presents the indication element pairs, type of interaction and color designations. The data is transferred in several stages:

1. Pre-processing of data by recalculation of the values obtained for the elements on the corresponding nodes and interaction surfaces.
2. Determination of correspondence between contact surfaces [21, 22].
3. Interpolation of values projection from the original surface onto the receiving surface.
4. Postprocessing of the interpolated data improving the convergence and cutting off any invalid variable values resulting from the interpolation.

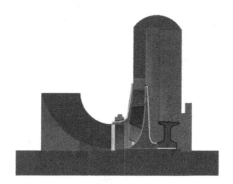

Fig. 3. The region interaction boundaries

The transmitted data are refined at each time step until the root mean square value of the residuals decreases to the set value.

The unsteady calculation was performed for a time interval from 0 to 0.02 s in increments of 50 μs, which corresponds to 400 time steps. The steady-state gas flow obtained in the calculation in the compressor path was used as the initial values of the gas-dynamic region parameters.

3   Analysis Methodology

The distributions of the gas-dynamic parameters in the compressor path and the displacements of the solid-state model, obtained from the results of non-stationary calculation, made it possible to set several control points at which the parameter changes in time were obtained. The time dependences made it possible to perform spectral analysis, according to the results of which the frequency response and phase response were obtained in the range from 0 to 10 kHz with a step of 50 Hz [23].

The structure dynamic state was assessed at several points.

Seven control points were selected in the gas-dynamic volume (Fig. 4, a):

– near the impeller blade;
– at the impeller inlet;
– at the impeller outlet;

Table 1. Color designation of the region interactions

No.	Element 1	Element 2	Interface	Color
1	Impeller flow volume	Impeller inlet	Fluid-Fluid	
2	Impeller flow volume	Impeller outlet	Fluid-Fluid	
3	Impeller flow volume	Labyrinth seal flow volume	Fluid-Fluid	
4	Impeller flow volume	Unloading device flow volume	Fluid-Fluid	
5	Impeller flow volume	Impeller blades	Fluid-Solid	
6	Impeller flow volume	Hub	Fluid-Solid	
7	Impeller flow volume	Shaft	Fluid-Solid	
8	Impeller flow volume	Shroud	Fluid-Solid	
9	Labyrinth seal flow volume	Labyrinth seal	Fluid-Solid	
10	Labyrinth seal flow volume	Shroud	Fluid-Solid	
11	Unloading device flow volume	Unloading device	Fluid-Solid	
12	Unloading device flow volume	Hub	Fluid-Solid	
13	Unloading device flow volume	Shaft	Fluid-Solid	
14	Hub	Shaft	Solid-Solid	
15	Hub	Unloading device	Solid-Solid	
16	Unloading device	Shaft	Solid-Solid	

– at the high-pressure area of the unloading device;
– at the low-pressure area of the unloading device;
– in the labyrinth seal of the unloading device;
– in the center of the contact surface with the unloading device.

In the solid-state model, 6 control points are defined (Fig. 4, b):

– on the impeller blade;
– near the labyrinth seal of the unloading device;
– in the center of the unloading device end section;
– in the center of the labyrinth seal contact surface with the labyrinth seal gas-dynamic volume;
– in the center of the shroud contact surface with the labyrinth seal gas-dynamic volume;
– on the shaft axis.

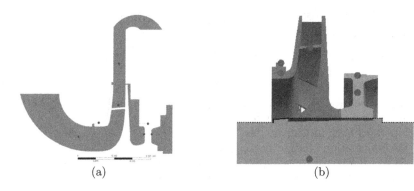

(a)   (b)

Fig. 4. Control points location: (a) gas-dynamic region; (b) solid state region

As a result, the following time dependences were obtained for each of 13 points:

1. Static pressure (from gas-dynamics).
2. Gas-dynamic forces along 3 axes (from gas-dynamics).
3. Structure displacement along 3 axes (from solid analysis).

The spectrograms obtained allow us to study the forced oscillations of the gas and structure. For real structures, the most interesting are the amplitudes of the maximum structure displacements. Exceeding the set values during the compressor operation can cause the rotor to touch the housing walls, which will lead to an emergency shutdown of the GTU. Based on these values, a critical area of the compressor operation can be determined.

4   Calculations

Determining the vibrational activity of the GTU rotor compressor in a 2FSI aeroelastic 3D statement using finite element methods and finite volumes is an extremely time-consuming task. This, in particular, is due to the complex geometry of the rotor design elements and gas-dynamic areas of the compressor path, as well as the small size of the labyrinth seal gaps (0.5 mm), which necessitates the development of detailed grid models that correctly resolve the gas-dynamic flow features.

Therefore, when creating the grid models, specialized ANSYS 18.0 modules were used and grid optimization was carried out. For example, to solve the 2FSI task with simultaneous account for the design of the shaft, bearings, impeller, labyrinth seal, unloading device and gas-dynamics of each element, the grid model is constructed with a relatively small number of finite elements ($\approx$10.2 million for gas and $\approx$0.2 million for the construction).

The calculations were performed for three compressor operation modes, which allowed us to determine the critical area of compressor operation.

When solving the 2FSI problem, the resources of the PNRPU HPC [24] and ANSYS 18.0 were used.

Below are the main specifications of the PNRPU HPC:

- 95 computational nodes;
- 128 four-core processors "Barcelona-3" (total number of cores is 512);
- 62 eight-core processors "Intel Xeon E5-2680" (total number of cores is 480);
- 24.096 TFlops peak performance;
- 78% performance in Linpack test packet;
- 27 TB information storage system capacity;
- 5888 GB RAM installed (32 GB/node with processors "Barcelona-3", 128 GB/node with processors "Intel Xeon E5-2680");
- 12 computing modules GTU NVIDIA Tesla M2090 (512 cores, 6 GB).

The results of the problem test runs on a different number of computational cores are shown in Fig. 5. It was found that the calculation of a deformable solid mechanics has the greatest influence on the productivity slowdown, the fraction of the computational time at one time step increases with an increase in the number of computational cores (Fig. 6).

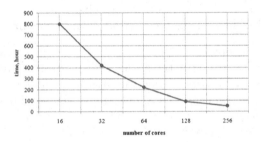

Fig. 5. Parallelization results for a different number of cores

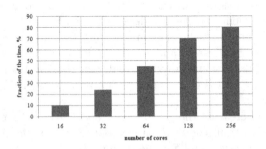

Fig. 6. The time spent to calculate the deformable solid mechanics for one iteration

The final program version was launched on 32 eight-core Intel Xeon E5-2680 processors. The results of each of the calculations took about 800 GB of disk space. The calculation time was 50 h.

5   Results

A spectral analysis [25] of the above dynamically changing parameters was performed. Some of the obtained spectrograms are presented in Fig. 7.   The results

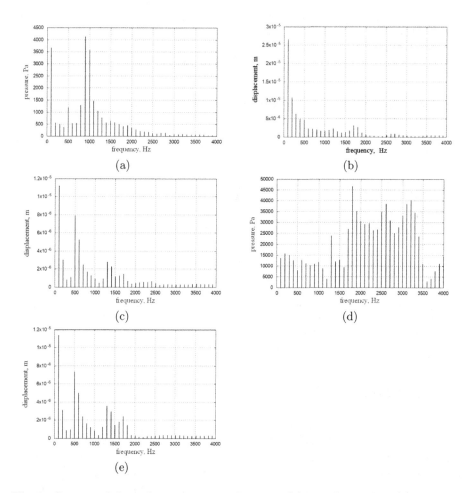

Fig. 7. Spectra of dependencies at control points: (a) impeller outlet; (b) impeller blade; (c) shaft axis; (d) labyrinth seal; (e) unloading device

show the presence of the blade frequency peak at the control point in the area of the output device (Fig. 7, a), which corresponds to the blade oscillation frequency (Fig. 7, b). The spectrogram for deformation at a point on the shaft axis (Fig. 7, c) shows the low-frequency component, which is also present on the spectrogram of pressure pulsations in the output device. The labyrinth compaction spectrogram show the vibrations (Fig. 7, d) in the high-frequency region; therefore, the labyrinth seal has a lesser effect on the vibration initiation. The unloading device spectrum (Fig. 7, e) also illustrated some peaks in the low-frequency region.

The critical area of the system operation with joint gas-dynamics consideration of the impeller, labyrinth seal and unloading device is limited by permissible displacement (up to 40 μm) [1–3]. Figure 8 shows histograms of maximum displacements at different GTU rotational speeds for various characteristic points in the rotor design with separate and combined accounting for gas-dynamics of the impeller, labyrinth seal and unloading device.

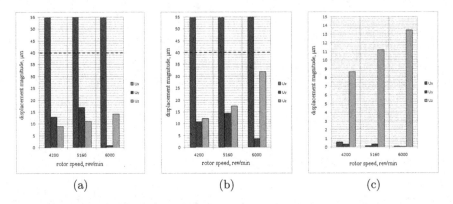

(a)   (b)   (c)

Fig. 8. Histograms of maximum displacements at various rotational speeds of the GTU rotor compressor (Ux, Uy - radial directions, Uz - axial direction): (a) on the contact boundary of the shroud with the labyrinth seal gap; (b) on the contact boundary of the unloading device with the labyrinth seal gap of the unloading device; (c) on the shaft axis.

6   Conclusion

The resources of the PNRPU high-performance computing complex made it possible to perform a coupled 2FSI calculation of gas-dynamics and the GTU centrifugal compressor stage. The required calculation time when using various core numbers is compared. It was found that with large core numbers, a significant portion of the time is spent on calculating the deformable solid mechanics. The solution to this problem will make it possible to extend the application of the 2FSI statement used to calculate the full compressor model consisting of several stages.

The calculation of the complex GTU compressor stage in the 2FSI statement shows the presence of peaks of pressure pulsations and displacements in the spectrograms of individual elements that are close in frequency. In certain modes of GTU operation, it can be assumed that the vibration frequencies of individual elements approach each other. This will lead to an increase in the displacement amplitudes and an emergency stop.

The analysis of the maximum displacement histograms shows the excessive radial displacements in the impeller labyrinth seals and unloading device labyrinth seals.

With an increase in the GTU rotational speed, the axial displacements increase. This is explained by an increase in the uncompensated axial force arising on the impeller. An increase in the angular velocity of the rotor leads to an increase in the pressure at the compressor stage outlet, which is transmitted to the high-pressure area of the unloading device. Different rates of changing the axial forces on the impeller and unloading device violate the axial rotor stability and increase loading action on the bearings.

The problem solution using the 2FSI approach allowed us to determine the frequencies of forced parameter oscillations in the gas and structure, taking into account their mutual influence, as well as to evaluate the amplitudes of the maximum structure displacements, which are of great importance in assessing the state of a working rotor.

Acknowledgements. The study was supported by the Russian Foundation for Basic Research (RFBR), project No. 17-47-590017.

References

1. Beloborodov, S.M., Tselmer, M.L.: Minimization of external disbalancing effects on the dynamics of the flexible rotor. J. Adv. Res. Nat. Sci. **3**, 30–32 (2017). (in Russian)
2. Beloborodov, S.M., Tsel'mer, M.L., Sviridov, E.V.: Precision balancing of impellers. Chem. Pet. Eng., 797–800 (2018). https://doi.org/10.1007/s10556-018-0424-0
3. Beloborodov, S.M., Tsimberov, D.M., Tsel'mer, M.L.: Experimental verification of the dynamic condition of the shafting. Bull. PNRPU. Eng. Mater. Sci. **19**(4) (2017). (in Russian)
4. Lokshtanov, E.A., Mikhaylov, V.M., Khorikov, A.A.: Statistical forecasting of the flutter of turbomachine blades. Naukova dumka **73–81** (1980). (in Ukrainian)
5. Govorov, A.A., Martirosov, M.I.: Self-oscillation processes of the aircraft engine compressor blades. Trudy MAI. (59), 10 (2012). (in Russian)
6. Galerkin, Y., Rekstin, A., Soldatova, K.: Aerodynamic designing of supersonic centrifugal compressor stages. WASET Int. J. Mec. Aerosp. Ind. Mechatron. Manuf. Eng. **9**(1), 123–127 (2015)
7. Galerkin, Y.B., Popova, E.Y., Soldatova, K.V.: Calculation analysis of an axial compressor supersonic stage impeller. WASET Int. J. Mech. Aerosp. Ind. Mechatron. Manuf. Eng. **9**(1), 118–122 (2015)
8. Makarov, A.A., Kas'yanov, S.V., Ezheva, I.V.: Analysis of the causes of increased vibration of the compressor rotor in operating conditions. Compress. Technol. **2**, 34–44 (2019). (in Russian)

9. Kalyulin, S.L., Modorskii, V.Ya., Shmakov, A.F.: Numerical coupled 2FSI analysis of gas-dynamic and deformation processes in the discharger of the model compressor of a gas transmittal unit. In: Fomin, V. (ed.) ICMAR 2018, AIP Conference Proceedings, vol. 2027, art. no. 030168 (2018). https://doi.org/10.1063/1.5065262

10. Hefeng, D., Chenxi, W., Shaobin, L., Zhen, S.X.: Numerical Research on segmented flexible airfoils considering fluid-structure interaction. Procedia Eng. **99**, 57–66 (2015). https://doi.org/10.1016/j.proeng.2014.12.508

11. Hussin, M.S., Ghorab, A., El-Samanoudy, M.A.: Computational analysis of two-dimensional wing aeroelastic flutter using Navier-Stokes model. Ain Shams Eng. J. **9**(4), 3459–3472 (2018). https://doi.org/10.1016/j.asej.2018.03.003

12. Rezaei, M.M., Zohoor, H., Haddadpour, H.: Aeroelastic modeling and dynamic analysis of a wind turbine rotor by considering geometric nonlinearities. J. Sound Vib. **432**, 653–679 (2018). https://doi.org/10.1016/j.jsv.2018.06.063

13. Beloborodov, S.M., Tselmer, M.L., Petrov, V.Yu., Modorsky, V.Ya.: Providing gas-dynamic tests for 2FSI subsystems. In: Fomin, V. (ed.) ICMAR 2018, AIP Conference Proceedings, vol. 2027, art. no. 040089 (2018). https://doi.org/10.1063/1.5065363

14. Modorskii, V.Ya., Mekhonoshina, E.V.: Numerical modeling of aeroelastic behavior of GTU centrifugal compressor rotor. Procedia Eng. **201**, 655–665 (2017). https://doi.org/10.1016/j.proeng.2017.09.6800

15. Butymova, L.N., Modorskii, V.Y., Petrov, V.Y.: Numerical modeling of interaction in the dynamic system "Gas-Structure" with harmonic motion of the piston in the variable section pipe. In: Fomin, V. (ed.) ICMAR 2016, AIP Conference Proceedings, vol. 1770, art. no. 030103 (2016). https://doi.org/10.1063/1.4964045

16. Modorskii, V.Ya., Butymova, L.N.: Numerical modeling of the labyrinth seal taking into account vibrations of the gas transmittal unit rotor in aeroelastic formulation. Procedia Eng. **201**, 666–676 (2017). https://doi.org/10.1016/j.proeng.2017.09.681

17. Modorskii, V.Ya., Mekhonoshina, E.V.: Impact of magnetic suspension stiffness on aeroelastic compressor rotor vibrations of gas pumping units. In: Fomin, V. (ed.) ICMAR 2016, AIP Conference Proceedings, vol. 1770, art. no. 030113 (2016). https://doi.org/10.1063/1.4964055

18. Koshelev, A.I., Narbut, M.A.: Mechanics of a deformable solid. SPb: SPbSU, 287 p. (2002). (in Russian)

19. Menter, F.R.: Review of the shear-stress transport turbulence model experience from an industrial perspective. Int. J. Compu. Fluid Dyn. **23**(4), 305–316 (2009). https://doi.org/10.1080/10618560902773387

20. Biesinger, T., Cornelius, C., Rube, C., Braune, A., Campregher, R., Godin, P.G., Schmid, G., Zori, L.: Unsteady CFD methods in a commercial solver for turbomachinery applications, pp. 2441–2452. ASME Paper. no. GT2010-22762 (2010). https://doi.org/10.1115/GT2010-22762

21. Jansen, K., Shakib, F., Hughes, T.J.R.: Fast projection algorithm for unstructured meshes. In: Computational Nonlinear Mechanics in Aerospace Engineering, vol. 146, no. 5, pp. 175–204 (1992). https://doi.org/10.2514/5.9781600866180.0175.0204

22. Galpin, P.F., Broberg, R.B., Hutchinson, B.R.: Three-dimensional Navier Stokes predictions of steady state rotor/stator interaction with pitch change. In: Proceedings of 3rd Annual Conference of the CFD Society of Canada, vol. 3 (1995)

23. Bendat, J.S., Piersol, A.G.: Random data analysis and measurement procedures. Measur. Sci. Technol. **11**(12), 1825–1826 (2011). https://doi.org/10.1088/0957-0233/11/12/702

24. Modorskii, V.Y., Shevelev, N.A.: Research of aerohydrodynamic and aeroelastic processes on PNRPU HPC system. In: Fomin, V. (ed.) ICMAR 2016, AIP Conference Proceedings, vol. 1770, art. no. 020001 (2016). https://doi.org/10.1063/1.4963924
25. Lyons, R.G.: Understanding Digital Signal Processing, 3rd edn. Pearson Education, London (2011)

Developing Cyber Infrastructure and a Model Climatic Wind Tunnel Based on the PNRPU High-Performance Computational Complex

Danila S. Maksimov$^{(\boxtimes)}$, Vladimir Ya. Modorskii, Grigory F. Masich,
Stanislav L. Kalyulin, Anton O. Mikryukov, Mikhail I. Sokolovskii,
and Ivan E. Cherepanov

Perm National Research Polytechnic University, Perm, Russia
{dsm,modorsky,ksl}@pstu.ru, cherepanovie@sbiw.ru

Abstract. The paper presents a promising direction of developing and using the cyberinfrastructure focused on pipelined/parallel data processing on a supercomputer of the experimental setup being developed at Perm National Research Polytechnic University, a small-sized wind tunnel (SSWT). The model climatic wind tunnel named Icing (MCWT) intensive data streams from which are transmitted on a high-speed optical network to the computer nodes of the supercomputer for their processing in real time created at PNRPU is described. The numerical gas-hydrodynamic results for the MCWT path obtained on this cyberinfrastructure are presented.

Keywords: Cyber infrastructure · Icing · Small-Sized Wind Tunnel · Numerical simulation · Supercomputer · Big data

1 Introduction

Technological base development for experimental research has led to the fact that modern experimental facilities generate continuously increasing volumes of data that need processing. In this area, we can distinguish the Dutch LOFAR project in astronomy and CERNLHC in high-energy physics. The later operates one of the most productive farms for source data processing. Experimental studies have also been actively carried out at Perm National Research Polytechnic University (PNRPU). This generated significant amounts of data [1,2]. To obtain reliable results in the study of complex non-stationary nonlinear interdisciplinary processes, it is necessary to conduct both computational and physical experiments. Their combination requires developing new paradigms of the distributed computing called the capacious concept of the cyberinfrastructure. Its essential infrastructural components include installations to connect to the high-speed optical networks, storage and visualization systems, computers and other installations for big data processing and generation. The article discusses the

© Springer Nature Switzerland AG 2020
L. Sokolinsky and M. Zymbler (Eds.): PCT 2020, CCIS 1263, pp. 336–350, 2020.
https://doi.org/10.1007/978-3-030-55326-5_24

aspects of developing and using the PNRPU cyberinfrastructure, which is part of the GIGA-URAL scientific and educational cyberinfrastructure. The Icing model climatic wind tunnel (MCWT) was created to test the cyberinfrastructure being developed at PNRPU, which is included in the unified scientific and educational network. At this stage, the MCWT was used to conduct physical and computational experiments to study the aircraft structure icing processes. Being formed on the aircraft wings, the icing leads to a significant decrease in the aircraft aerodynamic and control characteristics. The ice can also grow on aircraft engine nacelles, which can lead to an ice break off down in the engine. This would cause damage to the compressor elements and subsequent shutdown. The accident statistics of Army Aircraft Icing (2002), shows that between 1985 and 1999 there were 255 aircraft icing events; 12% of this number led to human losses; the material losses amounted to 28 million dollars [3]. According to the Aircraft Owners and Pilot Association (2007), there were 202 cases of aircraft icing in 1997–2007; 21% of the accidents led to human losses. With regard to the creation of a new family of aircraft engines such as PD, these issues should be paid particular attention. The article describes the solution to several problems. The first task was to create a common space of storage systems and calculators in order to ensure the possibility of high-performance processing of data received at the installation. For this, we used the GIGA-URAL scientific and educational network as a high-speed network. At that stage, it combined scientific and educational resources of Perm and Yekaterinburg and featured a speed of 30 Gbit/s [4–6]. One of the main sources of deterioration in the aggregate performance of geographically distributed highspeed applications is the low data transfer rate over the connection between terminal systems laid along long-distance high-speed communication lines [7]. The second task of the work described in this article was to include the MCWT [8] in the unified scientific and educational network. These systems are characterized by large and complex memory, processors, and input/output channels connected to the network with hierarchies and inherited parallelism in various subsystems. They require developing new methods and technologies for the efficient usage of the finite system architectures and resources. The third task was to expand the range of the MCWT reproducible flight conditions to increase the flow rate in the working area.

2   PNRPU Cyber Infrastructure

2.1   Cyber Infrastructure

The new paradigms of the distributed computing, according to the US National Science Foundation (NSF), are heavily focused on the term of cyberinfrastructure. The cyberinfrastructure combines the computing hardware, data and networks, digital sensors, observatories, experimental installations, an interactive set of software, firmware services and tools. The most important component of this infrastructure is a high-speed optical network built on a dark optical fiber and WDM [8] channel spectral multiplexing systems, which provide scalable petabyte speeds, which is shown in Fig. 1.

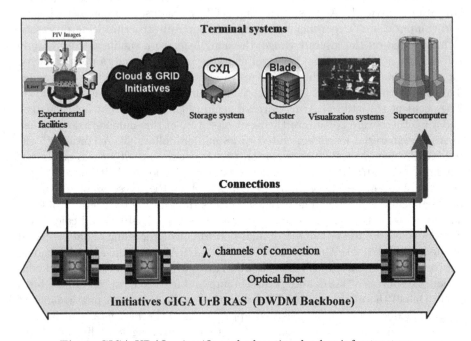

Fig. 1. GIGA URAL scientific and educational cyber infrastructure

Following the trends of building scientific and educational network infrastructures in the world, there was a problem of connecting the Center for High-Performance Computing Systems PNRPU and the Regional Center for Technical Competence "AMD-PNRPU", which host unique experimental facilities and computing resources, to the GIGA-URAL scientific and educational network fiber optic communication line (FOCL) at a speed of 10 Gbit/s. The system of discharged spectral channel multiplexing CWDW was used. This provided a speed increase up to hundreds of gigabits per second.

The purpose of connecting the PNRPU hosted computing cluster to a single space of storage systems and GIGA-URAL computers is to provide high-performance processing of the SSWT data generated in real time. The functional diagram of the created infrastructure is shown in Fig. 2.

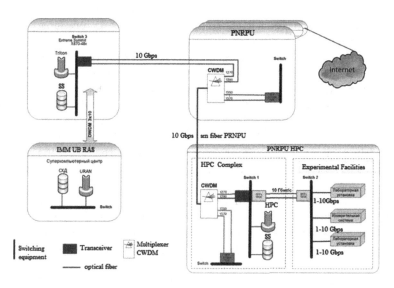

Fig. 2. Functional diagram of the created cyber infrastructure

Infrastructure reliability is ensured through the following solutions [9]. The article examines a complex approach to building a dependent corporate information and telecommunication infrastructure which covers the applications and services, servers and storage systems, data transmission networks and engineering support. It also shows the peculiarities of using the hyper-converged technology of virtualization for building fault-tolerant information systems.

The concept of cyberinfrastructure (CI) means comprehensive infrastructure according to the National Science Foundation (NSF) of the USA Cyberinfrastructure Vision for 21st Century Discovery NSF, March 2007 [1], that should combine all improvements in the field of information technology (IT). CI, as interpreted by NSF, combines hardware for computing, data and networks, digital sensors, observatories and experimental installations, an interactive set of software and services for firmware and tools. Out of the whole variety of CI aspects, the problem of incorporating the model icing wind tunnel into a single space of storage systems and computers was solved.

The high intensity of the generated data streams greatly complicates their processing. In addition, the required computing power is not always available at the experimental sites. Under these conditions, the transfer of calculations to remote supercomputers and the rejection of intermediate data storage in storage systems become a possible way of data processing a transition to the paradigm of distributed systems and in-memory computing. In-Memory Computing. The decrease in the cost of RAM DRAM (by 30% per year) has led to a significant increase in its volumes installed in computing systems. This growth has allowed placing more data directly in RAM to speed up the data processing. This approach has been called in-memory computing (IMC). In the forecast for 2013, Gartner analysts included the in-memory computing in the top ten "hottest" technologies, as the Big Data real-time operations had significantly accelerated and improved. As it was mentioned in a publication of Open Systems Publishing House, the rebirth of grids is explained by the ability of their architecture to meet the "three V" requirement for working with Big Data: velocity – high speed in memory, variability – the ability to store a wide variety of data in grids, volumes (the storage volumes in grids are theoretically endless). It is this approach (in-memory computing) that has been used to provide a dynamic analysis of the data generated by the small-sized wind tunnel (SSWT).

A lot of publications of foreign authors, for example, in the works of V. Vishwanath, noted that the current high-performance machines suffer from a significant imbalance between the computing power and the input/output speed. Their research is aimed at identifying the bottlenecks in the existing structure. The authors increased the input/output speed by asynchronous data feed and by scheduling the input/output using a request queue. Our approach is similar, but it is focused on high-speed long-distance communication channels (hundreds of km). The high value of the Bandwidth delay product (BDP) in which imposes specific requirements on the process of scheduling parallel data streams. In the distributed systems, the classical data processing approaches most commonly use such application protocols as CIFS and NFS to access storage and FTP, SCP, or GridFTP to transfer files. At the same time, the literature often notes insufficient performance of the transport protocols used at the lower levels of the OSI model with extended high-speed communication channels. It is generally recognized that one of the main degradation sources in the aggregate performance of geographically distributed high-speed applications is the poor end-to-end performance of the TCP protocol, which is present on all systems by default. Numerous TCP versions are aimed at increasing their efficiency (bandwidth) under various operating conditions. A huge number of works/publications are devoted to the development of strategies for controlling the congestion of the TCP versions in the presence of the competing flows (in fact, when using high-speed public Internet). Some of the works examine the TCP operation on the guaranteed high-speed communication channels of 10–100 Gbit/s, including using the remote direct memory access (RDMA network cards) mechanism. For example, the Large Data Access and Transport Proposal project, Linden Mercer (NRL), is being implemented on the created 100G experimental fragment of ESnet network. The data flow models accumulated by the organizer team and the experience in organizing the interaction of the computer installations [7] is shown in Fig. 3.

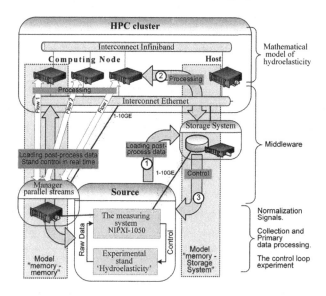

Fig. 3. Data processing models

The research variability of measured data and experimental control [8] is provided by two models of processing on a high-performance cluster. The memory-storage model has three stages:

1. Downloading data from the source to the storage system;
2. Processing data stored in the supercomputer storage system;
3. Uploading the processing results from the storage to the source.

Loading data go to storage system and subsequent processing (step 2) are performed either using file transfer protocols (FTP/GridFTP and SCP), or direct access to the data store using the file system protocols (CIFS and NFS/pNFS). This model is a classic scheme of the big data processing on a supercomputer. This scheme suggests breaking the measurement and counting processes in time and, in some cases, does not provide experimental control requirements.

The memory-memory model is focused on processing an intense data stream from the source in real time on a supercomputer. The idea of this model is based on the direct input of an intensive stream of structured data into the memory of the computational nodes of a supercomputer, bypassing an external data storage system. Using this approach is focused on the ability to control a full-scale experiment in the process of its implementation. Realized the ability to switch between automatic and manual control. The figure shows the smooth control systems and on/off relays required to control various elements. The connection diagram of the Icing MCWT experimental installation is shown in Fig. 4.

The "Managed level 2+ switch" SNR-S2990G-24FX-RPS was selected for connection to the experimental installation hardware and software complex of the system for measuring and recording fast processes based on Supercomputers.

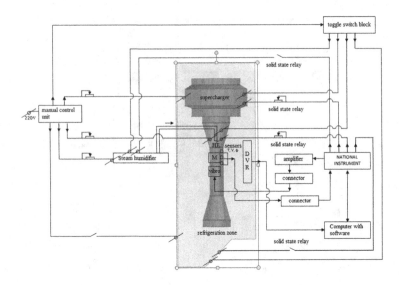

Fig. 4. Connection diagram of the experimental installation MCWT "Icing"

The Managed Level 2+ Switch has the following interfaces: 20 100/1000BaseX SFP ports are used to connect at a speed of 100/1000 Mbps installations, measuring systems and storage systems, a cluster and computers via copper or optics via SFP transceivers; 4 combo ports 10/100/1000BASE-T/100/1000BaseX SFP are used for connecting at a speed of 100/1000 Mbit/s installations, measuring systems and storage systems, a cluster and computers on copper 4 1/10G SFP+ ports are used to connect switches at 10 Gb/s and connections at 10 Gb/s of measurement systems, storage systems and a cluster. Fully hardware-based switching and ACL policy guarantee no traffic loss or delay; modern ASICs provide full-speed operation of all device ports. SNR-S2990G-24FX-RPS is equipped with a connector for connecting an uninterpretable power system with a voltage of 12 V to ensure fail-safe operation of the switch [10].

3   Conceptual Formulation of the Gas-Hydrodynamic Problem

3.1   Conceptual Physical Formulation

To perform a numerical simulation of the flow in the MCWT, the following conceptual formulation is done: - gas-dynamic processes are considered in a stationary setting;- the ideal gas flow is considered; - gas-dynamic flow is viscous, single-phase; - gravity is not taken into account; - at the same time, the flow in the working area, supercharger and in the internal cooled volume of the refrigerating chamber is investigated; - the walls of the working part are impermeable, adiabatic and rough (it is assumed that all roughness element protrusions lie inside the viscous layer), with the particles sticking; - to set the boundary conditions of input/output, in the zone behind the supercharger, there is assumed to be a gap.

The following solid-state 3D model of MCWT was built (Fig. 5):

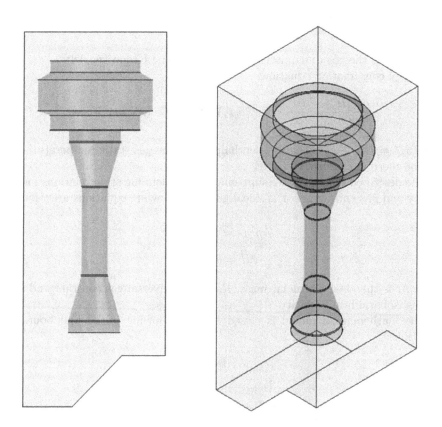

Fig. 5. Solid-state 3D model of MCWT "Icing"

3.2 Conceptual Mathematical Formulation

With the accepted physical model, a mathematical model has been developed that is based on the laws of conservation of mass, momentum, energy and is closed by the equations of state of an ideal compressible gas and turbulence, as well as by the initial and boundary conditions. It is required to solve a system of four independent equations, i.e. a Navier-Stokes equation system (1–4) [11]:

Mass conservation equation:

$$\frac{\partial \rho_{gas}}{\partial t} + \nabla \cdot (\rho_{gas} \boldsymbol{V}) = 0, \tag{1}$$

Momentum conservation equation:

$$\frac{\partial(\rho_{gas}H)}{\partial t} + \nabla \cdot (\rho_{gas}V \otimes V) = -\nabla P + \nabla((\mu + \mu_t) \cdot (\nabla V + (\nabla V)^T), \quad (2)$$

where P is the static pressure. The ratio for determining viscous stresses is:

$$\tau = \mu_g as \left(\nabla V + V\nabla - \frac{2}{3}\delta\nabla \cdot V\right), \quad (3)$$

where $\mu_g as$ is the gas dynamic viscosity and δ is the Kronecker delta.

Energy conservation equation:

$$\frac{\partial(\rho_{gas}H^*)}{\partial t} - \frac{\partial P}{\partial t} + \nabla \cdot (\rho_{gas}VH^*) = \nabla \cdot (\lambda\nabla T + V \cdot \tau), \quad (4)$$

where $H^* = H + \frac{|V|^2}{2}$ is the total enthalpy, T is the gas static temperature and λ is the thermal conductivity.

The described equations are supplemented by defining state relations for the density and gas enthalpy. For an ideal gas, the following equations are valid:

$$\rho_g as = \frac{PM}{R_0 T} \quad (5)$$

$$dH = c_p dT, \quad (6)$$

where M is the ideal gas molar mass, R_0 is the universal gas constant and c_p is the gas isobaric heat capacity.

The mathematical model is closed by the following initial and boundary conditions:

$$P_{total} = 1bar$$
$$T_{inlet} = -20\,°C$$
$$G_{outlet} = -0,65 \text{ kg/s}$$

The location of the gap zone of the solid-state model to set the boundary conditions is shown in Fig. 6:

To achieve a flow velocity of $70\,\text{m/s}$ in the working area, the mass flow rate corresponds to the maximum expected power of the supercharger. To solve the initial system of partial differential equations [12–15], the solid-state 3D model of the Icing MCWT was divided into cells. Figure 7 shows a grid model [16] built with ANSYS Mesher.

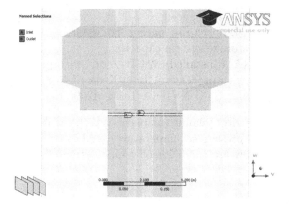

Fig. 6. Solid-state fracture zone for setting boundary conditions

Fig. 7. Grid model of the MCWT "Icing"

The main technical characteristics of PNRPU HPC: The number of finite elements in the MCWT amounted to 3.76 million. The grid model mainly consists of hexagonal elements. The maximum element size is 10 mm. As hardware for computational experiments, the resources of PNRPU HPC (peak performance 24 TFlops) were used. Technical characteristics of PNRPU HPC:

- 95 computational nodes;
- 128 four-core processors Barcelona-3 (total number of cores is 512);
- 62 eight-core processors Intel Xeon E5-2680 (total number of cores is 480);

- 24,096 TFlops peak performance;
- 78% performance in Linpack test packet ;
- 27 TB information storage system capacity;
- installed 5888 GB RAM installed (32 GB//node with processors "Barcelona-3" , 128 GB//node with processors "Intel Xeon E5-2680");
- 12 computing modules GTU NVIDIA Tesla M2090 (512 cores, 6 GB).

As a result of gas-hydrodynamic calculations [17], the fields of distribution of velocities, static, total pressures, temperatures, and other gas-dynamic parameters are obtained. Below are the fields of velocity streamlines (Fig. 8), and field of velocity (Fig. 9).

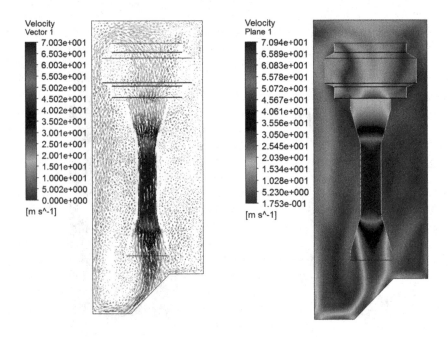

Fig. 8. Speed distribution lines and The distribution field in the MCWT

Based on the results of the gas-hydrodynamic calculations, ANSYS identified the possible overall dimensions of the supercharger and the working area of the Moscow Ring Road Icing, obtained a high uniformity of the gas-hydrodynamic flow in the working zone, and calculated the gas-dynamic parameters that ensure a uniform flow at the inlet to the supercharger.

4   Conducting Physical Experiments at the MCWT "Icing"

A series of physical experiments was carried out on the PNRPU Icing MCWT. Table 1 presents the plan of physical experiments:

Table 1. Plan for conducting physical experiments

No	$V_{gas}, m/s$	P_{gas}, Pa	T_{gas}, K	$\varphi, \%$	$\alpha, deg.$
1	70	101325	263.15	95	4
2	70	101325	253.15	95	4
3	70	101325	263.15	75	4

According to the calibration tests, the gas-dynamic flow velocity in the MCWT working area was constant and amounted to 70 m/s. The angle of attack was chosen equal to 4, as the most dangerous according to numerical experiments. The wing profile of NACA 0012 was made of aluminum [18–21]. The roughness of the sheet is 1.25 microns according to GOST 21631-76. The chord of the wing profile was 0.08 m. Figures 9, 10 and 11 show the contour-processed images from an IP camera for experiments No. 1, 2, 3.

Fig. 9. Icing with $V_{gas}, m/s = 70$ m/s, $T_{gas} = 263.15\,°$K, $\varphi = 95\%$ For timing: (a)–3 min., (b)–4 min, (c)–5 min

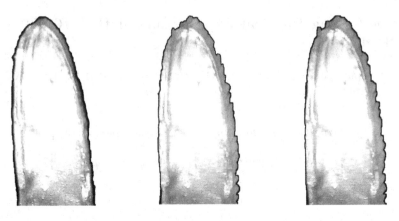

Fig. 10. Icing with $V_{gas}, m/s = 70$ m/s, $T_{gas} = 253.15\,°K$, $\varphi = 95\%$ For timing: (a)–3 min., (b)–4 min, (c)–5 min

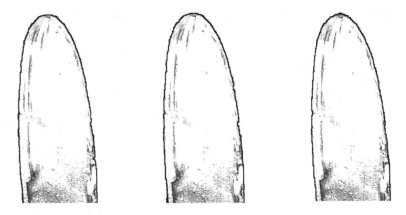

Fig. 11. Icing with $V_{gas}, m/s = 70$ m/s, $T_{air} = 263.15\,°K$, $\varphi = 95\%$ For timing: (a)–3 min., (b)–4 min, (c)–5 min

5   Conclusion

A component of cyberinfrastructure was created by connecting the experimental PNRPU setup, Icing MCWT, via a high-speed optical network to the computers, storage and visualization systems of the GIGA-URAL scientific and educational cyberinfrastructure. All data from MCWT was processed in the supercomputer memory, and the counting results were stored in the storage system. The developed solutions were tested to confirm their performance. Moreover the results of the computational experiment and the model physical experiment were obtained and verified. Based on the results of the verified gas-dynamic calculations of the modified MCWT design, no flow blocking was detected. Due to the decrease in the cross-section of the MCWT flow path, it was possible to achieve a speed of more than three times bigger than the speed of the base structure. From a maximum of 18.5 m/s on the base design to 70 m/s on the modified one.

Acknowledgements. The study was supported by the Russian Foundation for Basic Research (RFBR), project No. 17-47-590017.

References

1. Gaynutdinova, D.F., Modorsky, V.Y., Masich, G.F.: Infrastructure of data distributed processing in high-speed process research based on hydroelasticity tasks. Procedia Comput. Sci. **66**, 556–563 (2015). (in English)
2. Kalyulin, S.L., Modorskii, V.Ya., Maksimov, D.S.: Physical modeling of the influence of the gas-hydrodynamic flow parameters on the streamlined surface icing with vibrations. In: International Conference on the Methods of Aerophysical Research (ICMAR 2018), vol. 2027, Art. 040090, p. 5 (2018). https://doi.org/10.1063/1.5065364. (in English)
3. Runway Safety: International Civil Aviation IKAO, vol. 66, no. 2 (2011). (in Russian)
4. Masich, A.G., Masich, G.F.: Initiative GIGA UrB RAS. In: Computational Technologies, 2008, vol. 13. The Bulletin of KazNU al-Farabi. Mathematics, Mechanics and Informatics Issue, vol. 3, no. 58, part 2, pp. 413–418 (2008). (Joint issue). (in Russian)
5. Masich, A.G., Masich, G.F., Matveenko, V.P., Tiron, G.G.: Initiative GIGA UrB RAS: methodology of creation and architecture of the scientific and educational optical backbone of the Ural Branch of RAS. In: MIT-2011 Conference Proceedings, Belgrade, pp. 257–265 (2012). ISBN 978-86-83237-90-6 (AU). (in Russian)
6. Masich, A.G., Masich, G.F.: Problems of transport protocols in End-to-End connections over a high-speed long-range optical network and ways to solve them. In: Proceedings of the XX Baikal All-Russian Conference "Information and Mathematical Technologies in Science and Management", Part 1, pp. 224–232 (2015). (in Russian)
7. Masich, G.F., Masich, A.G.: From the "Initiative GIGA UrB RAS" to the cyber infrastructure of the Ural Branch of RAS. Perm Sci. Cent. J. UB RAS **4**, 41–56 (2009). (in Russian)
8. Klemenkov, G.P., Prihodko, Y.M., Puzyrev, L.N., Haritonov, A.M.: Modelling of aircraft icing processes in aeroclimatic tubes. Thermophys. Aeromeh. **15**(4), 563–572 (2008). (in Russian)
9. Masich, G.F., Latypov, S.R., Chugunov, D.P.: The aspects of information and telecommunication infrastructure's dependability. Perm Sci. Cent. J. UB RAS **3**, 25–40 (2018). https://doi.org/10.7242/1998-2097/2018.3.3. (in Russian)
10. Cyberinfrastructure Vision for 21st Century Discovery: National Science Foundation Cyberinfrastructure Council, March 2007. (in English)
11. Menter, F.R.: Review of the shear-stress transport turbulence model experience from an industrial perspective. Int. J. Comput. Fluid Dyn. **23**(4), 305–316 (2009). https://doi.org/10.1080/10618560902773387
12. Kalyulin, S.L., Modorskii, V.Ya., Cherepanov, I.E.: Numerical modeling of the influence of the gas-hydrodynamic flow parameters on streamlined surface icing. In: International Conference on the Methods of Aerophysical Research (ICMAR 2018), vol. 2027, art. no. 030180, p. 10 (2018). https://doi.org/10.1063/1.5065274. (in English)

13. Shchapov, V.A., Pavlinov, A.M., Popova, E.N., Sukhanovskii, A.N., Kalyulin, S.L., Modorskii, V.Y.: Supercomputer real-time experimental data processing: technology and applications. In: Voevodin, V., Sobolev, S. (eds.) RuSCDays 2018. CCIS, vol. 965, pp. 641–652. Springer, Cham (2019). https://doi.org/10.1007/978-3-030-05807-4_55

14. Kalyulin, S.L., Modorskii, V.Ya., Petrov, V.Y., Masich, G.F.: Computational and experimental modeling of icing processes by means of PNRPU high-performance computational complex. J. Phys. Conf. Ser. **965**, 012081 (2018). https://doi.org/10.1088/1742-6596/1096/1/012081

15. Kalyulin, S.L., Modorskii, V.Y., Paduchev, A.P.: Numerical design of the rectifying lattices in a small-sized wind tunnel. In: Fomin, V. (ed.) ICMAR 2016, AIP Conference Proceedings, vol. 1770, art. no. 030110 (2016). https://doi.org/10.1063/1.4964052

16. Jansen, K., Shakib, F., Hughes, T.J.R.: Fast projection algorithm for unstructured meshes. Comput. Nonlinear Mech. Aerosp. Eng. **146**(5), 175–204 (1992). https://doi.org/10.2514/5.9781600866180.0175.0204

17. Modorskii, V.Y., Shevelev, N.A.: Research of aerohydrodynamic and aeroelastic processes on PNRPU HPC system. In: Fomin, V. (ed.) ICMAR 2016, AIP Conference Proceedings, vol. 1770, art. no. 020001 (2016). https://doi.org/10.1063/1.4963924

18. Bendat, J.S., Piersol, A.G.: Random data analysis and measurement procedures. Meas. Sci. Technol. **11**(12), 1825–1826 (2011). https://doi.org/10.1088/0957-0233/11/12/702

19. Guffond, D., Hedde, T., Henry, R.: Overview of icing research at ONERA, advisory group for aerospace research and development. In: Fluid Dynamics Panel (AGARD/FDP) Joint International Conference on Aircraft Flight Safety - Actual Problems of Aircraft Development, Zhukovsky, Russia, p. 7 (1993)

20. Tran, P., Brahimi, M.T., Paraschivoiu, I.P., Tezok, F.: Ice accretion on aircraft wings with thermodynamic effects. In: 32nd Aerospace Sciences Meeting & Exhibit on American Institute of Aeronautics and Astronautics, Reno, Nevada, AIAA Paper, art. no. 0605 (1994)

21. Alekseenko, S.V., Prihodko, A.A.: Numerical simulation of icing cylinder and profile. Models review and calculation results. TsAGI Sci. J. **44**(6), 25–57 (2013). (in Russian)

Author Index

Printed in the United States
By Bookmasters